Enzymes as Targets for Drug Design

Enzymes as Targets for Drug Design

Edited by

Michael G. Palfreyman
Merrell Dow Research Institute
Cincinnati, Ohio

Peter P. McCann
Merrell Dow Research Institute
Cincinnati, Ohio

Walter Lovenberg
Merrell Dow Research Institute
Strasbourg, France

Joseph G. Temple, Jr.
Merrell Dow Pharmaceuticals, Inc.
Cincinnati, Ohio

Albert Sjoerdsma
Merrell Dow Research Institute
Cincinnati, Ohio

ACADEMIC PRESS, INC.
Harcourt Brace Jovanovich, Publishers
San Diego New York Boston
London Sydney Tokyo Toronto

Academic Press Rapid Manuscript Reproduction

This book is printed on acid-free paper. ∞

Academic Press, Inc.
San Diego, California 92101

United Kingdom Edition published by
Academic Press Limited
24–28 Oval Road, London NW1 7DX

Library of Congress Cataloging-in-Publication Data

Enzymes as targets for drug design.

1. Enzymes--Inhibitors. 2. Drugs--Design.
I. Palfreyman, Michael G. [DNLM: 1. Drug Design.
2. Enzyme Inhibitors--therapeutic use. QU 143 E6155].
QP601.5.E595 1989 615'.7 89-18313
ISBN 0-12-544030-8 (alk. paper)

Printed in the United States of America
90 91 92 9 8 7 6 5 4 3 2

CONTENTS

CONTRIBUTORS

Numbers in parentheses indicate the pages on which the authors' contributions begin.

Alan S. Bass (17), Committee on Clinical Pharmacology, University of Chicago, Chicago, Illinois 60637

Stephen B. Baylin (185), The Johns Hopkins Oncology Center, Johns Hopkins University School of Medicine, Baltimore, Maryland 21205

Philippe Bey (59), Merrell Dow Research Institute, Cincinnati, Ohio 45215

Robert A. Casero, Jr. (185), The Johns Hopkins Oncology Center, Johns Hopkins University School of Medicine, Baltimore, Maryland 21205

J. Richard Crout (1), Boehringer Mannheim Pharmaceuticals, Rockville, Maryland 20850

Wybren De Jong (31), Merrell Dow Research Institute, Strasbourg Center, 67084 Strasbourg Cedex, France

Fritz Gerhart (121), Merrell Dow Research Institute, Strasbourg Center, 67084 Strasbourg Cedex, France

Leon I. Goldberg (17), Committee on Clinical Pharmacology, University of Chicago, Chicago, Illinois 60637

Dennis A. Holt (85), Department of Medicinal Chemistry, Smith Kline and French Laboratories, King of Prussia, Pennsylvania 19406

Jean-Marie Hornsperger (121), Merrell Dow Research Institute, Strasbourg Center, 67084 Strasbourg Cedex, France

Michel J. Jung[1] (121), Merrell Dow Research Institute, Strasbourg Center, 67084 Strasbourg Cedex, France

M. Bashar Kahaleh (219), Division of Rheumatology and Immunology, Department of Medicine, Medical University of South Carolina, Charleston, South Carolina 29425

Manzoor M. Khan (45), Department of Medicine, Stanford University Medical Center, Stanford, California 94305

Ernest Kun (11), Department of Pharmacology, University of California, San Fransisco, California 94143

[1]Present address: Merrell Dow Research Institute, Lepetit Center, 21040 Gerenzano (Varese) Italy

E. Carwile LeRoy (219), Division of Rheumatology and Immunology, Department of Medicine, Medical University of South Carolina, Charleston, South Carolina 29425

Mark A. Levy (85), Department of Medicinal Chemistry, Smith Kline and French Laboratories, King of Prussia, Pennsylvania 19406

Walter Lovenberg (101), Merrell Dow Research Institute, Strasbourg Center, 67084 Strasbourg Cedex, France

Peter P. McCann (157), Merrell Dow Research Institute, Cincinnati, Ohio 45215

Ian A. McDonald (139), Merrell Dow Research Institute, Cincinnati, Ohio 45215

Kenneth L. Melmon (45), Departments of Medicine and Clinical Pharmacology, Stanford University Medical Center, Stanford, California 94305

Brian W. Metcalf (85), Smith Kline and French Laboratories, King of Prussia, Pennsylvania 19406

Michael B. Murphy (17), Committee on Clinical Pharmacology, University of Chicago, Chicago, Illinois 60637

John A. Oates (225), Department of Medicine, Vanderbilt University School of Medicine, Nashville, Tennessee 37232

Michael G. Palfreyman (139), Merrell Dow Research Institute, Cincinnati, Ohio 45215

Anthony E. Pegg (157), Departments of Cellular and Molecular Physiology and of Pharmacology, The Milton S. Hershey Medical Center, Pennsylvania State University, Hershey, Pennsylvania 17033

Lionel Resnick (211), Department of Dermatology, Mount Sinai Medical Center, Miami Beach, Florida 33140

Donald S. Robinson (233), Pharmaceutical Research and Development Division, Bristol-Myers Company, Wallingford, Connecticut 06492

Paul J. Schechter (201), Merrell Dow Research Institute, Cincinnati, Ohio 45215

Daniel Schirlin (121), Merrell Dow Research Institute, Strasbourg Center, 67084 Strasbourg Cedex, France

Paul Shapshak (211), Departments of Dermatology and Pathology, Mount Sinai Medical Center, Miami, Florida 33140

Elyse Singer (211), Department of Neurology, School of Medicine, University of California at Los Angeles, Los Angeles, California 90024, and Neurology Service, Veterans Administration, Medical Center-West Los Angeles, Wadsworth Division, Los Angeles, California 90073

Albert Sjoerdsma (157, 201), Merrell Dow Research Institute, Cincinnati, Ohio 45215

Karl Syndulko (211), Department of Neurology, School of Medicine, University of California at Los Angeles, Los Angeles, California 90024, and Neurology Service, Veterans Administration, Medical Center-West Los Angeles, Wadsworth Division, Los Angeles, California 90073

Wallace W. Tourtellotte (211), Department of Neurology, School of Medicine, University of California at Los Angeles, Los Angeles, California 90024, and Neurology Service, Veterans Administration, Medical Center-West Los Angeles, Wadsworth Division, Los Angeles, California 90073

Sidney Udenfriend (5), Roche Institute of Molecular Biology, Nutley, New Jersey 07110

Paul Van Giersbergen (31), Merrell Dow Research Institute, Cincinnati, Ohio 45215

Monique Zreika (139), Merrell Dow Research Institute, Strasbourg Center, 67084 Strasbourg Cedex, France

PREFACE

Enzyme inhibition frequently constitutes the molecular basis for the effectiveness of therapeutic agents. Progressively, new more powerful and more specific enzyme inhibitors are being designed, and with them new therapeutic agents are being discovered. These agents are often more specific in their effects and have the potential to treat a wider range of diseases. *Enzymes as Targets for Drug Design* presents a series of scientific discussions related to enzyme inhibitors that show the many facets of the drug discovery process from the basic sciences through clinical applications.

We want to stress that the choice of topics was a very personal one. From the outset, we planned a symposium, held on January 15–17, 1989, and this book, *Enzymes as Targets for Drug Design,* with the express intention of drawing together scientists who had worked at some stage with Albert Sjoerdsma. His long career has covered three phases: first at the University of Chicago, then at the National Institutes of Health, and since 1971 at the now Merrell Dow Research Institute. We have all had the pleasure of working together on many projects over the years. Walter Lovenberg and Peter McCann originally met Albert Sjoerdsma at the NIH, and Joseph Temple and Michael Palfreyman are later additions to his long list of colleagues and collaborators. This extended scientific interaction has been marked by two notable distinctions: first, a focused interest in enzymes and ways that their activity can be manipulated for therapeutic gain and, second, an outstanding group of co-workers whose research this book celebrates.

The chapters of this book cover three general areas: The first few chapters deal with examples of enzymes whose activities have recently been elucidated, or for which newer insights have been gleaned, but which do not yet have selective or potent inhibitors. The second part describes examples of enzymes where inhibitors have been identified but it is still not clear whether or not such an enzymatic blockade will be therapeutically beneficial. The final section describes clinical studies of newer, and not so new, enzyme inhibitors that are clearly of therapeutic importance.

The symposium on enzyme inhibitors provided an all too rare opportunity for basic scientists in biochemistry, chemistry, pharmacology, and cell biology to interact with clinicians and to share in the totality of the drug discovery process. We hope that the reader of this volume will gain a similar insight by appreciating the many facets that make up the mosaic of new knowledge needed to design molecules such as enzyme inhibitors which may have therapeutic applications.

THE EDITORS

ACKNOWLEDGMENTS

Of course neither the symposium nor this book could have happened without the dedication of many helpers. The editors extend their gratitude to these individuals and add a special thanks to Bj Hunt, Steven A. Biedenbach, Donna Gray-Williams, Kathy Lynn Rogers, and Elaine Semancik for making so much of this possible. We also thank the authors of the chapters, all renowned experts in their specialities, for the timely preparation of their manuscripts and Anne Watman and Beth Smith, who produced the final version under the direction of James M. Cretsos.

M.G.P., P.P.M., W.L., J.G.T., Jr., A.S.

1

ENZYME INHIBITORS AS DRUGS

J. Richard Crout

Boehringer Mannheim Pharmaceuticals Corporation
Rockville, MD 20850

This book is dedicated to a simple but powerful idea -- namely, that searching for enzyme inhibitors is a rational and efficient approach to drug discovery. That is not to say that this approach is necessarily easy.

In considering the topic of enzyme inhibitors as drugs, we should remind ourselves that it is one thing for a compound to inhibit an enzyme and quite another for that compound to become a marketed pharmaceutical product. For an enzyme inhibitor to become a practical drug, several criteria must be met:

1. The biochemical pathway that is inhibited must be related to a disease state in such a way that inhibition of that pathway in a patient is therapeutic.
2. The enzyme inhibitor must be specific so that unwanted inhibition of other pathways or receptors does not occur at therapeutic doses.
3. The compound must have the pharmacokinetic characteristics of a practical drug -- i.e., it must be absorbed, must penetrate to the site of action and must have a reasonably predictable dose response relationship and duration of action.
4. The compound must have an acceptable toxicological profile in animals, and the results of clinical studies in humans must demonstrate an appropriate balance between benefits and risks in therapeutic use.
5. The compound must survive a long and expensive clinical development process and ultimately be approved by regulatory agencies.
6. The compound must be economically viable in the marketplace and compete successfully with other therapeutic alternatives.

Given these obstacles, it is not surprising that relatively few of the large number of enzyme inhibitors known to enzymologists have become marketed drugs. Table I lists the drugs in standard medical sources that are currently marketed in the United States and are described as having enzyme inhibition as their mode of action (Gilman *et al.*, 1985; Physicians' Desk Reference, 1989).

TABLE I. Marketed Drugs That Are Enzyme Inhibitors

Drug	Enzyme Inhibited
Autonomic	
Physostigmine	Choline esterase
Neostigmine	
Organophosphates	
MAO Inhibitors	Monoamine oxidase
Metyrosine	Tyrosine hydroxylase
Cardiovascular	
Methyldopa	Aromatic L-amino acid decarboxylase
Captopril	Angiotensin converting enzyme (ACE)
Enalapril	
Lovastatin	Hydroxy-methylglutaryl (HMG)-CoA reductase
Antibacterials/Antivirals	
Sulfonamides	Dihydropteroic acid synthesis
Aminosalicylic acid	
Trimethoprim	Dihydrofolate reductase
Penicillins	Membrane transpeptidase
Cephalosporins	
Rifampin	DNA-dependent RNA polymerase
Acyclovir	DNA polymerase
Zidovudine (AZT)	Reverse transcriptase
Antineoplastics	
Methotrexate	Dihydrofolate reductase
Antimetabolites	
Fluorouacil (5-FU)	Thymidylate synthetase
Cytarabine (ARaC)	DNA synthesis
Mercaptopurine (6-MP)	Purine synthesis; nucleotide interconversions
Thioguanine	
Hydroxyurea	Ribonucleoside diphosphate reductase

(continued)

TABLE I. (continued)

Drug	Enzyme Inhibited
Endocrine	
Propylthiouracil	Thyroid peroxidase
Metyrapone	Adrenal 11β-hydroxylase
Aminoglutethimide	Adrenal 20α-hydroxylase
Miscellaneous	
Carbidopa	Aromatic *L*-amino acid decarboxylase
Disulfiram	Aldehyde dehydrogenase
Aspirin	Fatty acid cyclooxygenase
NSAID's	
Allopurinol	Xanthine oxidase

The autonomic agents include the choline esterase inhibitors, the monoamine oxidase inhibitors and metyrosine. The latter compound is used to inhibit norepinephrine production in patients with pheochromocytoma. Its availability as a therapeutic agent is a direct result of research done by Dr. Albert Sjoerdsma and his colleagues at the National Institutes of Health.

In the cardiovascular area two recently developed classes of enzyme inhibitors are proving to be true breakthroughs from both a medical and a marketing point of view. These are the ACE inhibitors used for the treatment of hypertension and congestive heart failure and the HMG CoA reductase inhibitors used for the management of lipid disorders. Methyldopa is also included on the list even though its mode of action is currently viewed as related to the formation of the false transmitter α-methyl norepinephrine. From the drug discovery point of view, it was originally evaluated as an inhibitor of norepinephrine synthesis in adrenergic neurons. This important antihypertensive drug is also the result of clinical investigation by Dr. Albert Sjoerdsma and his colleagues.

Enzyme inhibition is an important mechanism of action for a number of antibacterial agents. Sulfonamides and trimethoprim inhibit dihydrofolate synthesis at different steps, which is the rationale for their use together in a widely used combination product. Among the antibiotics the penicillins act by inhibition of bacterial cell wall synthesis. Rifampin, an important Merrell Dow drug used in the treatment of tuberculosis, acts by inhibiting DNA-dependent RNA polymerase activity in the microbial cell but not in mammalian tissues. The important antiviral drugs acyclovir and zidovudine (AZT) are also enzyme inhibitors. A number of antibiotics, e.g., the aminoglycosides, the tetracyclines and chloramphenicol, act by binding specific ribosomal subunits and thereby interfere with protein synthesis. These drugs have not been included on the list although arguably this mechanism might be called enzyme inhibition in a general sense.

The field of anticancer drugs presents a rich array of compounds that act on biochemical pathways. Those shown in Table I are those in which there is reasonable agreement in the literature on the enzymatic site of action. The purine and pyrimidine analogs are pro-drugs in the sense that they require phosphorylation before exerting their action as antimetabolites. Other antineoplastic drugs inhibit cellular proliferation by mechanisms that include the crosslinking of DNA, the inhibition of RNA synthesis, alkylation of amino acids in proteins, and/or binding to microtubules. These reactions affect a host of enzymatic pathways; on these grounds some might include all antineoplastic drugs in this table.

With respect to endocrine drugs, the antithyroid compound propylthiouracil has been shown to be an inhibitor of thyroid peroxidase. Two drugs are known to inhibit cortisol synthesis. Metyrapone inhibits the last step in this pathway and is used as a diagnostic test in patients with pituitary disease. Aminoglutethimide inhibits the first step in cortisol synthesis and is used to suppress hypercortisolism in patients with adrenal tumors.

The miscellaneous group of enzyme inhibitors includes several interesting drugs. Carbidopa is a dopa decarboxylase inhibitor that does not cross the blood brain barrier; it thereby potentiates the therapeutic effect of levodopa in patients with parkinsonism. Also in this group are disulfiram, a drug used to treat chronic alcoholism; allopurinol, an important uricosuric agent; and aspirin, the classic old drug that continues to find new uses and to reveal new mechanisms of action.

Table I is a reasonably comprehensive list representing current textbook knowledge. It is inevitable, however, that this list will grow in the future. Enzyme inhibition will undoubtedly be discovered as the mechanism of action of a number of drugs whose mechanisms are poorly understood today. More importantly, however, the list will grow because of contemporary scientific interest in enzyme inhibition as a route to drug discovery. The many important enzyme inhibitors currently in the research pipelines of pharmaceutical companies will become the therapeutic agents of tomorrow. This is the future that we celebrate at this Symposium.

REFERENCES

Gilman, A.G., Goodman, L.S., Rall, T.W., and Murad, F., eds. (1985). "Goodman and Gilman's The Pharmacological Basis of Therapeutics," 7th ed., New York, Macmillan.

"Physicians' Desk Reference," 43rd ed. (1989). Oradell, NJ, Medical Economics.

2

BIOGENESIS OF PHOSPHATIDYLINOSITOL GLYCOSYL MEMBRANE PROTEINS AND THEIR SIGNIFICANCE

Sidney Udenfriend

Roche Institute of Molecular Biology
Roche Research Center
Nutley, NJ 07110

INTRODUCTION

Membrane proteins are generally anchored through hydrophobic transmembrane spanning peptide sequences. Some proteins are known to be anchored by fatty acid residues that are covalently linked to the protein's carboxyl terminus. Another class of membrane proteins, discovered fairly recently, is anchored through the carboxyl terminus by a phosphatidylinositol glycosyl (PI-G) group (Figure 1). Alkaline phosphatase was the first membrane protein that was recognized as being PI-G tailed (Low and Finean, 1977; Ikezawa *et al.*, 1976). Over thirty PI-G tailed proteins are now known (Ferguson and Williams, 1988). However, only in the case of the variable surface glycoprotein

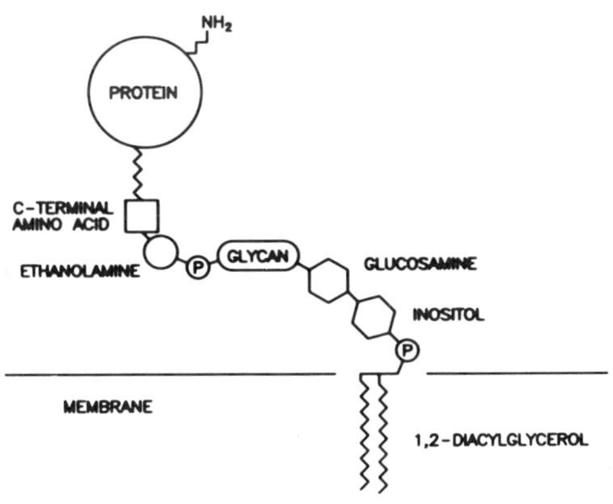

Fig. 1. *Generalized structure of PI-G tailed proteins (Reprinted by permission from Low et al., 1986).*

(VSG) of *Trypanosoma brucei* is the exact composition of the glycan known (Ferguson *et al.*, 1988). From these studies on trypanosome VSG and mammalian Thy-1 (Fatemi *et al.*, 1987; Fatemi and Tartakoff, 1986) it became apparent that the PI-G moiety is covalently attached to the C-terminus of the mature protein. However, this turned out not to be the C-terminus of the cDNA deduced protein, indicating that in each case a C-terminal peptide of 20-30 residues had been cleaved during the processing leading to PI-G tailing. One variant of VSG had its PI-G tail attached to an Asp and another to a Ser. Thy-1 was found to be linked to its PI-G tail through a C-terminal Cys.

We have utilized human placental alkaline phosphatase (PLAP) to investigate the biosynthesis of PI-G tailed proteins. To determine the attachment site of the PI-G moiety of human PLAP we treated the enzyme with chymotrypsin and utilized a site directed antibody directed to a peptide sequence at the C-terminus of mature PLAP (Bailey *et al.*, 1988) to isolate the PI-G tailed peptide. Upon sequencing, the chymotryptic peptide was found to end in an Asp, which represented Asp^{484} of cDNA deduced PLAP (Micanovic *et al.*, 1988), indicating a removal of a twenty-nine residue peptide during PI-G tailing. The C-terminal chymotryptic peptide cleaved from PLAP was also found to contain ethanolamine, glucosamine and neutral sugars, consistent with the known components of the PI-G moiety in VSG and Thy-1.

Several PI-G tailed proteins have now been examined and there is no consensus either in the nature of the amino acid residue to which the PI-G is attached nor the cDNA deduced sequences prior to or following the site of PI-G attachment (Figure 2). To investigate the requirements for PI-G tailing we produced a number of mutant forms of PLAP by transfecting COS cells with the corresponding cDNAs (Figure 3) (Berger *et al.*, 1988). In the case of PLAP, the C-terminal signal peptide could be shortened by six residues (from PLAP-513 to PLAP-507) without interfering with PI-G tailing or insertion of the product into the plasma membrane. However, shorter mutants (PLAP-502, 497, 493 and 489) were not PI-G tailed but instead were secreted into the medium. In fact the secreted forms of PLAP were produced by the cells in much larger amounts than were the membrane-bound forms. In another experiment PLAP-502, which is secreted, was extended by five residues to produce variants of PLAP-507.

PROTEIN	*C-TERMINAL SEQUENCE*
PLAP	-LAPPAGTTDAAHPGRSVVPALLPLLAGTLLLLETATAP
VSG 117A	--WENNACKDSSILVTKKFALTVVSAAFVALLF
VSG 221A	--NTNTTGSSNSFVISKTPLWLAVLLF
THY-1	-----KLVKCGGISLLVQNTSWLLLLLLSLSFLQATDFISL

Fig. 2. *C-terminal amino acid sequences of some nascent forms of proteins for which the site of PI-G tail attachment is known. Amino acid sequences are presented in the one-letter code. VSG is variable surface glycoprotein of Trypanosoma brucei; Thy-1 is the surface glycoprotein isolated from rat thymocytes.*

Wild Type	513--AGTTDAAHPGRSVVPALLPLLAGTLLLLETATAP
	507--AGTTDAAHPGRSVVPALLPLLAGTLLLL
Deletion	502--AGTTDAAHPGRSVVPALLPLLAG
Mutants	497--AGTTDAAHPGRSVVPALL
	493--AGTTDAAHPGRSVV
	489--AGTTDAAHPG
Cassette	502+(S14)-AGTTDAAHPGRSVVPALLPLLAGSIIII
Mutants	503+(SQ4)-AGTTDAAHPGRSVVPALLPLLAGSQQQQ
	518--AGTTDAAHPGRSVVPALLPLLAGYCCCCELLLLETATAP
	513--AGTTKAAHPGRSVVPALLPLLAGTLLLLETATAP
	513--AGTTWAAHPGRSVVPALLPLLAGTLLLLETATAP
Site-directed	513--AGTTSAAHPGRSVVPALLPLLAGTLLLLETATAP
Mutants	513--AGTTYAAHPGRSVVPALLPLLAGTLLLLETATAP
	513--AGTTLAAHPGRSVVPALLPLLAGTLLLLETATAP
	513--AGTTPAAHPGRSVVPALLPLLAGTLLLLETATAP
	VSV G
PLAP 489 G Hybrid	489--AGTTDAAHPG-SKL-[Memb. | Cytoplasm]

Fig. 3. *C-terminal amino acid sequences of nascent forms of PLAP encoded by wild type and mutant cDNAs. Amino acid sequences are presented in the one-letter code and are numbered starting from the first residue of the cDNA not including the NH_2-terminal signal peptide. cDNAs encoding various forms of nascent PLAP were transiently expressed by transfection into COS cells. Seventy-two hours post-transfection cell homogenates and media from 1×10^6 cells were routinely assayed for enzyme activity. When cells were incubated with (^{35}S)methionine production of labeled mutant forms of PLAP was determined by immunoprecipitation from cell homogenates and medium.*

The latter terminates in the sequence Thr-Leu-Leu-Leu-Leu. Replacement, with a different but physically similar sequence Ser-Ile-Ile-Ile-Ile yielded a mutant form of PLAP-507 that was PI-G tailed and membrane anchored. However, replacement of the hydrophobic sequence with the hydrophilic sequence Ser-Gln-Gln-Gln-Gln led to a form of PLAP-507 that was not PI-G tailed but was secreted into the medium. Extension of PLAP-502 with sixteen, mainly hydrophobic residues, yielded a PLAP-518 which was PI-G tailed and membrane bound even though it was longer than the wild-type PLAP-513. These studies show that the signal for PI-G tailing resides in the carboxyl terminus of the nascent protein. Production of hybrid molecules have also provided important information. When the nascent C-terminus of a PI-G tailed protein and the N-terminus of a normally secreted protein were combined the hybrid was PI-G tailed (Stroynowski *et al.*, 1987). When the amino terminus of PLAP was combined with the C-terminus of the G-protein of vesicular stomatis virus the hybrid was attached to the membrane by a membrane spanning peptide sequence, characteristic of the viral protein (Berger *et al.*, 1989).

Thus, we have begun to characterize the C-terminal structural requirements for PI-G tailing of a protein. In the case of PLAP the C-terminus of the nascent form of the wild-type protein is largely hydrophobic. More obvious comparisons can be made between N-terminal signal peptides and C-terminal peptides of nascent proteins destined for PI-G tailing. Both appear transiently in the prepro-protein and are cleaved during processing. They both are of similar size and are predominantly hydrophobic in nature. In addition, there is no single specific amino acid associated with the cleavage site of either type of signal peptide. Although there are smiliarities between the structures of the N-terminal and C-terminal signal peptides, there is one obvious difference that indicates that the mechanisms cannot be the same. The N-terminal signal peptide is cleaved from the nascent protein by a peptidase leaving a new N-terminus on the mature protein. In the case of PI-G tailed proteins removal of the C-terminal signal peptide from the nascent protein yields a mature protein with a substituted ethanolamine linked in amide linkage to the new C-terminus. Cleavage of an internal peptide bond and coupling in amide linkage requires activation of the carboxyl group, most probably through a transamidase or transpeptidase type of enzyme. From pulse chase experiments on Thy-1 in intact cells PI-G tailing occurs rapidly, conceivably cotranslationally or slightly later (Ferguson and Williams, 1988) and most likely occurs in the membrane of the endoplasmic reticulum. The hydrophobic C-terminal signal peptide may loosely anchor the nascent protein in the membrane in such a way as to make the residue destined to become the new C-terminus available for attack by a "transamidating" enzyme. To investigate this possibility requires studies in cell-free systems. Requirements for *in vitro* studies are labelled nascent protein as substrate, a PI-G generating system and the enzymes for PI-G tailing. We have generated nascent PLAP using a rabbit reticulocyte system and PLAP mRNA. Nascent PLAP was identified by its mass, as determined by SDS gel electrophoresis, and by the presence of both the N-terminal and C-terminal signal peptides. The latter were identified by specific site-directed antibodies (Bailey *et al.*, 1988). Since processing most likely takes place in the endoplasmic reticulum we investigated microsomal fractions from several different cell lines for their ability to process nascent PLAP to the PI-G tailed mature form. Evidence for processing would be a change in mass, through removal of the N-terminal and C-terminal signal peptides, and incorporation of at least some of the components of the PI-G moiety into the product. Of several cell lines that were investigated CHO cells were found to yield the most active microsomal fractions. Addition of CHO cell microsomal extracts to a rabbit reticulocyte translating system in the presence of ^{35}S-methionine and an ATP generating system yielded a product that fulfilled some of the criteria of a PI-G tailed protein. The ^{35}S-labeled product was precipitated by a polyclonal antibody to mature PLAP. The immunoprecipitate when subjected to SDS gel electrophoresis, migrated as a single product with the same mass as mature PLAP. The product had lost its C-terminal signal peptide but retained the C-terminus of mature PLAP. The N-terminal signal peptide had also been removed and the new N-terminus of mature PLAP appeared. Thus, fortified microsomal fractions contain, in addition to N-terminal signal peptidase, an enzyme(s) that can correctly cleave nascent PLAP at its C-terminus. When the incubations were carried out in the presence of labeled inositol, ethanolamine, or ^{32}P, only small amounts of radioactivity appeared in the mature PLAP, far less than stoichiometric amounts. Attempts to incorporate labeled ethanolamine into the putative PI-G tailed product have not been successful because the specific activity of commercially available ethanolamine is not sufficiently high to detect if it was incorporated. If a transamidase enzyme were responsible for the C-terminal cleavage then in our crude system, in the absence of the proper amine donor, the activated aspartate residue would instead condense with water. The result would be peptide cleavage to yield non PI-G tailed processed PLAP. The product we find is consistent with this.

During our studies it came to our attention that the PI-G moiety of PI-G tailed proteins was similar in structure to a newly characterized mediator of insulin action. Evidence for the existence of a low molecular weight, polar mediator(s) of insulin action on many intracellular enzymes had been presented by Larner and his associates several years ago (Larner *et al.*, 1979) and by Jarrett (Jarrett and Seals, 1979). Quite recently Saltiel and Cuatrecasas (1986) and Jarrett (Mato *et al.*, 1987) not only confirmed the existence of such a mediator but showed that it could be derived from a form of intracellular PI-G, that was not linked to protein. According to Saltiel and his colleagues active mediator contains no amino acid residues and no ethanolamine but contains similar glycosyl residues, unacetylated glucosamine and inositol phosphate. In other words, the inositol-glycosyl (I-G) mediator is the same as or similar to the I-G in PI-G. While this structural similarity may be a coincidence, in our opinion this does not seem likely. Larner (Romero *et al.*, 1988) has presented evidence that insulin brings about the release of PI-G tailed proteins on the surface of the cell. In fact he has shown that the release of AP from cells by inositol-specific phospholipase C parallels the appearance of insulin mediator. Larner has also shown that the interaction of insulin with its receptor leads to the simultaneous release of mediator, diacylglycerol and AP. In our laboratory we have shown that when insulin (at physiologic levels) is added to WISH cells (ATCC-CCL25) in culture AP is released (unpublished observations). Insulin has also been shown to release other PI-G tailed proteins from cells (Eckel *et al.*, 1978; Chan *et al.*, 1988). One must consider the purpose of the large and increasing number of recognized PI-G tailed proteins. It is conceivable that the elaborate PI-G moiety merely serves as another type of membrane anchor. However, there are good reasons to believe that PI-G tailed proteins may represent a new type of cellular signal transduction mechanism. Each PI-G tailed protein is capable of yielding several active moieties. When cells are treated with inositol-specific phospholipase C or insulin, release of the protein leaves diacylglycerol, a biologically active substance. Recently an inositol-specific phospholipase D has been demonstrated in animal tissus (Davitz *et al.*, 1987; Low and Prasad, 1988). Treatment of cells with phospholipase D liberates the protein and leaves phosphatidic acid in the membrane. The latter is also biologically active. Although many of the proteins that are PI-G tailed possess biological activities, their physiologic significance is generally not known. However, in the case of the PI-G tailed protein, heparin sulfate, the glycoprotein when released through insulin action, is taken up by the cells. The IG moiety, is one of several biologically active substances that can be produced by cleavage of PI-G tailed proteins from cells. All these biologically active products that can be derived from PI-G tailed proteins through enzyme or insulin action make it likely that the PI-G tail is more than an inert membrane anchor.

REFERENCES

Bailey, C.A., Howard, A., Micanovic, R., Berger, J., Heimer, E., Felix, A., Gerber, L., Brink, L., and Udenfriend, S. (1988). *Anal. Biochem.* **170**, 532-541.

Berger, J., Howard, A.D., Brink, L., Gerber, L., Hauber, J., Cullen, B.R., and Udenfriend, S. (1988). *J. Biol. Chem.* **263**, 10016-10021.

Berger, J., Micanovic, R., Greenspan, R.J., and Udenfriend, S. (1989). *Proc. Natl. Acad. Sci. U.S.A.* **86**, 1457-1460.

Chan, B.L., Lisanti, M.P., Rodriguez-Boulan, E., and Saltiel, A.R. (1988). *Science* **241**, 1670-1672.

Davitz, M.A., Hereld, D., Shak, S., Krakow, J., Englund, P.T., and Nussenzweig, V. (1987). *Science* **238**, 81-84.

Eckel, R.H., Fujimoto, W.Y., and Brunzell, J.D. (1978). *Biochem. Biophys. Res. Commun.* **84**, 1069-1075.

Fatemi, S.H. and Tartakoff, A.M. (1986). *Cell* **46**, 653-657.

Fatemi, S.H., Haas, R., Jentoft, N., Rosenberry, T.L., and Tartakoff, A.M. (1987). *J. Biol. Chem.* **262**, 4728-4732.

Ferguson, M.A.J. and Williams, A.F. (1988). *Annu. Rev. Biochem.* **57**, 285-320.

Ferguson, M.A.J., Homans, S.W., Dwek, R.A., and Rademacher, T.W. (1988). *Science* **239**, 753-759.

Ikezawa, H., Yamanegi, M., Taguchi, R., Myashita, T., and Ohyabu, T. (1976). *Biochim. Biophys. Acta* **450**, 154-164.

Jarett, L. and Seals, J.R. (1979). *Science* **206**, 1407-1408.

Larner, J., Galasko, G., Cheng, K., DePaoli-Roach, A.A., Huang, L., Daggy, P., and Kellogg, J. (1979). *Science* **206**, 1408-1409.

Low, M.G. and Finean, J.B. (1977). *Biochem. J.* **167**, 281-284.

Low, M.G. and Prasad, A.R. (1988). *Proc. Natl. Acad. Sci. U.S.A.* **85**, 980-984.

Low, M.G., Ferguson, M.A.J., Futerman, A.H., and Silman, I. (1986). *Trends Biochem. Sci.* **11**, 212-215.

Mato, J.M., Kelly, K.L., Abler, A., and Jarett, L. (1987). *J. Biol. Chem.* **262**, 2131-2137.

Micanovic, R., Bailey, C.A., Brink, L., Gerber, L., Pan, Y.C., Hulmes, J.D., and Udenfriend, S. (1988). *Proc. Natl. Acad. Sci. U.S.A.* **85**, 1398-1402.

Romero, G., Luttrell, L., Rogol, A., Zeller, K., Hewlett, E., and Larner, J. (1988). *Science* **240**, 509-511.

Saltiel, A.R. and Cuatrecasas, P. (1986). *Proc. Natl. Acad. Sci. U.S.A.* **83**, 5793-5797.

Stroynowski, I., Soloski, M., Low, M.G., and Hood, L. (1987). *Cell* **50**, 759-768.

3

STRUCTURE AND CATALYTIC FUNCTION OF ADP-RIBOSE POLYMERASE (ADPRT), A TARGET SITE FOR DRUGS INHIBITING MALIGNANT GROWTH

Ernest Kun[1, 2]

Department of Pharmacology and the Cardiovascular Research Institute
University of California
San Francisco, CA 94143

Laboratory for Environmental Toxicology and Chemistry
Romberg Tiburon Centers
San Francisco State University
Tiburon, CA 94920

I. INTRODUCTION

Effectivity of drug therapy of abnormal cell function can be assessed by the selectivity of action of a drug on a particular process and evaluated by the inverse relationship between drug concentration and the magnitude of the biological response to the drug. In other words, a highly effective drug induces a large or biologically significant, presumably curative cellular reaction at a low concentration with minimal side effects. In biophysical terms the phenomena described above contain a further parameter called "control strength," which means that the target or drug site should regulate rate limiting steps in cell function. The more rate determining the regulatory site, the greater is its control strength. This situation is analogous to the critical role of a light switch in regulating the power supply of a large building. Linear multistep processes are most susceptible to direct regulation, and many effective drugs (e.g., in the CNS, steroid synthesis, etc.) act in the above outlined manner as exemplified by papers presented in this Symposium.

[1]*Supported by a Research Grant of the United States Air Force of Scientific Research (AFOSR 890231).*

[2]*Research Career Award of USPH.*

Cellular regulation at the genetic level, comprising the highly complex domain of gene control in eukaryotes (Kun, 1988) belongs to a far more elusive territory, although in principle drug actions should follow the same rules set forth for other systems. The main experimental and theoretical difficulty is the complexity of macromolecular interactions, which are critical in gene regulations involving macromolecules such as specific DNA structures and molecules that bind to it. The present report deals with recent advances that identify molecular and catalytic activities of ADPRT. The self-regulatory and DNA topology modifying functions of ADPRT provide further evidence consistent with the participation of ADPRT in gene regulation. The exploration of ADPRT as a target site for drugs is based on enzymological principles and its application in cancer research is the subject of ongoing investigation.

II. RESULTS

A. The Primary Structure of ADPRT Binding Sites and Macromolecular Self-Association.

The biochemistry and biology of ADPRT is the subject of intensive research in many laboratories (for review see Althaus and Richter, 1987). What will be summarized here represents recent progress made in the author's laboratory and no attempt is made to review this field, except selected references will be given to closely related research.

ADPRT has been purified in several laboratories (see Althaus and Ritcher, 1987) and it is generally agreed that it is a nuclear protein with $M\gamma$=116-120 kDa. Purification has been laborious and for this reason we developed a method which requires only 2-3 working days to obtain the homogenous protein from calf thymus with high yield (Buki *et al.*, 1987). This technical advantage permitted the determination of the primary polypeptide structure of ADPRT (Buki and Kun, 1988), a work which was aided by the establishment of the cDNA sequence of ADPRT (Suzuki *et al.*, 1987; Kurosaki *et al.*, 1987; Cherney *et al.*, 1987; Alkhatib *et al.*, 1987).

The primary polypeptide structure, as reconstructed from plasmic peptides is illustrated in Figure 1. Two basic polypeptides (29 kDa and 36 kDa, a; b; see Buki and Kun, 1988) represent the DNA-binding (36 kDa) and histone-binding (29 kDa; 9) domains of the enzyme, respectively, which are cooperative, as determined by a nitrocellulose filter binding assay (Sastry and Kun, 1988). The histone binding terminal polypeptide assumes further significance by the fact that the enzyme-DNA complex, produced by the association of a synthetic octadeoxyribonucleotide duplex derived from an SV40 DNA sequence (Hakam *et al.*, 1987) is greatly dependent on certain histones as second cofactors for maximal polymerase enzymatic activity (Hakam *et al.*, 1989), without serving as ADP-ribose acceptors under initial velocity conditions. The histone binding site of ADPRT is of interest also in the mechanism of action of polyamines on nuclear processes, since polyamines appear to be able to compete with histones or replace them, a subject currently under investigation.

The 56 kDa polypeptide contains the NAD-binding domain, which appears to be specifically located in the 14-mer oligopeptide lost during plasmic digestion (see Buki and Kun, 1988). The darkened bars (Figure 1) indicate sequenced regions. The linear structure depicted in Figure 1 is to some extent misleading, because the ADPRT molecule exhibits 85% helicity (unpublished results), thus vicinity of various binding sites has to be assessed by a three dimensional structural model. Affinity labelling indicates that DNA- and NAD-binding sites are 8.5 A apart (McLic *et al.*, 1987).

A further anomaly of ADPRT is its propensity to form self-associated complexes (Bauer and Kun, 1989). Three independent techniques (molecular filtration, Ferguson analysis, and cross linkage) revealed that ADPRT exhibits a concentration-dependent self-association, and in all probability the enzymatically active species is the dimer, whereas polymers of ADPRT are inactive as enzymes. That the rate of the enzymatic reaction does not show a bimolecular kinetics is most probably explained by the relatively crude enzymatic assay customarily employed, consisting of determination of the "acid precipitable radioactive material" formed upon incubation of labelled NAD with the enzyme + DNA. This assay does not discriminate between initiator and elongation processes, thus is incapable of detecting the participation of ADPRT dimers, each enzyme molecule performing initially discrete reactions, as an initiator catalyst, ADPR acceptor and chain elongator.

B. Catalytic Activities of ADPRT.

Some degree of comprehension of the catalytic reactions performed by ADPRT depended on the recognition that this enzyme possesses two distinct, yet connected catalytic activities (see Sastry and Kun, 1988). These processes are schematically shown in Figure 2. The "metabolic" catalysis (reaction 1) consists of the DNA dependent polymerization of ADPR that is derived from the hydrolysis of NAD, when the ADPRT molecule serves also as the acceptor for covalently bound (ADPR)n polymer. This "metabolic" catalysis was until recently (Sastry and Kun, 1988) the only

Fig. 1. *Primary structure of ADPRT.*

(1) METABOLIC:

$$nNAD + ADPRT \longrightarrow ADPRT(ADPR)_n + nNA + nH^+$$

(2) DNA–TOPOLOGY:

$$ADPRT + DNA(REL.) \longrightarrow ADPRT - DNA(COND.)$$

A. (1) INHIBITS (2)

B. INHIBITORS OF (1) INDUCE (2)

Fig. 2. *Two self-regulatory activities of ADPRT.*

process known where ADPRT serves as an enzymatic catalyst, and all efforts to explain the biological function of ADPRT were necessarily confined to this metabolic process (see Althaus and Richter, 1987). Binding of ADPRT to certain closed circular relaxed DNAs (Sastry and Kun, 1988) produces condensation of DNA, as shown in reaction 2 of Figure 2. The binding of ADPRT to DNAs which have no termini produces enzymatically inactive complexes. However, if ADPRT binds to certain rare DNA structures, e.g., containing A-T rich regions and apparently containing DNA bends (Sastry *et al.*, 1989), the relaxed DNA condenses. DNA condensation is of considerable biological importance, and is critical in gene-regulation (Kun, 1988). The fascinating aspect of the two reactions shown in Figure 2 is that the "metabolic" reaction self-inhibits the DNA-topology regulating effect of ADPRT, and inhibiting the metabolic reaction with ADPRT <u>inhibitors</u> (drugs) re-converts ADPRT to perform reaction 2, i.e. altering DNA topology.

This binary self-regulation of dual activities appears unique for ADPRT and most probably represents the basis for the cellular control role of ADPRT, which is expressed in development and the regulation of the malignant phenotype (Tseng *et al.*, 1987).

Whereas the DNA topology regulatory effect of ADPRT depends on relatively rare and in its details unknown DNA structures (Sastry *et al.*, 1989) the coenzymic function of DNA on the polymerization of ADPR requires only non-specific DNA termini, as demonstrated by exonuclease protection experiments (Sastry *et al.*, 1989). It follows that the DNA binding 36 kDa polypeptide (Buki and Kun, 1988) must contain two types of DNA binding sites which have not been directly identified at present. The metabolic catalysis (Figure 2; <u>1</u>) consists of two types of processes: I. auto-ADP-ribosylation of the ADPRT molecule itself and II. the transfer of the ADPR groups to other nuclear protein acceptors. In the presence of μM concentrations of NAD at pH 8.0 ADPRT catalyzes at maximal rates poly-ADP-ribosylation by the mechanism schematically shown in I of Figure 3. The E-DNA complex is a submaximally functioning catalyst which becomes maximally active upon binding of histones.

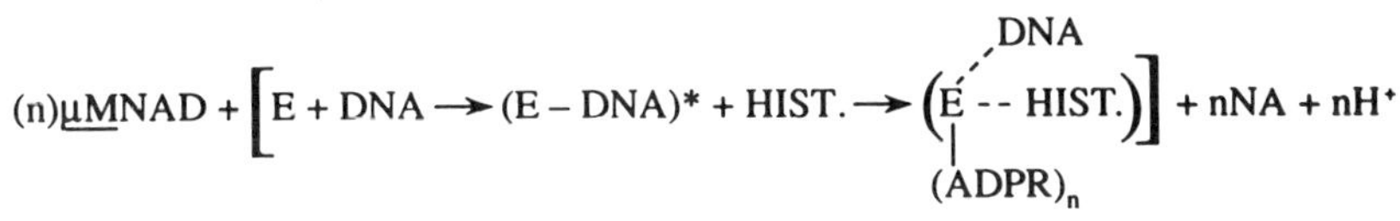

Fig. 3. (I). *Auto-ADP-ribosylation.*

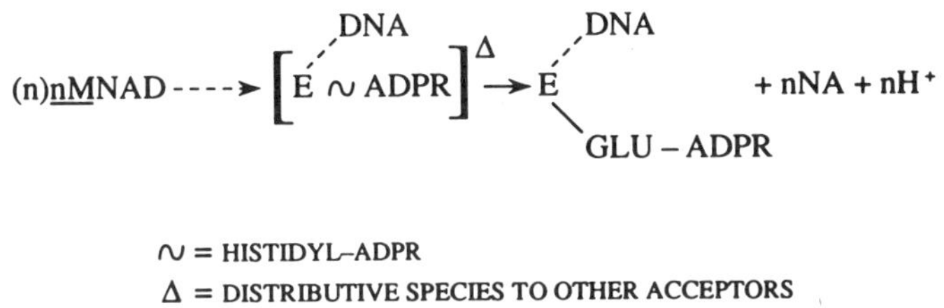

Fig. 3. (II). *Trans-ADP-ribosylation.*

However, the stable enzyme-glutamate-carboxy-esters of ADP-ribose (seeAlthaus and Richter, 1987) are not formed in one step, but require the generation of an enzyme-histidyl-ADP-ribose intermediate, which is unstable above pH 6.5 (Bauer *et al.*, 1986) and is the first or initiator adduct, formed on specific histidine residues of the ADPRP molecule itself (Bauer *et al.*, 1989). At pH above 6.5 this unstable intermediate (shown in brackets in II of Figure 3) migrates to glutamate carboxyl end groups, forming the established stable esters adducts (see Althaus and Richter, 1987).

When other acceptor proteins are associated with the ADPRT molecule (e.g., histones, topoisomerase I and II, etc.) the histidyl adduct can trans-ADP-ribosylate to other proteins.

It is clear from the above summary that new developments in the enzymatic and macromolecular properties of ADPRT and of DNA structures are likely to result in a more understandable role of ADPRT in cellular processes.

III. SUMMARY AND PROSPECTUS

Utilization of ADPRT sites as possible targets for drug action will depend on more in-depth molecular pharmacological research. Presently it seems predictable that the most outstanding pharmacological effects of ADPRT inhibitors which are not nicotinamide homologs (Hakam *et al.*, 1987; Tseng *et al.*, 1987) are in the realm of inhibition of malignant cell proliferation. Since drugs developed to perform these inhibitory effects on malignant proliferation are non-toxic (Tseng *et al.*, 1987), it seems feasible to pursue their synthesis and analysis of their modes of action at the molecular level. A possible synergism with other non-toxic drugs (e.g., interfering with polyamine pathways, see McCann *et al.*, 1987) seems another avenue of probable significance.

This paper is dedicated to Dr. Albert Sjoerdsma, a friend and colleague of long standing.

REFERENCES

Alkhatib, H.M., Chen, D., Cherney, B., Bhatia, K., Notario,V., Giri, C., Stein, G., Slattery, E., Roeder, R.G., and Smulson, M.E. (1987). *Proc. Natl. Acad. Sci. U.S.A.* **84**, 1224-1228.

Althaus, F.R., and Richter, C. (1987). "ADP-ribosylation of proteins: enzymology and biological significance" (Mol. Biol. Biochem. Biophys., Vol. 370), Berlin, Springer-Verlag.

Bauer, P.I. and Kun, E. (1989). *J. Cell Biol.* **107**, 829a.

Bauer, P.I., Hakam, A., and Kun, E. (1986). *FEBS Lett.* **195**, 331-338.

Bauer, P.I., Buki, K.G., and Kun, E. (1989). In preparation.

Buki, K.G. and Kun, E. (1988). *Biochemistry* **27**, 5990-5995.

Buki, K.G., Kirsten, E., and Kun, E. (1987). *Anal. Biochem.* **167**, 160-166.

Cherney, B.W., McBride, O.W., Chen, D., Alkhatib, H., Bhatia, K., Hensley, P., and Smulson, M.E. (1987). *Proc. Natl. Acad. Sci. U.S.A.* **84**, 8370-8374.

Hakam, A., McLick, J., Buki, K.G., and Kun, E. (1987). *FEBS Lett.* **212**, 73-78.

Hakam, A., McLick, J., Bauer, P.I., Niakan, F., and Kun, E. (1989). *J. Cell Biol.* **107**, 832a.

Kun, E. (1988). *Adv. Exp. Med. Biol.* **231**, 613-626.

Kurosaki, T., Ushiro, H., Mitsuuchi, Y., Suzuki, S., Matsuda, M., Matsuda, Y., Katunuma, N., Kangawa, K., Matsuo, H., Hirose, T., Inayama, S., and Shizuta, Y. (1987). *J. Biol. Chem.* **262**, 15990-15997.

McCann, P.P., Pegg, A.E., and Sjoerdsma, A. (eds.) (1987). "Inhibition of polyamine metabolism" Orlando, FL, Academic Press.

McLick, J., Bauer, P.I., Hakam, A., and Kun, E. (1987). *Biochemistry* **26**, 2226-2231.

Sastry, S.S. and Kun, E. (1988). *J. Biol. Chem.* **263**, 1505-1512.

Sastry, S.S., Buki, K.G., and Kun, E. (1989). *Biochemistry.* In press.

Suzuki, H., Uchida, K., Shima, H., Sato, T., Okamoto, T., Kimura, T., and Miwa, M. (1987). *Biochem. Biophys. Res. Commun.* **146**, 403-409.

Tseng, A., Jr., Lee, W.M.F., Kirsten, E., Hakam, A., McLick, J., Buki, K.G., and Kun, E. (1987). *Proc. Natl. Acad. Sci. U.S.A.* **84**, 1107-1111.

4

MODULATION OF THE DOPAMINERGIC SYSTEM IN CARDIOVASCULAR THERAPEUTICS

Michael B. Murphy
Alan S. Bass
Leon I. Goldberg[1]

Committee on Clinical Pharmacology
Departments of Medicine and Pharmacological and Physiological Sciences
The University of Chicago
Chicago, IL 60637

I. INTRODUCTION

This chapter focuses on the therapy of cardiovascular disease using agents that modulate biochemical processes linked to dopamine receptors. An outline is given of the role of dopamine, itself, as well as newer agonists and antagonists active at specific dopamine receptors. Finally, we describe the use of pro-drugs that rely on enzyme conversion to forms that are active at dopamine receptors and enzyme modulators which alter the concentration of dopamine at its active sites.

II. HISTORICAL PERSPECTIVE

Dopamine was synthesized in 1910 apparently simultaneously by Barger and Ewins at the Wellcome Laboratories in London and in Germany by Mannich and Jacobsohn (Goldberg, 1972). The London group felt that it was very similar to norepinephrine and epinephrine in its pharmacological effects because of the pressor response induced in several animal models (Goldberg, 1972). That it was certainly different from these sympathetic amines was first suggested by Holtz who demonstrated its hypotensive capacity in the guinea pig model (Holtz and Credner, 1942); however, this report went largely unnoticed for many years.

In the late 1940's an intense search was underway for sympathomimetic amines which might increase cardiac contractility without enhancing systemic vascular resis-

[1]*Died May 8, 1989.*

tance, for use in the treatment of heart failure. To quote Robert Walton in 1949, "Sympathomimetic compounds have been generally recognized as producing intense stimulation of cardiac contractile force. The effects, however, have ordinarily been considered too brief and too much associated with pressor effects to warrant consideration for use in conditions of heart failure. Modification of these drugs in the direction of that which has been shown for Arandhol would indicate better possibilities for such application in the more acute types of heart failure" (Walton and Brodie, 1949).

Shortly after his arrival at the National Institutes of Health in the laboratory of Dr. Albert Sjoerdsma, Dr. Leon Goldberg's interests were directed towards evaluating the pharmacology of dopamine. Within a short time it was clear that the hemodynamic effects of dopamine were quite different from those of norepinephrine in that its cardiovascular effects were exquisitely dose dependent. In the anesthetized dog model he demonstrated that low infusion rates were associated with a reduction in diastolic blood pressure accompanied by an increase in cardiac contractile force while higher doses led to an increase in peripheral vascular resistance and systemic blood pressure (Goldberg and Sjoerdsma, 1959). Shortly thereafter Horwitz, Fox and Goldberg (1962) administered dopamine to normal human volunteers and demonstrated hemodynamic changes identical to those seen in the dog. This led to the first administration of dopamine in a patient with heart failure. Serendipitously, this patient was studied in the metabolic ward where it was quickly realized that urine volume and sodium excretion increased significantly during the infusion of dopamine (Goldberg, 1977). Although initially attributed to improved cardiac output, a further study in normal volunteers demonstrated marked increases in renal blood flow and sodium excretion without change in arterial pressure (McDonald *et al.*, 1964). This suggested that dopamine might have specific effects on renal function, a suspicion that was rapidly confirmed in the anesthetized dog model where selective dilation of the renal vascular bed became the gold standard for identifying dopamine receptor agonists during the next 20 years. The fact that the renal vasodilator effect could be attenuated selectively by butyrophenones and phenothiazines, as they inhibit the effects of dopamine in the caudate nucleus, suggested that the renal effects were likely due to specific dopamine receptors (Goldberg and Yeh, 1971). During the ensuing years the complexity of the receptor effects of dopamine has been gradually unraveled.

III. RECEPTORS ACTED ON BY DOPAMINE

Dopamine acts on two dopamine receptor subtypes, DA_1 and DA_2 (Goldberg *et al.*, 1977; Goldberg and Rajfer, 1985). DA_1 receptors are located postsynaptically and mediate vasodilation in renal, mesenteric, coronary, and cerebral arterial blood vessels. DA_1-induced vasodilation is also seen in other vascular beds, but the responses differ among species and are usually less prominent. For example, it appears that greater numbers of DA_1 receptors are found in the human skeletal muscle vascular bed than in that of the dog (Goldberg *et al.*, 1977; Hughes *et al.*, 1987).

DA_2 receptors are located on postsynaptic sympathetic nerves and autonomic ganglia. These receptors inhibit release of norepinephrine from storage sites in sympathetic nerves. They also inhibit prolactin release from the anterior lobe of the pituitary gland and, in the emetic center of the medulla they induce nausea and vomiting. While these two receptor subtypes exhibit diverse functions, they also differ in the chemical structure activity relationships of their selective agonists and antagonists (Kohli and Goldberg, 1989). Currently fenoldopam and SCH 23390 are the most selective and specific DA_1 agonist and antagonist, respectively, while quinpirole and domperidone are the most selective DA_2 agonist and antagonist available.

Dopamine also acts on β- and α-adrenoceptors. It increases cardiac contractile force, heart rate, and atrioventricular conduction by directly stimulating β_1-adrenoceptors. Dopamine also produces these effects indirectly by releasing norepinephrine from myocardial sympathetic nerves. Although patients with severe congestive heart failure often exhibit reduced norepinephrine stores in the myocardium, dopamine still stimulates the myocardium by direct β_1 action (Carroll *et al.*, 1986). It is also a weak agonist at β_2-adrenoceptors (Goldberg *et al.*, 1977; Goldberg and Rajfer, 1985).

In addition to specific dopamine and β-adrenoceptors, dopamine also acts on postsynaptic α_1- and α_2-adrenoceptors (Goldberg and Rajfer, 1985), both of which induce vasoconstriction in arterial and venous vascular beds. In general, skeletal muscle, mesenteric and renal vessels are constricted more than those of cerebral and coronary tissues. Stimulation of presynaptic α_2-adrenoceptors by dopamine subserves similar functions as those of DA_2 receptors in that it inhibits release of norepinephrine from sympathetic nerve storage sites.

In addition to the cardiovascular and neuronal effects described above, dopamine also increases sodium, and to a lesser extent, potassium excretion (Meyer *et al.*, 1967). Although the natriuresis may be due entirely to increased renal blood flow (Frederickson *et al.*, 1985), there is growing evidence that it may result from activation of specific tubular dopamine receptors (Felder *et al.*, 1984; Bello-Reuss *et al.*, 1982).

IV. CLINICAL USE OF DOPAMINE

Dopamine produces different hemodynamic and renal effects in three infusion ranges which can be only approximately termed, low, medium, and high (Levinson *et al.*, 1985). Variations in the doses required to activate the different receptors make it difficult to predict the infusion rate required in individual patients. With the exception of hypotensive emergencies, where dopamine probably is not the ideal sole agent and where perfusion pressure must be rapidly increased, infusions should be begun at a low rate, approximately 0.5-1 μg/kg/min, and the dose should be titrated until the desired effect is achieved. In this setting the infusion rate can be doubled approximately every 5-10 minutes until the perfusion pressure is satisfactory.

A. Low Infusion Rates (0.1 to 2-4 μg/kg/min)

1. Activating DA_1 and DA_2 receptors

DA_2 receptors are activated at the lowest dose of dopamine, and as the dose is increased DA_1 receptors are recruited and hemodynamic changes may be observed (Meyer *et al.*, 1967; Beregovich *et al.*, 1974; D'Orio *et al.*, 1986). Initially, blood pressure may decrease slightly, but there is no change or a small decrease in heart rate.

These effects result from DA_1 mediated vasodilation and DA_2 mediated inhibition of the sympathetic nervous system. Diuresis and natriuresis usually are seen in this dose range, and renal blood flow may increase.

Low dose dopamine is also used to enhance renal perfusion and diuresis in diuretic-resistant patients (Henderson *et al.*, 1980; Lindner, 1983; Hilberman *et al.*, 1984; Parker *et al.*, 1981; Mann *et al.*, 1986). Additionally, it is employed during procedures which place renal blood flow at risk, including hepatic and renal transplant surgery (Polson *et al.*, 1987). There is some, albeit controversial, evidence that renal function may be preserved by the prophylactic use of dopamine in some patients (Paul *et al.*, 1986).

B. Medium Infusion Rates (2-4 to 6 μg/kg/min)

1. Recruitment of β_1-adrenoceptors

Medium infusion rates are used predominantly in the treatment of heart failure (Goldberg *et al.*, 1977; Meyer *et al.*, 1967; Beregovich *et al.*, 1974). Cardiac output usually increases, renal blood flow is enhanced, and systemic vascular resistance is decreased. Heart rate may not change, may increase or decrease, according to the predominance of DA_2 and β_1-adrenoceptor effects in the individual patient. Pulmonary wedge pressure usually remains unchanged. It is important that the infusion rate be maintained at a level which does not cause vasoconstriction and, consequently, increase afterload.

Since infused dopamine may increase venous tone (Butterworth *et al.*, 1987), as it does not dilate the capacitance vessels, dopamine is often used in combination with nitroprusside or nitroglycerin which dilate the venous beds (Miller *et al.*, 1977; Keung *et al.*, 1980; Loeb *et al.*, 1983). It also may be used in combination with dobutamine, which does not act on DA_1 receptors but which permits greater increments in cardiac output than dopamine due to its effect on β-adrenoceptors (Richard *et al.*, 1983; Pedersen and Mortensen, 1987). To achieve the desired hemodynamic and renal actions, the infusion rates of these two amines must be carefully titrated.

C. High Infusion Rates (greater than 6 μg/kg/min)

1. Recruitment of α-adrenoceptors

At high infusion rates blood pressure and systemic vascular resistance increase with the recruitment of α_1- and α_2-adrenoceptors. In normal subjects blood pressure may increase at infusion rates ranging from 2 to 10 μg/kg/min (Goldberg, 1977); on the other hand, some patients in shock require more than 50 μg/kg/min (Sprung *et al.*, 1984). It is important to correct plasma volume deficits prior to starting dopamine since cardiac output and renal blood flow may not increase because of poor venous return, and excessive vasoconstriction with peripheral necrosis may occur. In many patients in whom adequate perfusion pressure cannot be achieved with dopamine, the addition or substitution of norepinephrine may be helpful (Goldberg *et al.*, 1969; Schaer *et al.*, 1985).

D. Novel Dopamine Agonists for Congestive Heart Failure

Propylbutyl dopamine (PBDA) exhibits both DA_1 and DA_2 activity but has little β_1-adrenoceptor activity (Kohli *et al.*, 1980). Its α-adrenoceptor agonist effects are also negligible. Consequently, while DA_1 and DA_2 receptors are activated, there is little stimulation of the heart or vasoconstriction. Administration of PBDA to 11 patients with congestive heart failure at rates of 5, 10, and 20 µg/kg/min led to dose-dependent reductions in arterial pressure (Fennell *et al.*, 1983). Left ventricular filling pressure and pulmonary and systemic vascular resistance were reduced, while cardiac index increased without a change in stroke work or heart rate. On the other hand, PBDA caused some nausea and vomiting, although these were not recorded at infusion rates of 20 µg/kg/min, the likely effective dose in the treatment of heart failure.

Dopexamine activates DA_1 and DA_2 receptors, but differs from dopamine in that it is a potent agonist at β_2-adrenoceptors with minimal α- or β_1-adrenoceptor activity (Brown *et al.*, 1985a; Brown *et al.*, 1985b). It also blocks the uptake of norepinephrine into sympathetic nerve terminals (Bass *et al.*, 1987). In patients with congestive heart failure dopexamine increases cardiac index, stroke volume index, and heart rate while reducing systolic and pulmonary vascular resistance at infusion rates of 1 to 6 µg/kg/min (Dawson *et al.*, 1985; Svensson *et al.*, 1986; Bayliss *et al.*, 1987; Tan *et al.*, 1987). It is unclear whether these effects of dopexamine are the results of its vasodilator effects or whether a direct positive inotropic effect occurs. On the one hand, it is possible that dopexamine activates myocardial β_2-adrenoceptors to increase cardiac contractility and heart rate; however, this is controversial since cardiac contractility traditionally has been assumed to result from β_1- and not β_2-adrenoceptor activation (Bayliss *et al.*, 1987). On the other hand, blockade of norepinephrine uptake might produce a greater increase in contractile force and heart rate in patients with heart failure than in normal subjects (Bass *et al.*, 1987). The latter effect of dopexamine may result in higher concentrations of norepinephrine at the synapse, particularly since patients with congestive heart failure exhibit higher than usual levels of circulating norepinephrine. Finally, whether fenoldopam, a selective DA_1 receptor agonist which has been reported to increase cardiac output and decrease systemic resistance, is of value in the treatment of heart failure following oral administration, remains uncertain (Young *et al.*, 1985).

E. Selective Dopamine Agonists in the Treatment of Hypertension

The hypotensive effects of DA_1 and DA_2 receptor activation led to the suggestion more than 20 years ago that dopamine agonists might be useful antihypertensive drugs (McNay *et al.*, 1966). Because of its α-adrenoceptor activity, dopamine, itself, is not a useful blood pressure-lowering agent. Early studies in animal models and in patients with severe hypertension demonstrated that dopamine was an effective hypertensive agent following the administration of the α-adrenoceptor blocking agent, phenoxybenzamine (McNay *et al.*, 1966). Blood pressure reduction was accompanied by improved renal blood flow and sodium excretion.

Fenoldopam, a benzazepine derivative, is a selective DA_1 receptor agonist devoid of β- or α-adrenoceptor activity at therapeutic concentrations (Hahn *et al.*, 1982). A number of studies have demonstrated the ability of oral fenoldopam to lower blood pressure in hypertensive patients (Ventura *et al.*, 1984; Carey *et al.*, 1984), although the hypotensive effect is short lived (Carey *et al.*, 1984) due to its short half-life and its poor oral bioavailability (Weber *et al.*, 1988). Studies of parenteral fenoldopam in patients with mild hypertension have demonstrated linear log dose hypotensive response relationships beginning at 0.05 μg/kg/min, accompanied by an increase in heart rate which is baroreflex induced (Murphy *et al.*, 1987). Fenoldopam also has been shown to increase renal blood flow and sodium excretion while lowering systemic blood pressure (Murphy *et al.*, 1987). Following termination of infusion, blood pressure and heart rate return rapidly to baseline.

The short plasma half-life of fenoldopam (approximately 7-9 minutes) (Weber *et al.*, 1988) may make it a useful agent in the treatment of emergency hypertension. Comparison of fenoldopam with nitroprusside, currently the standard therapy in patients with severe hypertension, reveals that both agents are equally effective in lowering blood pressure (Elliott *et al.*, 1989). On the other hand, while nitroprusside has no significant beneficial effects on renal function, fenoldopam increases sodium and water excretion while increasing creatinine clearance in these patients. These renal effects of fenoldopam may offer advantages over current vasodilator therapy during the acute reduction of severe hypertension, particularly in patients with preexisting renal impairment. Fenoldopam also may offer advantages over nitroprusside in preserving renal blood flow during controlled intra-operative hypotension (Aronson *et al.*, 1987).

F. DA_2 Receptor Agonists in Hypertension

Several studies suggest that selective DA_2 receptor agonists may have useful hemodynamic effects in the treatment of hypertension since they inhibit sympathetic activity. Bromocriptine has been shown to lower blood pressure in hypertensive patients and, indeed, postural hypotension is often an unwanted side effect of its use in the treatment of Parkinson's disease or hyperprolactinemia (Mercuro *et al.*, 1985; Lombardi *et al.*, 1987; Whitworth *et al.*, 1987). Mercuro *et al.* (1985) have shown that it reduces blood pressure in both normal and hypertensive individuals. They have shown that this effect is mediated by peripheral, rather than central nervous, DA_2 receptor activation since the hypotensive effects could be prevented by prior administration of the DA_2 receptor antagonist, domperidone, at doses that do not cross the blood brain barrier (Mercuro *et al.*, 1984). On the other hand, McNay *et al.* (1987) reported that the selective DA_2 receptor agonist, quinpirole, raised arterial blood pressure in some hypertensive patients, perhaps as a result of centrally-mediated increased sympathetic outflow or vasopressin release (Nagahama *et al.*, 1986). Additionally, Whitworth and her colleagues (1987) have reported that the side effects of bromocriptine, predominantly nausea and orthostatic hypotension, as well as its high cost preclude its widespread use in the treatment of hypertension.

G. Dopaminergic Pro-drugs

Since dopamine has poor oral bioavailability, it must be administered intravenously, and this limits the use of dopamine for chronic therapy. On the other hand, several compounds which are orally bioavailable and which are enzymatically converted to dopamine have been investigated for their potential in the treatment of cardiovascular disease.

1. Levodopa

Levodopa is the immediate precursor of dopamine and is decarboxylated by aromatic amino acid decarboxylase to yield dopamine. In the early 1970's Whitsett and Goldberg (1972) investigated the β_1-adrenoceptor activity of levodopa using systolic time intervals measured noninvasively. Single doses of levodopa, 1-1.5 grams orally, produced the same shortening of the systolic time interval as infusions of dopamine at 2 and 4 μg/kg/min. These effects were blocked by propranolol. They also showed that levodopa increased renal blood flow, inulin clearance, and sodium excretion.

Later, use of levodopa in a patient with severe heart failure (Goldberg *et al.*, 1977), which permitted the weaning of intravenous dopamine, led to the controlled evaluation of levodopa in ten patients with severe congestive heart failure (Rajfer *et al.*, 1984). Levodopa, in doses of 1.5-2 grams per day, increased cardiac index and stroke volume index while reducing systemic vascular resistance. There were no changes in mean systemic pressure, heart rate, left ventricular filling pressure, or mean right atrial pressure. These beneficial effects lasted for 4-6 hours following each oral dose. Rajfer and his colleagues (1987) showed that these beneficial hemodynamic effects also were maintained for up to 12 weeks of treatment.

The most common adverse effects of levodopa, nausea and vomiting, can be minimized by gradual increments in the dose. Patients should be commenced at 250 mg four times a day and the dose increased to 1-2 grams over a 5-7 day period. Pyridoxine, 50 mg, must be coadministered since it is a cofactor for decarboxylation of levodopa.

2. Ibopamine

Ibopamine is the diisobuteric ester of N-methyl dopamine (epinine). Following hydrolysis by plasma esterases, epinine is released and activates DA_1 receptors in addition to α-adrenoceptors (Visioli, 1986). Epinine is also a potent β_2-adrenoceptor agonist.

Several invasive studies have shown that oral doses of 50-300 mg of ibopamine increase cardiac index and diminish systemic vascular resistance (Visioli, 1986; Itoh *et al.*, 1985; Dei Cas *et al.*, 1983; Nakano *et al.*, 1986). However, one study (Rajfer *et al.*, 1986) suggests that continued therapy is associated with attenuation of the hemodynamic effects. Ibopamine administration also results in diuresis and natriuresis. Since ibopamine does not penetrate the chemoreceptor trigger zone in the area postrema, it does not cause nausea and vomiting. On the other hand, large doses of ibopamine have been shown to elevate pulmonary wedge pressure transiently in some patients with severe congestive heart failure (Ren *et al.*, 1984), but this does not appear to be the case when smaller doses are used.

3. γ-Glutamyl-dopa

In 1976 Orlowski and Wilk reported that γ-glutamyl derivatives of amino acids were concentrated and metabolized as free amino acid, primarily in the kidney which contains high concentrations of γ-glutamyl transpeptidase. In a study in anesthetized rats they showed that parenteral administration of γ-glutamyl-dopa (glu-dopa) led to the production of high concentrations of dopamine in the kidney with substantially smaller amounts in other tissues, such as brain, heart, liver, lung, muscle, and spleen (Wilk *et al.*, 1978). They also demonstrated concomitant increases in renal blood flow and arterial blood pressure, the latter suggesting that plasma dopamine levels are elevated sufficiently to activate peripheral α-adrenoceptors. Subsequent studies in man have shown that the compound is natriuretic and diuretic and promotes an increase in glomerular filtration rate with decreased plasma renin activity (Worth *et al.*, 1985). These changes have been accompanied by little or no change in heart rate or systemic blood pressure. The renal, hemodynamic, and tubular effects of glu-dopa in man have been blunted by prior administration of the dopamine receptor antagonist, (R)-sulpiride (MacDonald *et al.*, 1988). Additional studies have shown that the administration of glu-dopa reduces the histological, renal tubular, and renal hemodynamic consequences of glycerol administered to produce experimental acute renal failure in rats (Casson *et al.*, 1983). However, whether glu-dopa offers significant therapeutic possibilities in man remains to be determined.

H. Modulators of Synthetic and Metabolic Enzymes of Dopamine

The enzymes involved in dopamine synthesis and metabolism have been targets of research seeking better therapy for heart failure or hypertension. Inhibition of aromatic amino acid decarboxylase to lower sympathetic activity and inhibition of dopamine β-hydroxylase to increase concentrations of dopamine while, at the same time inhibiting norepinephrine production, have been the main focus of this research.

I. Dopa Decarboxylase Inhibitors

Dopa decarboxylase inhibitors (AADC) were introduced as adjunctive treatment in the management of patients with Parkinson's disease (Rinne and Molsa, 1979). Carbidopa, coadministered with levodopa, prevents the peripheral conversion of levodopa to dopamine, enhancing its access to the central nervous system. In the laboratory AADC inhibitors have been used largely to evaluate the endogenous role of dopamine in the control of urinary sodium excretion. Based on earlier suggestions that dopamine may play a regulatory role in the tubular handling of sodium, Ball and Lee (1977) observed that the treatment of normal volunteers on a normal sodium diet with carbidopa reduced urinary dopamine and sodium excretion. Sowers *et al.* (1984) have reported blunting of natriuresis and of the increments in dopamine excretion in conscious dogs subjected to saline infusion following carbidopa administration. More recently, evidence has been presented that the natriuresis and increment in urinary dopamine excretion produced by a protein meal are also suppressed by pretreatment with carbidopa (Williams *et al.*, 1986). These data suggest that, under physiological conditions, the renal excretion of sodium may be directly related to the endogenous production of dopamine.

Experiments with the competitive AADC inhibitors have not demonstrated direct cardiovascular changes resulting from inhibition of dopamine synthesis. However, studies with newer, highly efficacious, irreversible AADC inhibitors have reported interesting effects on arterial blood pressure. Monofluoromethyl dopa, which inhibited the conversion of dopa to dopamine to an extent that tissue concentrations of dopamine and norepinephrine were depressed, showed a reduction in mean arterial pressure in spontaneously hypertensive rats of about 50 mmHG (Fozard *et al.*, 1981). However, similar experiments in Wistar-Kyoto normotensive animals had little effect on blood pressure. While these agents are useful tools in investigating the physiological role of dopamine, it is unlikely that they will have a therapeutic role in cardiovascular disease. The tissue depletion of norepinephrine required to lower systemic blood pressure probably would be associated with significant central depressive effects as well as unacceptable postural hypotension.

J. Dopamine β-Hydroxylase Inhibition

Another enzymatic approach to the modulation of the system is the inhibition of dopamine β-hydroxylase (DBH) (EC 1.14.17.1, a copper-containing mixed function oxidase) which would be expected to decrease norepinephrine synthesis and enhance dopamine levels. As early as 1969, Hidaka and colleagues reported that fusaric acid produced this biochemical effect and resulted in blood pressure reduction in spontaneously hypertensive rats. This finding was confirmed in later human studies where blood pressure reduction was sometimes accompanied by early tachycardia and enhanced urinary norepinephrine excretion, probably due to direct neuronal norepinephrine release by the inhibitor (Velasco *et al.*, 1975). In the 1970's, other compounds, A32390A, BRL8242, and the precursor of fusaric acid, bupicomide, provoked blood pressure reduction in several models of hypertension while significantly increasing blood plasma and tissue dopamine/norepinephrine ratios. Blood pressure reduction was accompanied by diuresis, natriuresis, and increased renal blood flow with many of these compounds. However, the most widely studied, fusaric acid, was abandoned because of hepatotoxicity.

Most recently, scientists at Smith Kline & French have reported data with the difluorobenzyl-imidazone, SK&F 102698, which produces a long-lived hypotensive effect with once daily dosing in the spontaneously hypertensive rat. Interestingly, central injection of this compound also reduces blood pressure, an effect that is blocked by SCH 23390, indicating centrally-mediated dopaminergic controlled blood pressure (Ohlstein *et al.*, 1988). However, one must be cautious in evaluating this approach to antihypertensive therapy. Sympatholytic agents, in general, are associated with significant side effects, such as drowsiness, dry mouth, and postural hypotension, and consequently, are losing favor in current antihypertensive practice. Since SK&F 102698 crosses the blood brain barrier, one might expect reduction in brain norepinephrine levels with associated lethargy, drowsiness, or depression. Furthermore, the recent demonstration of profound postural hypotension in two patients with DBH deficiency (Man In't Veld *et al.*, 1987) lends support to the suspicion that pharmacological inhibition of the enzyme may have similar effects. However, these issues remain to be examined in sufficiently large clinical trials in hypertensive patients.

V. SUMMARY

During the past 30 years enormous advances have been made in our understanding of the pharmacological effects of dopamine. The identification of specific dopamine receptors has opened new therapeutic possibilities. Full exploitation of the system will be achieved only with the development of orally bioavailable agonists and antagonists with adequate biological half-lives. Evaluation of pro-drugs and enzyme modulators may prove to be a useful alternative; however, they all await adequate evaluation in clinical studies.

REFERENCES

Aronson, S., Roth, S., Glock, D., and Goldberg, L.I. (1987). *Clin. Res.* **35**, 881A.

Ball, S.G. and Lee, M.R. (1977). *Br. J. Clin. Pharmacol.* **4**, 115-119.

Bass, A.S., Kohli, J.D., Lubbers, N.L., and Goldberg, L.I. (1987). *J. Pharmacol. Exp. Ther.* **242**, 940-944.

Bayliss, J., Thomas, L., and Poole-Wilson, P. (1987). *J. Cardiovasc. Pharmacol.* **9**, 551-554.

Bello-Reuss, E., Higashi, Y., and Kanega, Y. (1982). *Am. J. Physiol.* **242**, F634-F640.

Beregovich, J., Bianchi, C., Rubler, S., Lomnitz, E., Cagin, N., and Levitt, B. (1974). *Am. Heart J.* **87**, 550-557.

Brown, R.A., Dixon, J., Farmer, J.B., Hall, J.C., Humphries, R.G., Ince, F., O'Connor, S.E., Simpson, W.T., and Smith, G. (1985a). *Br. J. Pharmacol.* **85**, 599-608.

Brown, R.A., Farmer, J.B., Hall, J.C., Humphries, R.G., O'Connor, S.E., and Smith, G.W. (1985b). *Br. J. Pharmacol.* **85**, 609-619.

Butterworth, J.F., Austin, J.C., Johnson, M.D., Berrizbeitia, L.D., Dance, G.R., Howard, G., and Cohn, L.H. (1987). *Anesth. Analg.* **66**, 209-214.

Carey, R.M., Stote, R.M., Dubb, J.W., Townsend, L.H., Rose, C.E., and Kaiser, D.L. (1984). *J. Clin. Invest.* **74**, 2198-2207.

Carroll, J.D., Lang, R.M., Neumann, A.L., Borow, K.M., and Rajfer, S.I. (1986). *Circulation* **74**, 815-825.

Casson, I.F., Clayden, D.A., Cope, G.F., and Lee, M.R. (1983). *Clin. Sci.* **65**, 159-164.

Dawson, J.R., Thompson, D.S., Signy, M., Juul, S.M., Turnbull, P., Jenkins, B.S., and Webb-Peploe, M.M. (1985). *Br. Heart J.* **54**, 313-320.

Dei Cas, L., Bolognesi, R., Cucchini, F., Fappani, A., Riva, S., and Visioli, O. (1983). *J. Cardiovasc. Pharmacol.* **5**, 249-253.

D'Orio, V., El Allaf, D., Juchmes, J., and Marcelle, R. (1986). *Arch. Int. Physiol. Biochim.* **92**, S11-S20.

Elliott, W.J., Weber, R.R., Nelson, K.S., Oliner, C.W., and Murphy, M.B. (1989). Fourth International Interdisciplinary Conference on Hypertension in Blacks, Nairobi, Kenya, June 28-July 2. (Abstr.). In press.

Felder, R.A., Blecher, M., Calcagno, P.L., and Jose, P.A. (1984). *Am. J. Physiol.* **247**, F499-F505.

Fennell, W.H., Taylor, A.A., Young, J.B., Brandon, T.A., Ginos, J.Z., Goldberg, L.I., and Mitchell, J.R. (1983). *Circulation* **67**, 829-836.

Fozard, J.R., Mohring, J., Palfreyman, M.G., and Koch-Weser, J. (1981). *J. Cardiovasc. Pharmacol.* **3**, 1038-1049.

Frederickson, E.D., Bradley, T.J., and Goldberg, L.I. (1985). *Am. J. Physiol.* **249**, F236-F240.

Goldberg, L.I. (1972). *Pharmacol. Rev.* **241**, 1-29.

Goldberg, L.I. (1977). *Proc. R. Soc. Med.* **70** (Suppl. 2), 7-15.

Goldberg, L.I. and Rajfer, S.I. (1985). *Circulation* **72**, 245-248.

Goldberg, L.I. and Sjoerdsma, A. (1959). *J. Pharmacol. Exp. Ther.* **127**, 212-218.

Goldberg, L.I. and Yeh, B.K. (1971). *Eur. J. Pharmacol.* **15**, 36-40.

Goldberg, L.I., Talley, R.C., and McNay, J.L. (1969). *Prog. Cardiovasc. Dis.* **12**, 40-51.

Goldberg, L.I., Hsieh, Y.Y., and Resnekov, L. (1977). *Prog. Cardiovasc. Dis.* **19**, 327-340.

Hahn, R.A., Wardell, J.R., Sarau, H.M., and Ridley, P.T. (1982). *J. Pharmacol. Exp. Ther.* **223**, 305-313.

Henderson, I.S., Beattie, T.J., and Kennedy, A.C. (1980). *Lancet* **2**, 827-828.

Hidaka, H., Nagatsu, T., Takaeya, K., Takeuchi, T., Suda, H., Kojairi, K., Matsuzaki, M., and Umezawa, H. (1969). *J. Antibiot.* (Tokyo) **22**, 228-230.

Hilberman, M., Maseda, J., Stinson, E.B., Derby, G.C., Spencer, R.J., Miller, D.C., Oyer, P.E., and Myers, B.D. (1984). *Anesthesiology* **61**, 489-494.

Holtz, P. and Credner, K. (1942). *Naunyn-Schmiedeberg's Arch. Pharmacol. Exp. Pathol.* **200**, 356-388.

Horwitz, D., Fox, S.M., and Goldberg, L.I. (1962). *Circ. Res.* **10**, 237-245.

Hughes, A.D., Thom, S.A., Woodall, N.M., Redman, D., and Sever, P. (1987). *J. Hypertens.* **5**, 337-340.

Itoh, H., Kohli, J.D., Rajfer, S.I., and Goldberg, L.I. (1985). *J. Pharmacol. Exp. Ther.* **233**, 87-93.

Keung, E.C., Ribner, H.S., Schwartz, W., Sonnenblick, E.H., and LeJemtel, T.H. (1980). *J. Cardiovasc. Pharmacol.* **2**, 113-119.

Kohli, J.D. and Goldberg, L.I. (1989). In "Cardiovascular function of peripheral dopamine receptors" (J.P. Hieble, ed.). New York, Marcel Dekker. In press.

Kohli, J.D., Weder, A.B., Goldberg, L.I., and Ginos, J.Z. (1980). *J. Pharmacol. Exp. Ther.* **213**, 370-374.

Levinson, P.D., Goldstein, D.S., Munson, P.J., Gill, J.R., Jr., and Keiser, H.R. (1985). *J. Clin. Endocrin. Metab.* **60**, 821-826.

Lindner, A. (1983). *Nephron* **33**, 121-126.

Loeb, H.S., Ostrenga, J.P., Gaul, W., Witt, J., Freeman, G., Scanlon, P., and Gunnar, R.M. (1983). *Circulation* **68**, 813-820.

Lombardi, C., De Cotiis, R., Spedini, C., Missale, C., Memo, M., and Spano, P.F. (1987). *Eur. J. Clin. Pharmacol.* **33**, 519-521.

McDonald, R.H., Jr., Goldberg, L.I., McNay, J.L., and Tuttle, E.P., Jr. (1964). *J. Clin. Invest.* **43**, 1116-1124.

MacDonald, T.M., Jeffrey, R.F., Freestone, S., and Lee, M.R. (1988). *Br. J. Clin. Pharmacol.* **25**, 203-212.

McNay, J.L., MacCannell, K.L., Meyer, M.B., and Goldberg, L.I. (1966). *J. Clin. Invest.* **45**, 1045-1046.

McNay, J.L., Henry, D.P., and DeLong, A.F. (1987). *Clin. Exp. Hypertens.* **9**, 1098.

Man In't Veld, A.J., Boomsma, F., Molenman, P., and Schalekamp, M.A. (1987). *Lancet* **1**, 183-184.

Mann, H.J., Fuhs, D.W., and Hemstrom, C.A. (1986). *Drug Intell. Clin. Pharm.* **20**, 421-438.

Mercuro, G., Rossetti, Z.L., Tocco, L., Rivano, A.C., and Cherchi, A. (1984). *Eur. Heart J.* **5** (Suppl. 1), 168 (Abstr.).

Mercuro, G., Rossetti, Z.L., Tocco, L., Rivano, C.A., Cherchi, A., and Gessa, G.L. (1985). *Eur. J. Clin. Pharmacol.* **27**, 671.

Meyer, M.B., McNay, J.L., and Goldberg, L.I. (1967). *J. Pharmacol. Exp. Ther.* **156** 186-192.

Miller, R.R., Awan, N.A., Joye, J.A., Maxwell, K.S., DeMaria, A.N., Amsterdam, E.A., and Mason, D.T. (1977). *Circulation* **55**, 881-884.

Murphy, M.B., McCoy, C.E., Weber, R.R., Frederickson, E.D., Douglas, F.L., and Goldberg, L.I. (1987). *Circulation* **76**, 1312-1318.

Nagahama, S., Chen, Y-F., Lindheimer, M.D., and Oparil, S. (1986). *J. Pharmacol. Exp. Ther.* **236**, 735-742.

Nakano, T., Morimoto, Y., Kakuta, Y., Konishi, T., Kodera, T., Kanamaru, M., and Takezawa, H. (1986). *Arzneimittelforschung* **36**, 1829-1834.

Ohlstein, E.H., Kruse, L.I., Ezekiel, M., Sherman, S.S., Erickson, R., DeWolf, W.E., and Berkowitz, B.A. (1987). *J. Pharmacol. Exp. Ther.* **241**, 554-559.

Ohlstein, E.H., Kruse, L.I., Ezekiel, M., and Willette, R.N. (1988). *J. Pharmacol. Exp. Ther.* **247**, 186-195.

Orlowski, M. and Wilk, S. (1976). *Eur. J. Biochem.* **71**, 549-555.

Parker, S., Carlon, G.C., Isaacs, M., Howland, W.S., and Kahn, R.C. (1981). *Crit. Care Med.* **9**, 630-632.

Paul, M.D., Mazer, C.D., Byrick, R.J., Rose, D.K., and Goldstein, M.B. (1986). *Am. J. Nephrol.* **6**, 427-434.

Pedersen, J.E. and Mortensen, S.A. (1987). *Acta Cardiol.* (Brux) **42**, 295-298.

Polson, R.J., Park, G.R., Lindop, M.M., Farman, J.V., Calne, R.Y., and Williams, R. (1987). *Anaesthesia* **42**, 15-19.

Rajfer, S.I., Anton, A.H., Rossen, J.D., and Goldberg, L.I. (1984). *New Engl. J. Med.* **310**, 1357-1362.

Rajfer, S.I., Rossen, J.D., Douglas, F.L., Goldberg, L.I., and Karrison, T. (1986). *Circulation* **73**, 740-748.

Rajfer, S.I., Rossen, J.D., Nemanich, J.W., Douglas, F.L., Davis, F., and Osinski, J. (1987). *J. Am. Coll. Cardiol.* **10**, 1286-1293.
Ren, J.H., Unverferth, D.V., and Leier, C.V. (1984). *J. Cardiovasc. Pharmacol.* **6**, 748-755.
Richard, C., Ricome, J.L., Rimailho, J.A., Bottineau, G., and Auzepy, P. (1983). *Circulation* **67**, 620-626.
Rinne, U.K. and Molsa, P. (1979). *Neurology* **29**, 1584-1589.
Schaer, G.L., Fink, M.P., and Parrillo, J.E. (1985). *Crit. Care Med.* **13**, 492-496.
Sowers, J.R., Crane, P.D., Beck, F.W.J., McClanahan, M., King, M.E., and Mohanty, P.K. (1984). *Endocrinology* **115**, 2085-2090.
Sprung, C.L., Caralis, P.V., Marcial, E.H., Pierce, M., Gelbard, M.A., Long, W.M., Duncan, R.C., Tendler, M.D., and Karpf, M. (1984). *New Engl. J. Med.* **311**, 1137-1143.
Svensson, G., Sjogren, A., and Erhardt, L. (1986). *Eur. Heart J.* **7**, 697-703.
Tan, L.B., Littler, W.A., and Murray, R.G. (1987). *J. Cardiovasc. Pharmacol.* **10**, 280-286.
Velasco, M., Gilbert, C.A., Rutledge, C.O., and McNay, J.L. (1975). *Clin. Pharmacol. Ther.* **18**, 145-153.
Ventura, H.O., Messerli, F.H., Frohlich, E.D., Kobrin, I., Oigman, W., Dunn, F.G., and Carey, R.M. (1984). *Circulation* **69**, 1142-1145.
Visioli, O. (1986). *Arzneimittelforschung* **36**, 285-286.
Walton, R.P. and Brodie, O.J. (1949). *J. Pharmacol. Exp. Ther.* **96**, 343-351.
Weber, R.R., McCoy, C.E., Ziemniak, J.A., Frederickson, E.D., Goldberg, L.I., and Murphy, M.B. (1988). *Br. J. Clin. Pharmacol.* **25**, 17-21.
Whitsett, T.L. and Goldberg, L.I. (1972). *Circulation* **45**, 97-106.
Whitworth, J.A., Macdonald, I., and Kincaid-Smith, P. (1987). *Aust. N.Z. J. Med.* **17**, 457-458.
Wilk, S., Mizoguchi, H. and Orlowski, M. (1978). *J. Pharmacol. Exp. Ther.* **206**, 227-232.
Williams, M., Young, J.B., Rosa, R.M., Gunn, S., Epstein, F.H., and Landsberg, L. (1986). *J. Clin. Invest.* **78**, 1687-1693.
Worth, D.P., Harvey, J.N., Brown, J., and Lee, M.R. (1985). *Clin. Sci.* **69**, 207-214.
Young, J.B., Leon, C.A., Pratt, C.M., Suarez, J.M., Aronoff, R.D., and Roberts, R. (1985). *J. Am. Coll. Cardiol.* **6**, 792-796.

5

ALPHA-METHYLDOPA

Wybren De Jong

Merrell Dow Research Institute
Strasbourg Research Center
67084 Strasbourg Cedex, France

Paul Van Giersbergen

Rudolf Magnus Institute for Pharmacology
University of Utrecht
3521 GD Utrecht, The Netherlands

α-Methyldopa (L-α-methyl-3,4-dihydroxyphenylalanine) was first synthesized in 1951 by Merck in the USA (Pfister and Stein, 1959; Sjoerdsma, 1982; Henning, 1984; Stein *et al.*, 1954). A few years later, Sourkes (1954) reported that it was a powerful inhibitor of dopa decarboxylase (1-aromatic amino acid-1-decarboxylase (E.C. 4.1.1.28)) *in vitro.* Subsequent observations *in vivo* in several animal species confirmed this inhibition (Dengler and Reichel, 1958; Westermann *et al.*, 1958). In these experiments the generation of vasoactive products from administered precursors like 1-DOPA was inhibited. No effects of α-methyldopa on basal blood pressure were apparent in these studies (Sjoerdsma, 1982; Dengler and Reichel, 1958; Westermann *et al.*, 1958). In the fall of 1959 pronounced hypotensive activity of α-methyldopa was unexpectedly discovered in human hypertensive subjects by Sjoerdsma and associates (Sjoerdsma, 1982; Oates *et al.*, 1960). This finding initiated the use of α-methyldopa and in particular its active 1-isomer (Methyldopa, Aldomet®) as a major therapeutic agent in hypertension (Henning, 1984; Gifford, 1972). Early attempts to demonstrate a hypotensive action in dogs were not successful, except for the study of Goldberg and associates who reported a slowly developing fall in blood pressure in conscious dogs (Sjoerdsma, 1982; Goldberg *et al.*, 1960). In subsequent years the hypotensive activity was reported in a variety of animal species under various experimental conditions (Henning, 1984; Nijkamp and De Jong, 1977).

It is now generally accepted that α-methyldopa has to be taken up by catecholaminergic neurons and analogous to the endogenous precursor of noradrenaline, 1-DOPA, it is processed to α-methylnoradrenaline (Figure 1). The early investigations of Carlsson and Lindqvist (1962) revealed that the conversion of α-methyldopa to α-methylnoradrenaline in brain is associated with a drop in endogenous noradrenaline.

Enzymes as Targets for Drug Design

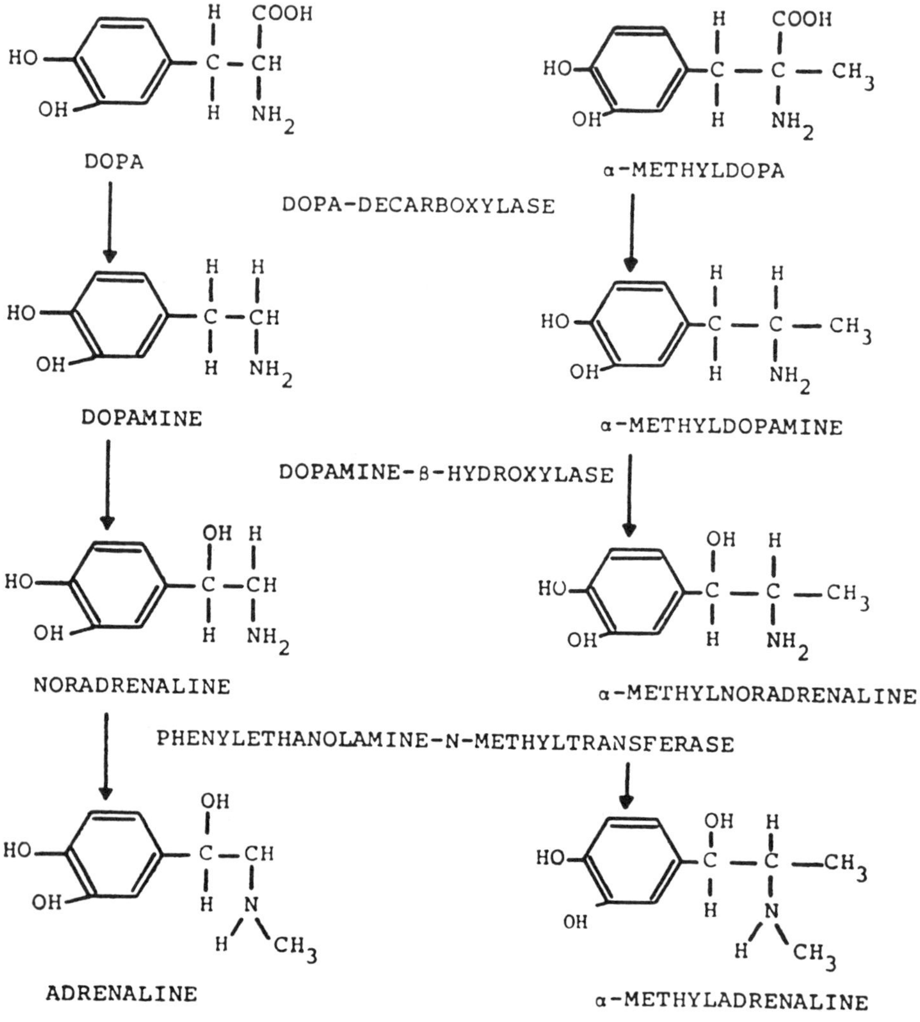

Fig. 1. *Biosynthetic pathway of 1-DOPA and α-methyldopa both in the central nervous system and peripheral sympathetic nerves.*

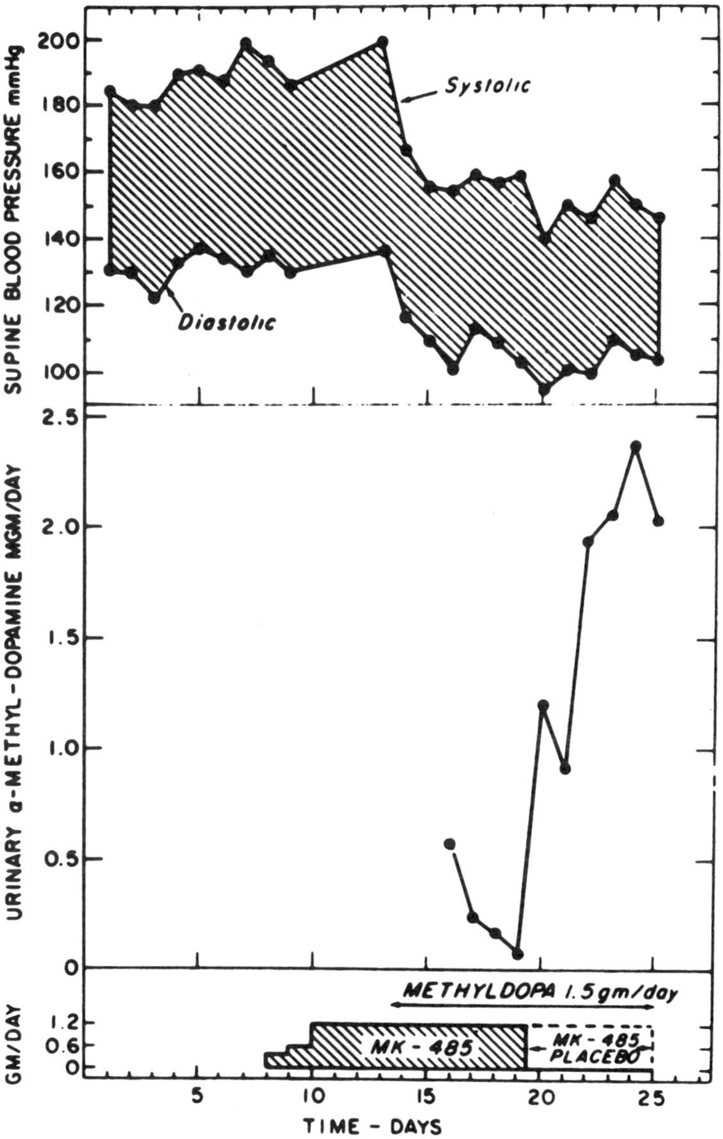

Fig. 2. *Failure of α-methyldopa hydrazine (MK 485) to alter the blood pressure response to methyldopa despite apparent inhibition of decarboxylation to primary α-methyldopamine. (Reproduced by permission of the American Heart Association, Inc. (Sjoerdsma et al., 1963)).*

The basic approach of Sjoerdsma and associates in the National Institutes of Health in Bethesda, with respect to amine metabolism in hypertensive patients centered on inhibition of enzymatic pathways in relation to blood pressure regulation (Sjoerdsma, 1982). In this line of research α-methyldopa was studied. An interesting early observation was the finding that inhibition of dopa decarboxylase activity by a hydrazino analog of 1-DOPA (MK 485) failed to affect the hypotensive action of α-methyldopa although decarboxylase activity appeared to be effectively inhibited (Figure 2).

This observation, together with the transient sedation observed in man, led to the first suggestion that central effects may be important for the hypotension (Sjoerdsma *et al.*, 1963; Gillespie *et al.*, 1962). Other explanations for the fall in blood pressure focused upon more direct interference with the function of the peripheral sympathetic nervous system (Table I).

Carlsson and Lindqvist (1962) suggested that α-methylnoradrenaline may substitute for and take over the functions of noradrenaline following its depletion. Functional studies demonstrated the release of α-methylnoradrenaline from tissues of animals treated with α-methyldopa and showed efficacy of α-methylnoradrenaline in the same dose range as noradrenaline in different animal preparations (for review see Henning, 1984; Nijkamp and De Jong, 1977). However, Day and Rand postulated the "false transmitter hypothesis" indicating that α-methylnoradrenaline may act as a less efficient transmitter compared to endogenous noradrenaline (Day and Rand, 1963; Day and Rand, 1964). This hypothesis was mainly based upon the lesser pressor potency of α-methylnoradrenaline relative to noradrenaline in different animal preparations and the earlier finding of α–methylnoradrenaline in brain tissue (Carlsson and Lindqvist, 1962).

TABLE I.

Proposed Mechanisms of Action of α-Methyldopa

- Inhibition of sympathetic nervous system
- Inhibition of dopa decarboxylase
- Inhibition of related enzymes
- Depletion of norepinephrine (Sjoerdsma , 1963)
- Via amine metabolites of α-methyldopa (Sjoerdsma *et al.*, 1963)
- False transmitter (Day and Rand, 1963, 1964)
- Central nervous system (Gillespie *et al.*, 1962; Henning, 1984; Nijkamp and DeJong, 1977)
- α-Methylnoradrenaline is a more potent transmitter in the brain and is less effective in the periphery as compared to endogenous noradrenaline (De Jong and Nijkamp, 1976; Nijkamp and De Jong, 1977)

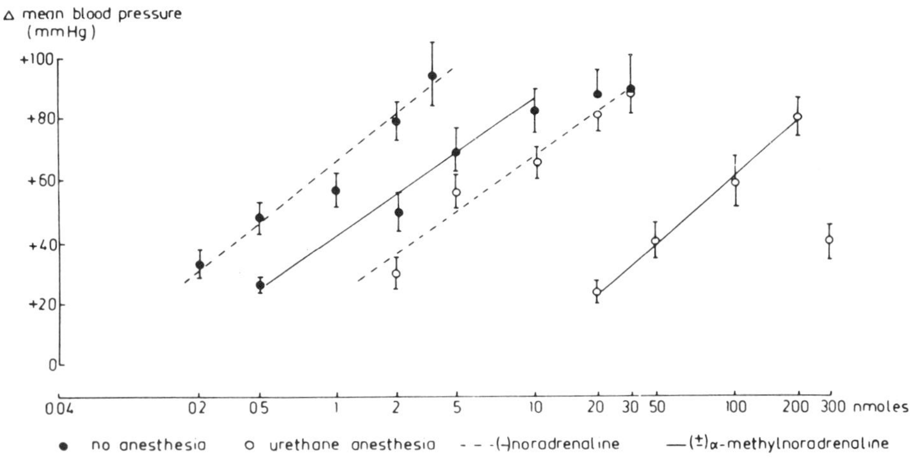

Fig. 3. *Dose-response effect of α-methylnoradrenaline and noradrenaline on mean blood pressure after intravenous administration to anesthetized and conscious normotensive rats. Data are presented as the mean ± S.E.M. of 6-10 animals. (Reproduced with kind permission from Nijkamp and De Jong, 1977).*

As shown in Figure 3 in anesthetized and in conscious rats α-methylnoradrenaline is not as potent as noradrenaline at inducing a pressor response upon intravenous administration. There are numerous observations which are in agreement with these findings, however, a considerable number of reports are not, but this discrepancy has been reviewed elsewhere (Henning, 1984; Nijkamp and De Jong, 1977). Interestingly, no major impairment or blockade of peripheral sympathetic nervous function appears to occur following administration of α-methyldopa in animals and man (Henning, 1984; Nijkamp and De Jong, 1977; Gifford, 1972).

Direct evidence for a central site of action of α-methyldopa came from experiments in which small amounts of the drug were administered into the vertebral artery, the ventricular system of the brain or locally into specific areas of the brain (Henning, 1984; Nijkamp and De Jong, 1977). Henning and Van Zwieten (1968) observed hypotension and bradycardia in anesthetized cats which received α-methyldopa via the vertebral artery in a dose having no effect upon blood pressure following systemic administration.

As illustrated in Figure 4, hypotension was evident with a low dose of α-methyldopa given by infusion into a lateral brain ventricle of conscious renal hypertensive rats. Similar results were obtained by other investigators under a variety of conditions (for review see Henning, 1984; Nijkamp and De Jong, 1977). Another approach was the differential inhibition of the catecholamine biosynthetic pathway both within the brain and peripherally. Inhibition of dopa decarboxylase in the central nervous system

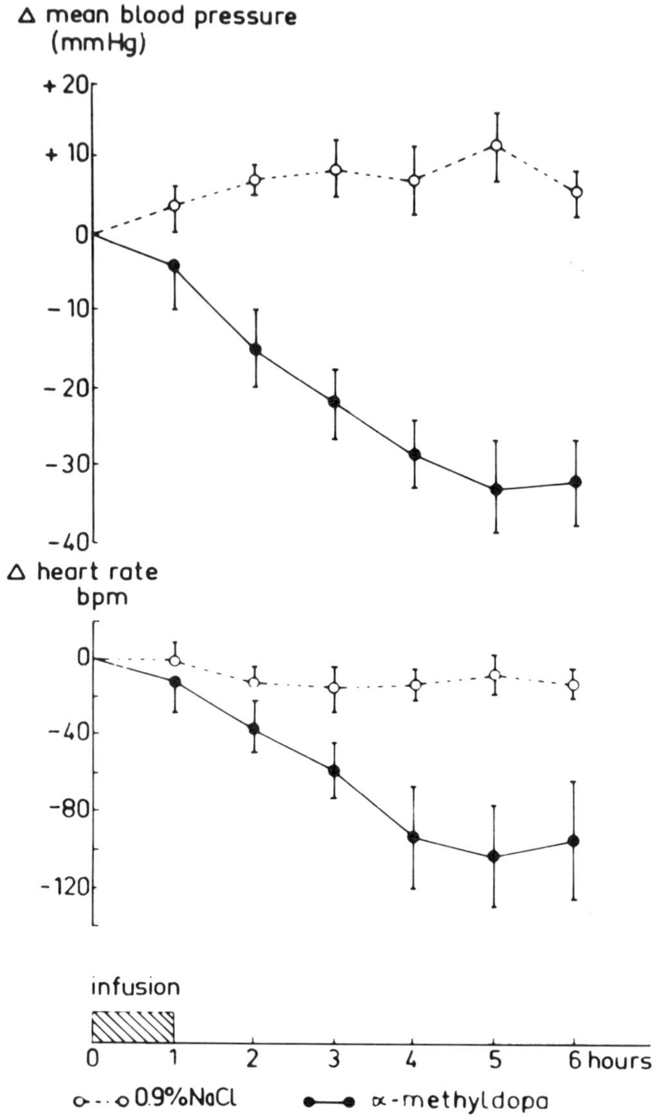

Fig. 4. *Effect of α-methyldopa infusion (10 mg/kg/hr) into the lateral ventricle on mean blood pressure and heart rate of conscious renal hypertensive rats. Data are expressed as the mean ± S.E.M. of 7-10 animals. (Reproduced with kind permission from Nijkamp and De Jong, 1977).*

interfered with the action of α-methyldopa, but peripheral inhibition of the enzyme was without effect (Henning, 1984; Nijkamp and De Jong, 1977). The results shown in Figure 2 are similarly explained since MK 485 does not cross the blood brain barrier. Thus, the initial idea that peripheral inhibition of dopa decarboxylase explained the hypotensive action was ended and instead, central conversion of α-methyldopa by this enzyme as a requirement for the fall in blood pressure was introduced. Further studies involving inhibition of dopamine-β-hydroxylase demonstrated that the generation of α-methylnoradrenaline is necessary to induce the hypotensive and bradycardiac effects of α-methyldopa (Henning, 1984; Nijkamp and De Jong, 1977), and blockade of brain α_2-adrenoceptors was found to attenuate the effects of the drug (Henning, 1984; Nijkamp and De Jong, 1977). Taken together these data provide conclusive evidence for a central site of action of α-methyldopa. After entering the brain and subsequent uptake into catecholaminergic neurons it is converted to firstly α-methyldopamine and then to α-methylnoradrenaline, the active compound. As recently discussed by Henning (1984), a role of α-methyladrenaline is not likely, but cannot be excluded.

A number of brain areas may be involved in the α-methyldopa-induced hypotension. However, mid-collicular transection does not abolish the hypotensive response in rats (Nijkamp and De Jong, 1977). In cats the vertebral artery is an effective route of administration (Henning, 1984; Henning and Van Zwieten, 1968). This artery mainly supplies the pons-medulla oblongata region in this species. In the rat, however, administration of a low dose of α-methyldopa via this route did not induce a fall in blood pressure (Nijkamp and De Jong, 1977). This was later explained by the fact that the vertebral artery in the rat mainly supplies blood to the neck musculature (Wellens *et al.*, 1976). These observations and other literature reports (Henning, 1984; Nijkamp and De Jong, 1977) indicate that caudal brainstem areas, including the nucleus of the solitary tract (NTS) are necessary for the hypotensive action of α-methyldopa. The NTS contains the first central relay of the baroreceptor pathways and is of importance for the integration of cardiovascular responses. The medio-caudal part of the NTS contains the baroreceptor terminals and is densely innervated by catecholaminergic fibers. In fact, noradrenaline is thought to exert an inhibitory control on blood pressure in this region (De Jong *et al.*, 1975). This area of the NTS is sensitive to applied catecholamines where they induce hypotension and bradycardia (De Jong, 1974; Versteeg *et al.*, 1984). Similarly, microinjection of α-methylnoradrenaline into this area of the NTS also decreases blood pressure and heart rate (Nijkamp and De Jong, 1977; De Jong and Nijkamp, 1976).

As depicted in Figure 5, α-methylnoradrenaline is more potent than noradrenaline following application in the NTS, and has a longer duration of action. This may be due to the fact that α-methylnoradrenaline is not a substrate of monoamine oxidase (Henning, 1984). Thus, instead of functioning as a less potent or less efficient transmitter, α-methylnoradrenaline may replace noradrenaline in the brain and act as a more potent transmitter and in this way exert its antihypertensive action. The diminished peripheral pressor action relative to noradrenaline may well contribute to its therapeutic effect (Table I).

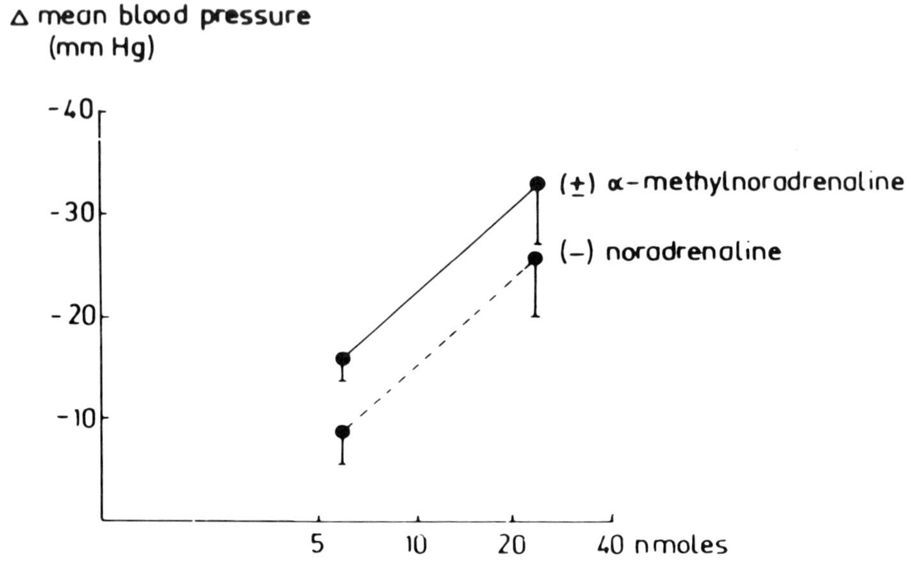

Fig. 5. *Effect of α-methylnoradrenaline and noradrenaline on mean blood pressure of anesthetized rats after bilateral microinjection in the area of the nucleus tractus solitarii (data are presented as the mean ± S.E.M. of 6 animals).*

A number of observations point to the involvement of central opioid systems in hypertension and in the hypotensive response induced by central α_2-adrenoceptor agonists like clonidine and α-methyldopa (Petty and De Jong, 1984; Kunos *et al.*, 1981; Naranjo *et al.*, 1985; Sitsen and De Jong, 1983; Sitsen and De Jong, 1984; Van Giersbergen and De Jong, 1988; Van Giersbergen *et al.*, 1989a; Van Giersbergen *et al.*, 1989b). Anesthesia may be a confounding factor in experiments addressing the involvement of endogenous opioid systems in cardiovascular responses (Van Giersbergen and De Jong, 1988). Therefore, conscious rats with chronic intracisternal cannulas were used to study this putative role of brain opioids in the action of α-methyldopa (Figure 6).

A chronic indwelling femoral artery catheter was employed to monitor cardiovascular parameters (Van Giersbergen and De Jong, 1988). This approach permits the study of α-methyldopa effects in the conscious state and in addition, the intracisternal route of administration allows the use of low doses of the drug without interference from direct peripheral effects (Van Giersbergen and De Jong, 1988; Van Giersbergen *et al.*, 1989a; Van Giersbergen *et al.*, 1989b; Van der Maas *et al.*, 1986). As shown in Figure 7, α-methyldopa, administered intracisternally, dose-dependently decreased the blood pressure and heart rate of conscious normotensive rats. Pretreatment with a single dose (30 µg) of the opiate receptor antagonist naltrexone via the same cannula caused a parallel shift to the right of the dose-response curve; the effects on the blood

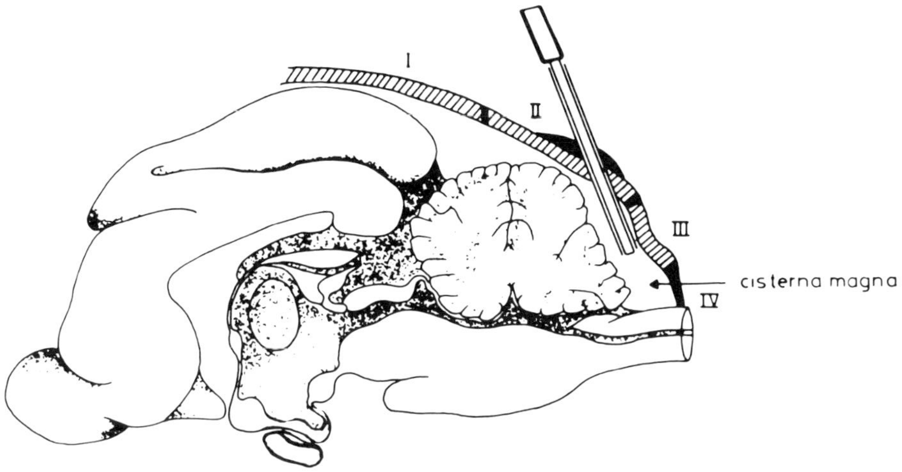

Fig. 6. *A sagittal cross-section of the brain and skull, showing the localization of the cisterna magna cannula. I parietal bone; II interparietal bone; III occipital bone; IV atlanto-occipital membrane. (Reproduced with kind permission from Van Giersbergen et al., 1988).*

pressure curve being more pronounced than on the heart rate curve. Similar results were obtained with a slightly different protocol in which one dose of α-methyldopa was used with pretreatment consisting of different doses of the opioid antagonist (Van Giersbergen and De Jong, 1988). In an additional series of experiments interference with the opioid system was obtained by employing antisera raised to opioid peptides in rabbits (Van Giersbergen *et al.*, 1989a).

Figure 8 depicts the effect of pretreatment with the antiserum B4. This antiserum binds β-endorphin but shows no cross-reactivity with enkephalins. It caused a similar displacement to the right of the α-methyldopa dose-response curve as naltrexone. Antisera raised to met-enkephalin or γ- and α-endorphin had no effect (Van Giersbergen *et al.*, 1989a). Furthermore, the hypotensive action of the peripherally acting vasodilator hydralazine was not affected by the B4 antiserum (Van Giersbergen *et al.*, 1989a). These results indicate that opioid peptides play a role in the hypotensive mechanism of action of α-methyldopa. Arguments for a role of the NTS in this mechanism were presented by Petty and de Jong who in anesthetized rats could inhibit the hypotensive action of α-methylnoradrenaline applied locally into this nucleus using an antagonist and also using a β-endorphin antiserum (Petty and De Jong, 1984).

The protocol of these studies necessitated pretreatment with the opioid antagonist or antiserum. As shown in Figure 9, treatment with naltrexone (30 μg) at different time

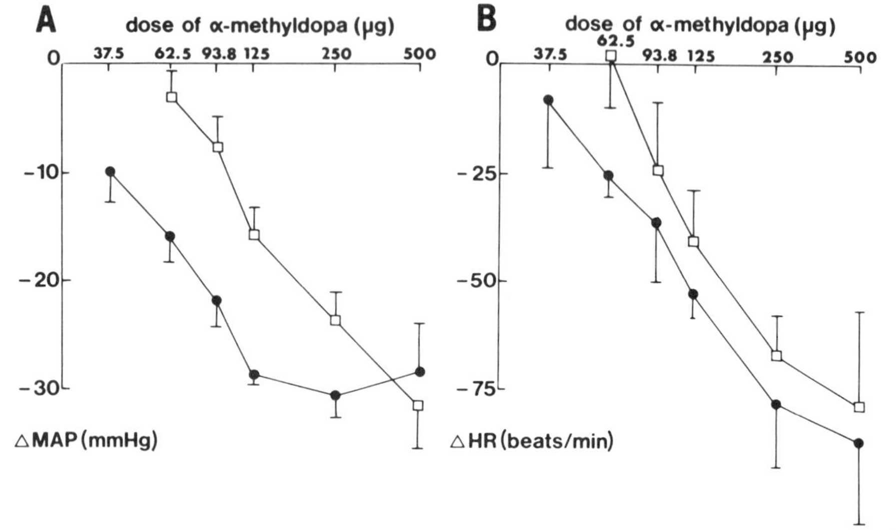

Fig. 7. *Effect of i.c. pretreatment with naltrexone (30 μg) on the decrease in mean arterial pressure (MAP; mmHg; A) and heart rate (HR, beats/min; B) induced by increasing doses of α-methyldopa (37.5-500 μg) given by the same route in conscious normotensive rats. Naltrexone (□) or saline (●) was injected 20 min before α-methyldopa. MAP and HR were determined 4 hr after injection of α-methyldopa (n=5-12 for the various doses). The dose-response curves for α-methyldopa were shifted to the right, p< .001 (A) and p< .05 (B). (Reproduced with kind permission from Van Giersbergen and De Jong, 1988).*

intervals after α-methyldopa was not effective. This dose of naltrexone was, however, effective when administered 20 min before the injection of α-methyldopa (Figure 7). Thus, it appears that stimulation of an opioid receptor during the first hour following the administration of α-methyldopa is critical for the induction of hypotension and bradycardia. Kunos and associates (1981) reported that α-adrenoceptor agonists could release β-endorphin-like immunoreactivity from brainstem slices of spontaneously hypertensive rats. Furthermore, Petty and De Jong (1982) demonstrated that β-endorphin in low doses (pg range) induced hypotension and bradycardia when injected into the NTS. Overall, these findings indicate a role of β-endorphin or a closely related peptide in the α-methyldopa-induced hypotension. The fact that treatment with naltrexone or an antiserum became ineffective one hour or later after the administration of the agonist may reflect a more complicated interaction with the opioid system (Van Giersbergen and De Jong, 1988; Van Giersbergen *et al.*, 1989a). It might be speculated that the opioid peptide effect is essential during the first hour for its subsequent continued release, or that the synthesis of α-methylnoradrenaline and/or its release is affected. It had been suggested that the generation of the latter compound and its subsequent release in the NTS correlates with the hypotensive effect of α-methyldopa (Louis *et al.*, 1984).

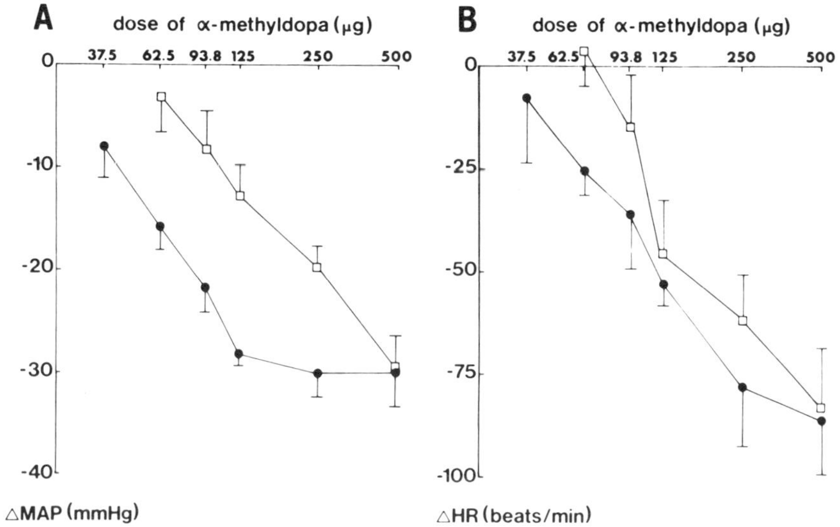

Fig. 8. *Effect of intracisternal pretreatment with the β-endorphin antiserum B4 (0.1 μl) on the α-methyldopa (37.5-500 μg i.c.) dose-response curve regarding decreases in mean arterial pressure (MAP; A) and heart rate (HR) (B) in conscious normotensive rats. The antiserum (❑) or saline (●) was injected 20 min before α-methyldopa. MAP and HR were determined 4 hr after injection of α-methyldopa (n=5-14 for the various doses). The dose-response curves for α-methyldopa were shifted to the right, p <0.001 (A) and p <0.05 (B). (Reproduced with kind permission from S. Karger AG, Basel (Van Giersbergen et al., 1989)).*

SUMMARY

α-Methyldopa was initially used for its ability to inhibit the enzyme dopa decarboxylase. The finding of its hypotensive effect in man was a surprise (Sjoerdsma, 1982; Oates *et al.*, 1960). It is a potent antihypertensive agent which has to be converted to its principal active metabolite α-methylnoradrenaline in the central nervous system in order to induce hypotension and bradycardia. The enzymes involved in the conversion are dopa decarboxylase and dopamine-β-hydroxylase. In the brain α-methylnoradrenaline substitutes for endogenous noradrenaline and it appears to act as a more potent neurotransmitter. In contrast, its peripheral pressor action in general is less pronounced than that of noradrenaline, which may contribute to the therapeutic efficacy of α-methyldopa.

Lower brainstem structures appear to mediate the hypotensive action, in particular the nucleus of the solitary tract has been implicated as a site of action of α-methylnoradrenaline. Recent data indicate that interference with central opioid receptor stimulation abolishes the hypotensive action of α-methyldopa. There are indications that β-endorphin or a closely related peptide plays a key role in these processes, however, the precise mechanism remains to be elucidated. The role of brain

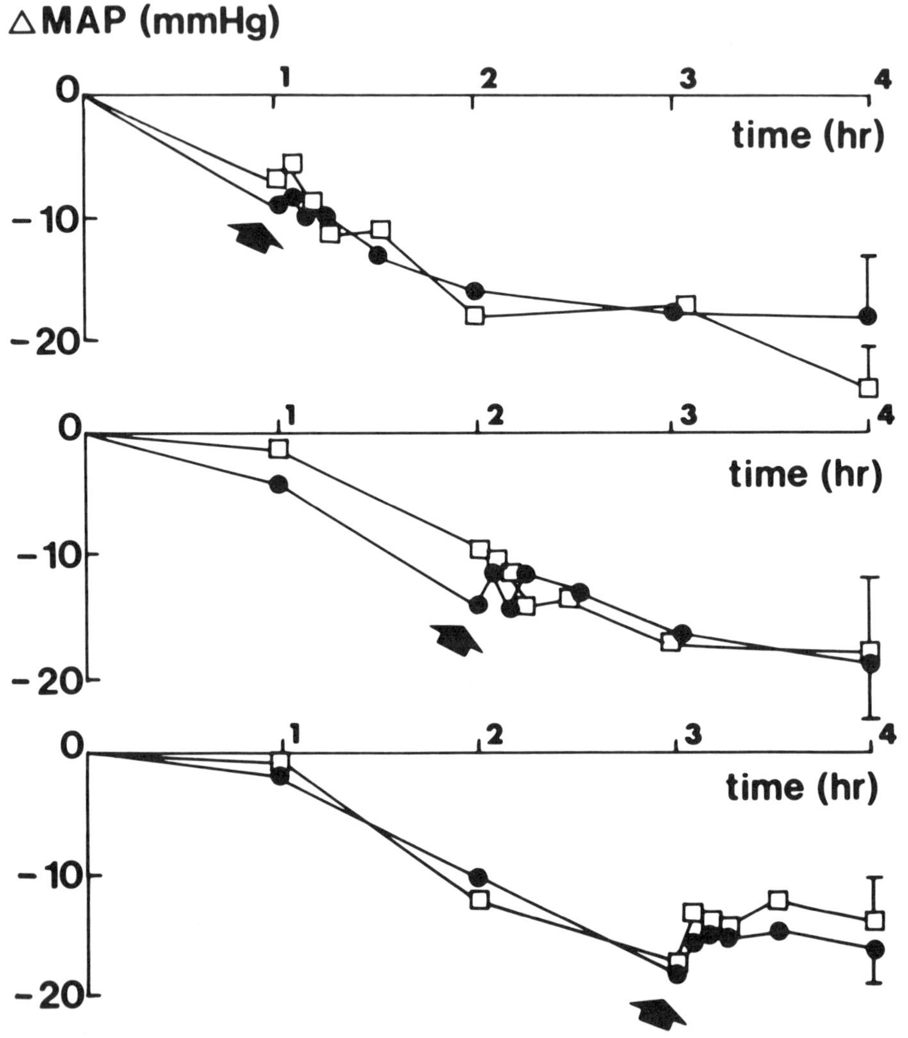

Fig. 9. *Effect of naltrexone (30 μg i.c.) administered 1, 2 or 3 hr after injection of α-methyldopa (93.8 μg) on the cardiovascular responses induced by α-methyldopa administration (n=5). Arrows indicate points of administration of naltrexone (❑) or saline (●). Mean arterial pressure (MAP; mmHg) was recorded continuously. (Reproduced with kind permission from Van Giersbergen and De Jong, 1988).*

opioid receptor stimulation in the hypotensive action of α-methyldopa again emphasizes the central site action of this antihypertensive agent.

REFERENCES

Carlsson, A. and Lindqvist, M. (1962). *Acta Physiol. Scand.* **54**, 87-94.
Day, M.D. and Rand, M.J. (1963). *J. Pharm. Pharmacol.* **15**, 221-224.
Day, M.D. and Rand, M.J. (1964). *Int. J. Neuropharmacol.* **3**, 173-175.
De Jong, W. (1974). *Eur. J. Pharmacol.* **29**, 179-181.
De Jong, W. and Nijkamp, F.P. (1976). *Br. J. Pharmacol.* **58**, 593-598.
De Jong, W., Zandberg, P., and Bohus, B. (1975). *Prog. Brain Res.* **42**, 285-298.
Dengler, H. and Reichel, G. (1958). *Naunyn Schmiedebergs Arch. Exp. Pathol. Pharmakol.* **234**, 275-281.
Gifford, R.W., Jr., ed. (1972). Methyldopa in the management of hypertension. West Point, PA, Merck, Sharp & Dohme.
Gillespie, L., Oates, J.A., Crout, J.R., and Sjoerdsma, A. (1962). *Circulation* **25**, 281-291.
Goldberg, L.I., DaCosta, F.M., and Ozaki, M. (1960). *Nature* **188**, 502-504.
Henning, M. (1984). In "Handbook of hypertension. Vol. 3: Pharmacology of antihypertensive drugs" (P.A. van Zwieten, ed.), pp. 154-193. Amsterdam, Elsevier.
Henning, M. and Van Zwieten, P.A. (1968). *J. Pharm. Pharmacol.* **20**, 409-417.
Kunos, G., Farsang, C., and Ramirez-Gonzales, M.D. (1981). *Science* **211**, 82-84.
Louis, W.J., Conway, E., Summers, R., Beart, P., and Jarrott, B. (1984). *Hypertension* **6** (Suppl. 2), 40-44.
Naranjo, J.R., Fernandez-Roman, M., Urdin, M.C., and Fuentes, J.A. (1985). *Gen. Pharmacol.* **16**, 287-290.
Nijkamp, F.P. and De Jong, W. (1977). *Prog. Brain Res.* **47**, 349-368.
Oates, J.A., Gillespie, L., Udenfriend, S., and Sjoerdsma, A. (1960). *Science* **131**,1890-1891.
Petty, M.A. and De Jong, W. (1982). *Eur. J. Pharmacol.* **81**, 449-457.
Petty, M.A. and De Jong, W. (1984). *Neuropharmacology* **23**, 643-648.
Pfister, K., III and Stein, G.A. (1959) U.S. Patent 2,868,818.
Sitsen, J.M.A. and De Jong, W. (1983). *Hypertension* **5**, 185-190.
Sitsen, J.M.A. and De Jong, W. (1984). *Clin. Exp. Hypertens.* [A] **A6**:1345-1356.
Sjoerdsma, A. (1963). *Heart Bull.* **12**, 1-4.
Sjoerdsma, A. (1982). *Br. J. Clin. Pharmacol.* **13**, 45-49.
Sjoerdsma, A., Vendsalu, A., and Engelman, K. (1963). *Circulation* **28**, 492-502.
Sourkes, T.L. (1954). *Arch. Biochem. Biophys.* **51**, 444-456.
Stein, G.A., Bronner, H.A., and Pfister, K. (1954). *J. Am. Chem. Soc.* **77**, 700-703.
Van der Maas, M.L., Head, G.A., and De Jong, W. (1986). *J. Cardiovasc. Pharmacol.* **8**, 1014-1019.
Van Giersbergen, P.L.M. and De Jong, W. (1988). *J. Pharmacol. Exp. Ther.* **244**, 341-347.

Van Giersbergen, P.L.M., Head, G.A., and De Jong, W. (1988). In "Opioid peptides and blood pressure control" (K.O. Stumpe, K. Kraft, A.J. Faden, eds.), pp. 174-182. Berlin, New York, Springer-Verlag.

Van Giersbergen, P.L.M., Wiegant, V.M., and De Jong, W. (1989a). *Neuroendocrinology* **49**, 71-79.

Van Giersbergen, P.L.M., Tierney, S.A.V., Wiegant, V.M., and De Jong, W. (1989b). *Hypertension* **13**, 83-90.

Versteeg, D.H.G., Petty, M.A., Bohus, B., and De Jong, W. (1984). In "Handbook of hypertension, Vol. 4: Experimental and genetic models of hypertension" (W. De Jong, ed.), pp. 397-430. Amsterdam, Elsevier.

Wellens, D., Wouters, L., Nijkamp, F.P., and De Jong, W. (1976). *Experientia* **32**, 85-87.

Westermann, E., Balzer, H., and Knell, J. (1958). *Naunyn Schmiedebergs Arch. Exp. Pathol. Pharmakol.* **234**, 194-205.

6

CONSIDERING RECEPTOR NON-SPECIFICITY AS A FACTOR FOR DEVELOPING EFFECT, TISSUE AND RECEPTOR SPECIFIC AGENTS[1]

Kenneth L. Melmon
Manzoor M. Khan

Division of Clinical Pharmacology
Departments of Medicine and Pharmacology
Stanford University
Stanford, CA 94305

I. INTRODUCTION

Lessons learned from the study of structure activity relationships and molecular profiles of receptors for catecholamines and other cell surface agonists often have been used to seek analogous enzyme-substrate relationships. We have learned about the heterogeneity of similar receptors. In fact heterogeneity of receptors is more the rule than exception. The variety in receptors has been described by various investigators in terms of subclasses, families, types and subtypes. The subclasses of adrenergic (both α and β), muscarinic, cholinergic, histaminergic, dopaminergic, serotoninergic and adenosinergic receptors have been well characterized. The more we learn about the heterogeneity of the receptors, the transducers that different receptors employ, and the selective effects that different agonists have on a specific subclass of a receptor type, the more we can justify looking for parallels in enzyme-substrate relationships. The ongoing work that defines the properties of receptors and their agonists and antagonists might also apply to enzymes and their substrates. Perhaps allosteric forms of enzymes are to substrates as some subclasses of α- and β-adrenergic receptors or histaminergic systems are to respective agonists or antagonists.

Recently we have developed similar congener derivatives and conjugates of four chemically unrelated pharmacophores. The compounds used for derivatization included isoproterenol (a β-agonist), propranolol, practolol (β-antagonists), and histamine. These low molecular weight agonists and antagonists of two separate receptor

[1]*Work supported by NIH-NIAID AI-23463.*

Enzymes as Targets for Drug Design

systems were derivatized to determine whether adding pharmacologically inert aliphatic or aromatic groups (tails) to the molecules might preserve their pharmacologic properties. These structural modifications imparted net chemical changes in the new molecules. We had postulated that the chemical changes would influence the distribution and pharmacokinetics of the compound. The congeners and conjugates were expected to behave as their progenitor did *in vitro*, binding to classical receptor subtypes and inducing the same second messenger as the prototype drug had. The only pharmacologic novelty we guessed could occur, would have been via kinetic and pharmaco-dynamic properties *in vivo*. We believed that the changes in *in vivo* properties would have been explained by altered kinetics.

Our observations differed from our expectations. The most pronounced skews in the pharmacologic properties of the derivatives were found in structurally similar congener derivatives and conjugates of isoproterenol, propranolol, practolol and histamine. Yet, the pharmacologically inert tail was altering the pharmacologic properties of each ligand in analogous ways. The changes in pharmacologic properties did not appear to be based on gross differences in the pharmacokinetics of the new chemicals versus their progenitors (Ferraiolo *et al.*, 1987). Rather the changes appear to be based on congener-induced alterations in binding properties of the new chemicals, at/or nearby but not exclusively on the receptor (Schramm *et al.*, 1986; Eimerl *et al.*, 1987; Marki *et al.*, 1988; Eigenmann *et al.*, 1989).

The pharmacologically inert tails altered the potency, efficacy and tissue and affect specificity of the pharmacophores by altering their affinity for the microenvironment of the respective receptors. The free tails themselves did not affect the pharmacologic function of the prototype agonists or antagonists in *in vitro* experiments. Because similar derivatives of unrelated chemicals had similar alterations in potency, receptor affinity and tissue and effect specificity, we postulate that the microenvironments of different receptors in different tissues may be chemically similar. We also wonder whether other agonists or antagonists not yet derivatized will behave as catecholamine, histamine, propranolol, and practolol derivatives. Preliminary data so indicate with the derivatives of antagonists of adenosine (Jacobson *et al.*, 1986).

Go one more step. Assume these chemical tails have generalizable, and ultimately predictable affects on the pharmacologic properties of the various agonists. Would such observations be used and extrapolated as a model strategy for understanding substrate-enzyme relationships? That is, could there be a non-specific microenvironment on or near active sites of enzymes that could be determinants of substrate specificity? If so, the scheme we have used to develop tissue and effect specific agents for adrenergic and histaminergic receptors might be used to do the same for defined enzyme systems.

II. METHODS

A. Synthesis of Congener Derivatives and Conjugates of Norepinephrine, Propranolol, Practolol, and Histamine

1. General Schemes

Figure 1 illustrates the general approach to constructing the congener derivatives and conjugates of each pharmacophore. The complete methods for the synthesis, purification and characterization of the catecholamines (up to octa peptides) (Jacobson *et al.*, 1983a; Jacobson *et al.*, 1983b; Reitz *et al.*, 1985a; Reitz *et al.*, 1985b; Fakuda *et al.*, 1987); propranolol and practolol (Khan *et al.*, 1987a); and histamine (up to dipeptides) (Khan *et al.*, 1987b), their purification and characterization have been fully described.

B. Pharmacologic Testing of Congeners and Conjugates

The pharmacologic testing for each of the derivatives of the β-adrenergic agonists and antagonists included binding and displacement assays on whole cell and membrane preparations of S-49 mouse lymphoma cells (a source of β_2-receptors), turkey erythrocytes (β_1-receptors), rat adipocytes (β_1-receptors), rat heart (predominantly β_1-receptors), and rat lung (predominantly β_2-receptors). Biologic activity of the progenitors and their derivatives were first measured by either cAMP accumulation (agonists) or inhibition of isoproterenol-induced cyclic cAMP generation. Later more complex effects were assessed in *in vitro* preparations of intact heart, papillary muscle from the heart, and bronchial tissue (Marki *et al.*, 1988; Eigenmann *et al.*, 1989; Rosenkranz *et al.*, 1983a; Rosenkranz *et al.*, 1983b). *In vivo* tests ranged from simple measurements of heart rate and blood pressure in anesthetized rats, to complex measurements of cardiac contractility, chronotropic actions, coronary vasodilation, and

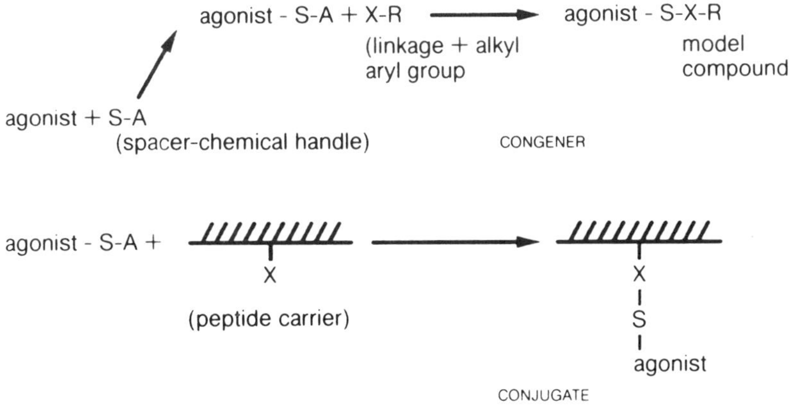

Fig. 1. *Congener approach to drug conjugate design. (Reprinted by permission from Melmon and Khan, 1987.)*

systemic metabolic and hormonal effects ordinarily expected of catecholamines in rats, rabbits, and/or dogs.

The biologic functions of the derivatives of histamine were tested by various *in vitro* assays including the accumulation of intracellular cAMP in murine lymphocytes (H_2-receptor effect), assessing the suppressor cell activity in mixed leukocyte reaction (MLR) preparations (H_1-receptor effect), measuring the adenylate cyclase activation in the guinea pig myocardium (H_2-receptor effect) and vasoconstriction of the rabbit aorta (H_1-receptor effect). The immunomodulatory effects of two of the derivatives of histamine were tested in murine *in vivo* models. The models included the induction and assay of cytolytic T cell activity, antibody response to sheep red blood cells and oxazolone-induced delayed type hypersensitivity response.

III. RESULTS

A. Catecholamines

About 100 derivatives were made of norepinephrine to determine which congener derivatives and conjugates were active on S-49 lymphoma cells and on other *in vitro* models mentioned above (Jacobson *et al.*, 1983a; Jacobson *et al.*, 1983b; Reitz *et al.*, 1985a; Reitz *et al.*, 1985b; Fakuda *et al.*, 1987). All the active compounds were made from substitutions on the amino end of the amine. Since the pharmacologically most interesting congeners developed from branched chain substitution of norepinephrine, we discuss our congeners as being derived from isoproterenol. As can be seen in Table I, showing a representative sample of congeners' actions on S-49 cells, relatively small changes in chemical composition made remarkable differences in potency when compared to isoproterenol. The pharmacologic properties of the congeners differed from the progenitor in far more than potency. Effect and receptor selectivity as well as pharmacodynamics were altered.

For instance, compound 3 in Table I was orally active (Eigenmann *et al.*, 1989). (Isoproterenol is not orally active). It had clear β_1-selectivity in radioligand studies (Marki *et al.*, 1988; Eigenmann *et al.*, 1989; Rosenkranz *et al.*, 1983a). It exerted extremely tight binding to receptors in that its effects could not be terminated readily by washing tissues (Marki *et al.*, 1988; Eigenmann *et al.*, 1989). Some of the altered potency appeared related to the molecule's interaction within the lipid microenvironment of the receptor, although all of its potency changes could not be explained on that basis (Schramm *et al.*, 1986). All of the effects it exhibited were based on its β-mimetic properties. Blockade, for example, of its inotropic effects on isolated cardiac tissues was competitive with propranolol. When the cardiac tissue was washed clear of blocker, the effects of the agonist returned without adding additional agonist. Thus, part of the affinity of the compound to tissues was in addition to its binding to receptor.

As a full β-agonist, it elicited positive inotropic effects with 10-360 times the potency of isoproterenol and an extremely long duration of action (Marki *et al.*, 1988; Eigenmann *et al.*, 1989). Electrophysiological studies in isolated guinea pig papillary muscles confirmed the β_1-receptor mediated actions -- all of which would be expected of a drug that had β_1-agonistic action (Marki *et al.*, 1988; Eigenmann *et al.*, 1989;

Table I. *In vitro* biological activity of congeners and model derivatives of isoproterenol. (Reprinted by permission from Melmon *et al.*, 1987.)

Biological activity was measured by cyclic AMP accumulation in S-49 cells. Relative activity is expressed as the ratio of the K_a for isoproterenol to K_a for the test compound.

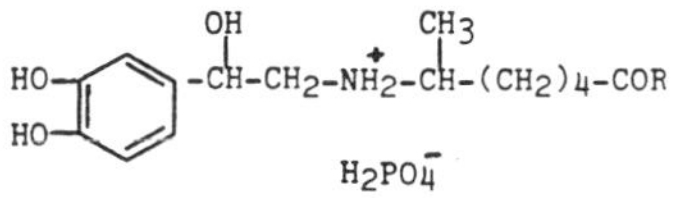

Compound	R	n[a]	Activity relative to isoproterenol
1. Carboxylic acid	$-OH$	4	1.6×10^{-4}
2. p-Toluide	$-NH-C_6H_4-CH_3$	4	5.4×10^{1}
3. p-Trifluoromethylanilide	$-NH-C_6H_4-CF_3$	4	8.6×10^{2}
4. m-Trifluoromethylanilide	$-NH-C_6H_4-CF_3$ (meta)	4	9.8×10^{-1}
5. o-Trifluoromethylanilide	$-NH-C_6H_4-CF_3$ (ortho)	4	2.5×10^{-4}
6. p-Butylanilide	$-NH-C_6H_4-(CH_2)_3CH_3$	4	3.2×10^{3}
7. p-Methoxylanilide	$-NH-C_6H_4-OCH_3$	4	3.1
8. p-Toluide	$-NH-C_6H_4-CH_3$	2	7.1×10^{-3}
9. p-Toluide	$-NH-C_6H_4-CH_3$	3	3.6×10^{-1}
10. p-Toluide	$-NH-C_6H_4-CH_3$	5	1.9×10^{-1}

[a]n=Number of methylene groups in the spacer group.

Rosenkranz *et al.*, 1983a). What seemed peculiar at the time, was that the isoproterenol derivatives did not have a number of actions that isoproterenol did. For instance, the drug was not able to reverse histamine-induced bronchoconstriction. It did not increase O_2 consumption in the periphery, and although it had profound inotropic effects, its chronotropic actions were far less than caused by isoproterenol (Marki *et al.*, 1988; Eigenmann *et al.*, 1989). In essence the agonist was not only selective to β_1-actions, it also became tissue selective (Rosenkranz *et al.*, 1983a).

Testing of stereospecific diasteromers indicated that almost all of the activity was in the RR isomer. What was striking was the fact that alteration of the configuration to S at either site led to remarkably less pharmacologic activity (Schramm *et al.*, 1986; Marki *et al.*, 1988). These studies indicated that configuration of the asymmetrical site of the pharmacologically inert tail profoundly influenced the activity of the ligand. The case was being made that non-specific sites, on or near the receptor, were playing a role in receptor selectivity and tissue specific actions of these congener derivatives.

B. β-Adrenergic Antagonist Derivatives: Propranolol and Practolol

We next asked whether the potential value of making tissue and effect specific β-agonists could also apply to β-antagonists and whether the ligands that had been used to create the most interesting skews in potency and effect with the agonist would do the same with antagonists.

Approximately 50 derivatives of propranolol and practolol were constructed. Early pharmacologic screening demonstrated that the compounds that resembled the isoproterenol congener derivatives and conjugates were those with the most interesting changes in potency, tissue and effect specificity. Table II shows a sampling of propranolol derivatives that were analogous to the isoproterenol derivatives. The analogous compounds in the two series had similarities, as well as differences, in their structure activity relationships. As shown in Table II, compounds 4 and 6 correspond with compounds 2 and 3 of the agonist series and reveal the remarkable tissue and effect specificity of the two antagonist derivatives. Derivatives of propranolol were much more potent β_1-antagonists than practolol or any of its derivatives. Furthermore, they developed tissue specificity far different from their progenitors. In fact compound 6 and a dipeptide of propranolol were the first competitive inhibitors of the atypical β_1-receptors on adipocytes (Khan *et al.*, 1987a; Zhu *et al.*, 1989). The same dipeptide was one of the most potent selective β_1-antagonists of the series and was far more potent in this selective effect than practolol or any of its congeners and conjugates (Khan *et al.*, 1987a).

The β-adrenergic antagonist activities of the *p*-toluidide and a *p*-trifluoromethylanilide derivative of propranolol (compounds 4 and 6, Table II) were tested in intact rats to determine whether the unusual *in vitro* profiles on myocardium and adipose tissue were found *in vivo*. The relative potency of *p*-toluidide derivatives studied in

Table II. Relative potencies of β-andrenergic antagonist derivatives. (Reprinted by permission from Melmon *et al.*, 1987).

	S-49/Lung	Fat cells/Heart	S-49/Fat cells	Lung/Heart**
Propranol	.54	1.0	0.1	.77
1 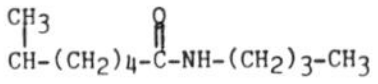	0.1	25	.003	10
4 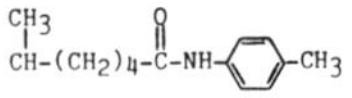	.0054	15	.0005	1.3
6 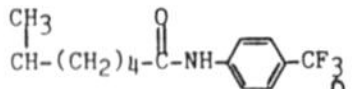	.05	414	.0003	2.5
9 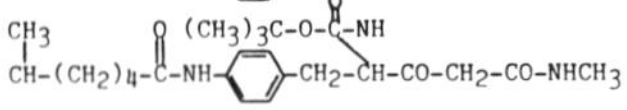	.26	3.0	.13	1.5
Practolol	.25	.81	.91	14.5
14 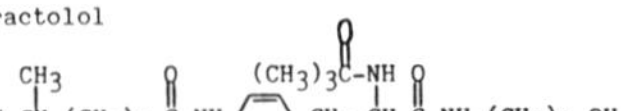	.048	1.45	1.39	42.0

*Taken from inhibition constants (Ki).

**Rat lung and heart contain 15-20% beta 1 and beta 2 receptors respectively.

pithed rats was 1.3 for dP/dt and 4.6 for heart rate compared to propranolol. That for *p*-trifluoromethylanilide was 2.5 dP/dt and 4.6 for heart rate. The SR-isomer of the *p*-toluidide derivatives was 81 times more potent than propranolol in inhibiting effects on dP/dt and 27 times more effective than propranolol on heart rate while the SS-isomer was 0.17 times on dP/dt and 0.16 times as potent as the parent compound on heart rate (Zhu *et al.*, 1989).

In fasted rats, infusion of isoproterenol resulted in an increase of 10.6 ± 3.1 mg/dl in plasma non-esterified fatty acid (NEFA) and an increase of 46 ± 9 mg/dl in glucose. Unlike propranolol, neither the *p*-toluidide nor the *p*-trifluoromethylanilide derivatives blocked the increase in plasma NEFA, although they both blocked the increase in plasma glucose (Zhu *et al.*, 1989). The *in vitro* tests had predicted the *in vivo* selectivity for the β_1-adrenergic receptors on the heart as opposed to the β_1-like adrenoceptors on adipose tissue for the *p*-toluidide and *p*-trifluoromethylanilide derivatives. These findings were qualitatively and generally quantitatively in agreement with our previous findings *in vitro*.

C. Histamine

After testing the congener derivatives and conjugates of β-adrenergic agonists and antagonists, and observing the most interesting skews in the pharmacologic effects of similar congener derivatives and conjugates of both series, our focus changed. The chemical tails we used for derivatization had not changed the pharmacokinetics and thereby the pharmacodynamics of congener derivatives and conjugates. Instead, they were influencing the tissue and effect specificity of the compounds by some effects they seemed to have on the microenvironment of the receptors. Key questions were what molecular interactions in the microenvironment of the receptor allowed the effect and tissue specificity and were microenvironments of the β-adrenergic receptors common to other cell surface receptors. We have not answered the first of the questions. But by making only 19 congeners and conjugates of histamine, we have found that at least two pharmacologically unrelated receptor systems whose agonists are not chemically related behave similarly as congener derivatives and conjugates.

For purposes of this discussion, five of the 19 derivatives of histamine had the most interesting differences in their pharmacologic effects from histamine. They are listed in Table III and by now, not surprisingly, the list, includes the most notable compounds in the β-agonist and antagonist series. All of the compounds were active in only one of the three tissues tested (Khan *et al.*, 1986), whereas histamine stimulated all three. Methylene chain length had been a critical factor in potency of the catecholamine and antagonist derivatives. Therefore, the effects of the changes in the methylene chain length, on the pharmacologic activity of the congener derivatives and conjugates of histamines were studied. Compounds 1 and 2 differed from each other by one methylene chain length. Both were active only on lymphocytes (the natural suppressor cells). But the one methylene group difference changed compound 1, a very potent H_1- and H_2-receptor stimulator of lymphocyte receptors, to an exclusive stimulator of H_1-receptors on lymphocytes (Khan *et al.*, 1986). The same one methylene group difference between the two dipeptides (compounds 4 and 5) was sufficient for com-

Table III. Responses to congener derivatives and conjugates of histamine. (Reprinted by permission from Melmon and Khan, 1987.)

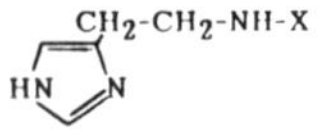

	Relative Potencies		Effects on	
	Natural Suppressor Cells (H_2)	Guinea Pig Myocardium (H_2)	Natural Suppressor Cells (H_1)	Rabbit Aorta (H_1)
1) 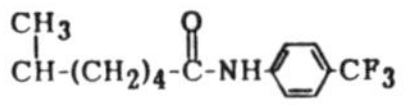	4.3×10^4	Inactive	+	Inactive
2) 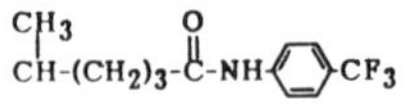	Inactive	Inactive	+	Inactive
3) 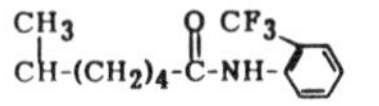	Inactive	Inactive	Inactive	Inactive
4) 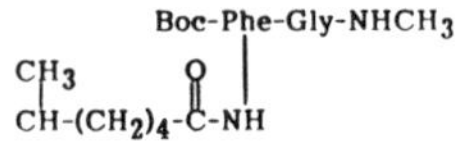	Inactive	0.7×10^{-3}	Inactive	Inactive
5) 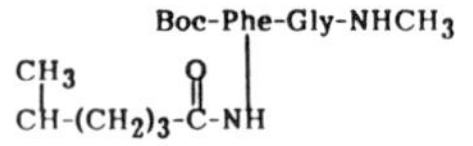	Inactive	Inactive	+	Inactive

H_2 activity on natural suppressor cell was measured by cAMP accumulation assays, and for guinea pig myocardium (H_2) by adenylate cyclase stimulation assay. H_1 activity on natural suppressor cells was measured by the effects of the congeners and conjugates of histamine on the suppressive capacity of NS cells in mixed leukocyte reaction. Relative potency is expressed as the ratio of K_a for histamine to K_a for the compound. K_A for histamine on natural suppressor cells was 1.8×10^{-5} and on guinea pig myocardium was 1.7×10^{-7}.

pound 4 to be an H_2-receptor agonist on myocardium and for compound 5 to be an H_1-receptor agonist on the natural suppressor cells (Khan *et al.*, 1986). Recent data from Dr. J.C. Schwartz' laboratory in Paris indicates that both compounds 1 and 2 (Table III) were blockers of H_3 receptors in CNS tissue with Ki values of 3 μM each (data not reported).

Finally, comparison of the potencies of the two series of agonist congener derivatives and conjugates generally shows the similarities of like congener derivatives of isoproterenol and histamine (Table IV). It is evident that the pharmacologic activity of various congener derivatives is based upon several factors of modification. These criteria include the sensitivity of each series to chain length, the requirement for presence of a branched chain, the effects of a toluidide group, and the dependence on para methyl or trifluoromethyl substitution on the group. Also note the dramatic drop in potency when the trifluoromethyl group is placed in the ortho position (Table IV).

In *in vivo* experiments, compound 1 suppressed the generation of humoral and cell mediated cytolytic cell responses but did not affect the generation of DTH response (manuscript in preparation). Compound 2 had no effect on generation of any of the *in vivo* immune responses studied. The mechanisms of the suppressive effect of compound 1 are not known but its value as a potential immune modulator is evident (data not reported).

Table IV. Comparative potencies of similar isoproterenol and histamine congener derivatives on S-49 lymphoma and natural suppressor cells. (Reprinted by permission from Melmon *et al.*, 1987.)

	Activity Relative to Isoproterenol	Activity Relative To Histamine
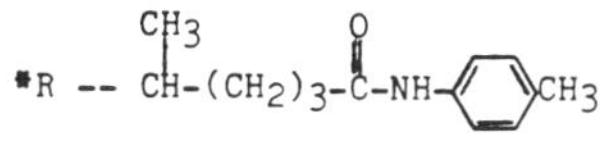	3.11×10^{-1}	3.0×10^{3}
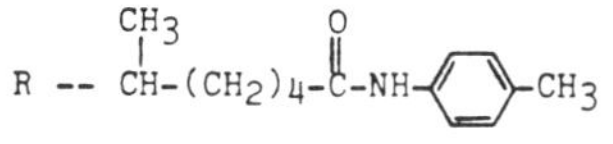	4.40×10^{1}	$.30 \times 10^{1}$
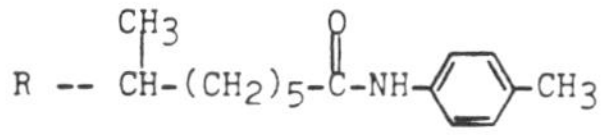	1.11×10^{-1}	$.20 \times 10^{1}$
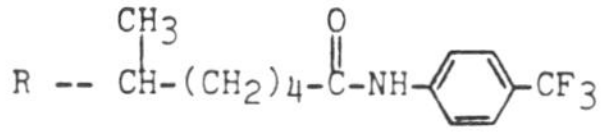	8.60×10^{2}	4.2×10^{4}
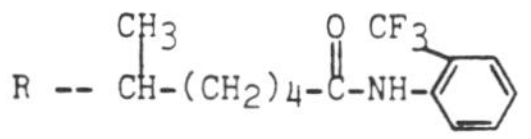	2.6×10^{-4}	Inactive
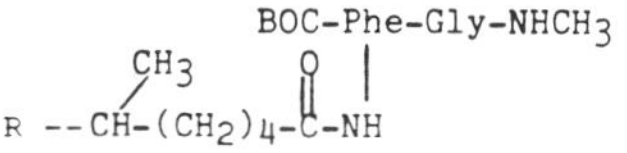	0.4×10^{1}	Inactive
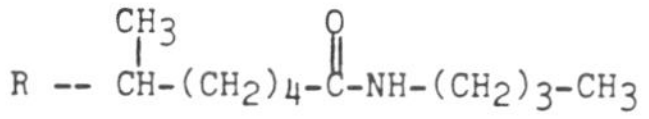	2.5×10^{-1}	Inactive

Biological activity was measured by cAMP accumulation in a mouse natural suppressor cell and in S-49 cells. Relative activity is expressed as the ratio of Ka for histamine or isoproterenol to Ka for the compound.

*R= 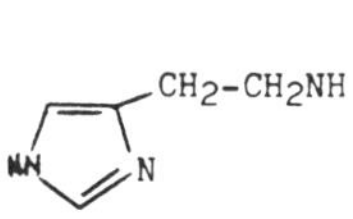 or 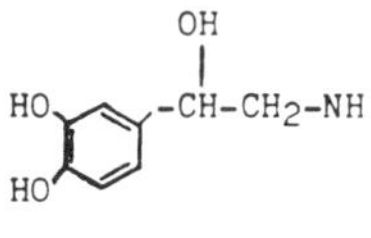

To qualitatively summarize the parallels in pharmacologic properties of each series of derived pharmacophores, a review of Table V shows that similar derivatives of the pharmacophores had similar changes in their potency as well as their tissue selectivity profiles. Furthermore, it is interesting to note that Jacobson *et al.* (1986) found analogous changes in pharmacologic activities of a series of adenosine antagonists. The next obvious and most interesting experiments would be to derivatize adenosine receptor agonists and histamine receptor antagonists. This strategy could be used to further test the generalizability of the value of our chemical tails which utilize the receptor microenvironment to alter the potency and restrict the effects of membrane active pharmacophores. Such experiments are in progress.

IV. DISCUSSION AND CONCLUSIONS

There is nothing new about the legitimacy of derivatizing pharmacologic ligands to alter the potency, tissue and effect specificity of the ligand. Most of the time, however, schemes for derivatization are stylized to fit a given molecule. Little can be said about the value of a derivatization scheme for one molecule when it is applied to another. What is different from other work in this regard is that our work deliberately used the same derivatization scheme for unrelated molecules and came out with what appears to be generalizable results. In summary, when five different molecules active in three different pharmacologic receptor systems, were similarly derivatized, the congener derivatives developed similar types of skews in their pharmacologic profile. The quality of the skews was not predictable for any given series, but that a skew would be found became predictable as witnessed by the ability to limit the number of tested compounds in successive series (Melmon *et al.*, 1987). These observations probably are not proof of the hypothesis that focuses on receptor non-specificity as a way to develop tissue and effect specific compounds from a progenitor with otherwise wide-spread effects. If these examples are not sufficient to prove that the approach is valid, they are at least provocative. Clearly there is justification for extending the approach to other molecules. For both pharmacologic and therapeutic sake, we suggest that antihistamines and adenosine be subjected to the same congener and conjugate technique.

The mechanisms of the altered pharmacology of these derivatives are not known, but could be assessed in future experiments. So far we can say that the changes in pharmacologic profile are not related entirely to the receptor recognition site. Instead, they seem dependent on some elements of the non-specific areas of receptor as opposed to the portion of receptor that recognizes agonist or antagonist. The adrenergic receptors that mediate the physiologic effects of catecholamines belong to a family of plasma membrane receptors that are coupled to G proteins. A number of neurotransmitters are included in that family and histamine is likely to be one of those. Experiments have partially defined the molecular mechanisms of binding of agonists and antagonists specifically to their receptors in this family. Among the approaches that have helped to delineate the domains of a receptor that are involved in its specificity and mechanisms of linking receptor occupation with G protein coupling are the construction of chimeras of α_2- and β_2-adrenergic receptors (Dixon *et al.*, 1986; Kobilka *et al.*, 1987; Cerione *et al.*, 1985; Weiss *et al.*, 1988; Kobilka *et al.*, 1988). In these experiments, about seven hydrophobic membrane spanning domains of the receptors have been identified, but only a couple (usually the 6th and 7th domains) were involved with recognition of agonists and antagonists (Dixon *et al.*, 1986; Kobilka *et al.*, 1987; Cerione *et al.*, 1985; Weiss *et al.*, 1988; Kobilka *et al.*, 1988). Perhaps the specificity of our congener derivatives is dependent on the non-specificity of one or some combination of several other domains of the receptor molecule. Direct binding studies using our reagents on various chimeras could verify or refute this concept. In

Table V. Structure predicts pharmacologic novelty. (Reprinted by permission from Melmon *et al.*, 1987.)

STRUCTURE	ISOPROTERENOL (Type I)	PROPRANOLOL (β_2) (Type II)	PRACTOLOL (β_1)	HISTAMINE (Type III)
METHYLENE CHAIN				
C_2	⬇	⬇	—	↓
C_3	↓	⬇	↓	⬆ H_1-Lymph
C_4	⬆	⬆	⬆	⬆ H_1,H_2-Lymph
C_5	↓	↓	↓	↓
BRANCHED METHYLENE CHAIN	↑	↑	↑	↑
TOLUIDIDE MOIETIES				
P-methyl	↑ β_1 (Heart vs Lung-TS)	↑ β_1 (Fat vs Heart) RS & TS	↑ β_1 RS & TS	⬆ RS-H_1 & TS-lymph
P-trifluoromethyl	⬆ OA: TS [Peripheral vs Coronary Metabolic, Skeletal Muscle (Glycogen)]	↓ β_1 (Fat vs Heart)	↑ β_2,RS & TS	⬆ RS-H_1 & TS-lymph
Ortho	↓	—	—	↓
Meta	↔	—	—	—
AMINO ACID	↑; β_1 (TS, RS)	↑; β_1 (TS, RS)	↑; β_1 TS, RS	—
DIPEPTIDE	↑; OA; (TS, RS)	↓ Fat vs Heart	—	TS-Heart vs lymph RS-H_2
LONGER PEPTIDE	↔ or ↓	—	—	—

* Histamine derivatives not active on heart (H_2) can be vasoconstrictive (H_1). III-1 or vasorelaxing with H_2 III-6 & 9

OA = Oral activity, TS = Tissue selectivity, NT = No Tachyphylaxis in heart

RS = Receptor selectivity, —— = Not tested

addition, perhaps the generalization of the concept, we propose, also could be tested by derivatizing ligands for receptors not included in the catecholamine receptor family. If the strategy worked, and if their domains were analogous to the neurotransmitter family, the mechanisms could be proven. If the new derivatives for agonists or antagonists of receptors unrelated to catecholamine and histamine receptors did not develop tissue or effect specificity, examination of the differences between the "non-specific" domains of the receptor systems might provide an explanation.

Extrapolation of the concepts discussed above can also be applied to enzyme-substrate relationships even without a molecular explanation for why our compounds developed specificity. If we were to design substrates with chemical variability not in the region of substrate recognition sites, but in the "less important" domains of the enzyme, we might begin taking advantage of enzyme non-specificity in the same way we may take advantage of receptor non-specificity, to develop receptor or tissue specific drugs. Could the derivatization concept be applied to the decarboxylases, hydroxylases, reductases, proteases and oxidases discussed in this Symposium? It might be time to find out.

Successful application of the hypothesis could even have medical value. Having catecholamine-based drugs that were orally active, inotropically active but chronotropically inactive would not be trivial (Marki *et al.*, 1988; Eigenmann *et al.*, 1989; Rosenkranz *et al.*, 1983a). Having β-blockers that did not share the LDL-cholesterol and triglyceride raising effects of other β-blockers is clinically interesting (Rosenkranz *et al.*, 1983b). Could the same be said about brain specific inhibitors of decarboxylase, or MAO, etc.? Developing a histamine-based immunomodulating drug has its appeal (Melmon and Khan, 1987). Our data makes each of these possibilities a little less of a dream than before.

ACKNOWLEDGEMENTS

We thank Drs. Murray Goodman, Quan Ming Zhu, Roberto Rosenkranz, and Kenneth Jacobson for participating in these projects and are grateful to Ms. Sara Fisher for her fine secretarial assistance.

REFERENCES

Cerione, R.A., Staniszewski, C., Benovic, J.L., Lefkowitz, R.J., Caron, M.G., Gierschik, P., Somers, R., Spiegal, A.M., Codina, J., and Birnbaumer, L. (1985). *J. Biol. Chem.* **260**, 1493-1500.

Dixon, R.A., Kobilka, B.K., Strader, D.J., Benovic, J.L., Dohlman, H.G., Frielle, T., Bolanowski, M.A., Bennett, C.D., Rands, E., Diehl, R.E., Mumford, R.A., Slater, E.E., Sigal, I.S., Caron, M.G., Lefkowitz, R.J., and Strader, C.D. (1986). *Nature* **321**, 75-79.

Eigenmann, R., Burkard, W.P., Holck, M., Marki, H.P., Osterrieder, W., Goodman, M., and Melmon, K.L. (1989). *Arzneimittelforschung*. In press.

Eimerl, S., Schramm, M., Lok, S., Goodman, M., Khan, M.M., and Melmon, K.L. (1987). *Biochem. Pharmacol.* **36**, 3523-3527.

Ferraiolo, B.L., Halldin, M.M., Asscher, Y., Akita, Y., Melmon, K.L., Benet, L.Z., and Castagnoli, N., Jr. (1987). *Pharmacology* **34**, 157-166.

Fukuda, M., Beechan, C.M., Verlander, M., Goodman, M., Khan, M.M., and Melmon, K.L. (1987). *Int. J. Pept. Protein Res.* **29**, 300-317.

Jacobson, K.A., Marr-Leisy, D., Rosenkranz, R.P., Verlander, M.S., Melmon, K.L., and Goodman, M. (1983a). *J. Med. Chem.* **26**, 492-499.

Jacobson, K.A., Verlander, M.S., Rosenkranz, R.P., Melmon, K.L., and Goodman, M. (1983b). *Int. J. Pept. Protein Res.* **22**, 284-304.

Jacobson, K.A., Kirk, K.L., Padgett, W.L., and Daly, J.W. (1986). *Mol. Pharmacol.* **29**, 126-153.

Khan, M.M., Marr-Leisy, D., Verlander, M., Bristow, M.R., Strober, S., Goodman, M., and Melmon, K.L. (1986). *J. Immunol.* **137**, 308-314.

Khan, M.M., Marr-Leisy, D., Rosenkranz, R.P., Hoffman, B.B., Verlander, M.S., Goodman, M., and Melmon, K.L. (1987a). *Biochem. Pharmacol.* **36**, 2175-2182.

Khan, M.M., Melmon, K.L., Marr-Leisy, D., Verlander, M.S., Egli, M., Lok, S., and Goodman, M. (1987b). *J. Med. Chem.* **30**, 2115-2120.

Kobilka, B.K., Matsui, H., Kobilka, T.S., Yang-feng, T.L., Francke, U., Caron, M.G., Lefkowitz, R.J., and Regan, J.W. (1987). *Science* **238**, 650-656.

Kobilka, B.K., Kobilka, T.S., Daniel, K., Regan, J.W., Caron, M.G., and Lefkowitz, R.J. (1988). *Science* **240**, 1310-1316.

Marki, H.P., Crameri, Y., Eigenmann, R., Krasso, A., Ramuz, H., Bernauer, K., Goodman, M., and Melmon, K.L. (1988). *Helv. Chim. Acta* **71**, 320-336.

Melmon, K.L. and Khan, M.M. (1987). *Trends Pharmacol. Sci.* **8**, 437-441.

Melmon, K.L., Khan, M.M., and Goodman, M. (1987). *UCLA Symp. Mol. Cell. Biol., New Ser.* **65**, 113-129.

Reitz, A.B., Sonveaux, E., Rosenkranz, R.P., Verlander, M.S., Melmon, K.L., Hoffman, B.B., Akita, Y., Castagnoli, N., and Goodman, M. (1985a). *J. Med. Chem.* **28**, 634-642.

Reitz, A.B., Avery, M.A., Rosenkranz, R.P., Verlander, M.S., Melmon, K.L., Hoffman, B.B., Akita, Y., Castagnoli, N., and Goodman, M. (1985b). *J. Med. Chem.* **28**, 642-647.

Rosenkranz, R.P., Hoffman, B.B., Jacobson, K.A., Verlander, M.S., Klevans, L., O'Donnell, M., Goodman, M., and Melmon, K.L. (1983a). *Mol. Pharmacol.*, **227**, 429-435.

Rosenkranz, R.P., Jacobson, K.A., Verlander, M.S., Klevans, L., O'Donnell, M., Goodman, M., and Melmon, K.L. (1983b). *J. Pharmacol. Exp. Ther.* **227**, 267-273.

Schramm, M., Eimerl, S., Goodman, M., Verlander, M.S., Khan, M.M., and Melmon, K.L. (1986). *Biochem. Pharmacol.* **35**, 2805-2809.

Weiss, E.R., Kelleher, D.J., Woon, C.W., Soparkar, S., Osawa, S., Heasley, L.E., and Johnson, G.L. (1988). *FASEB J.* **2**, 2841-2848.

Zhu, Q.M., Lee, C., Laughton, C., Khan, M.M., and Melmon, K.L. (1989). *Biochem. Pharmacol.* **38**, 1563-1567.

7

MECHANISM-BASED ENZYME INHIBITORS AS AN APPROACH TO DRUG DESIGN

Philippe Bey

Merrell Dow Research Institute
2110 E. Galbraith Road
Cincinnati, OH 45215

I. INTRODUCTION

Enzyme inhibitors acting in the enzyme's active site are classically categorized as reversible or irreversible. From a functional and kinetic standpoint, reversible inhibition is usually associated with a rapid noncovalent, time-independent interaction between the enzyme and the inhibitor, whereas irreversible inhibition is characterized by a time-dependent component, a chemical modification of the inhibitor, and the formation of a stable covalent bond between the enzyme and the inhibitor. Prior to 1970 the prototype of the irreversible inhibitor was the so-called affinity labeling agent. These agents usually are constructed by attaching to a reversible ligand of the target enzyme a chemically reactive group which eventually alkylates a nucleophilic residue within or nearby the active site. Although very useful for the study of enzymes *in vitro*, affinity labeling agents are generally not suitable as drugs owing to their intrinsic chemical reactivity.

In the late 1960's, Jencks and Wolfenden revived the concept formulated in 1946 by Pauling that inhibitors resembling the structure of the transition state should bind more tightly to the active site than analogs of the substrate in the ground state. Jencks (1966) identified many such inhibitors in the literature for which he coined the term transition-state analogs while Wolfenden (1969) first developed a thermodynamic analysis that gave the concept a theoretical basis and then put it to the experimental test by designing transition-state analog inhibitors for triose phosphate isomerase. At about the same time, the serendipitous discovery by Bloch *et al.* (Endo *et al.*, 1970) that β-hydroxydecanoyl thioester dehydrase is irreversibly inactivated by 3-decynoyl-*N*-acetylcysteamine, a chemically inert acetylenic analog of its substrate, laid the groundwork for the development of the new concept of "suicide" enzyme inhibition (Abeles and Maycock, 1976). These two concepts were to change fundamentally the research on enzyme inhibition as they opened the possibility to design rationally specific reversible and irreversible inhibitors for enzymes whose mechanism of action is established. Extensive exploitation over the last 15 years of these two concepts has led to the development of a wealth of new specific enzyme inhibitors (Silverman, 1988; Wolfenden and Frick, 1987). In this paper I present selected examples of suicide and transition-state inhibitors of therapeutic potential or relevance which for the most part were developed at the Merrell Dow Research Institute under the tenure of Al Sjoerdsma. I also illustrate that a functional distinction between the various types of suicide and transition-state inhibitors is a matter of degree and that, from a mechanistic standpoint, these inhibitors form a continuum which will be referred to as mechanism-based inhibitors.

II. DESIGN OF MECHANISM-BASED INHIBITORS BASED ON THE CONCEPT OF SUICIDE INHIBITION

Suicide inhibitors were originally defined as chemically unreactive pseudo-substrates that are transformed by the catalytic machinery of the target enzymes into chemically reactive species which eventually alkylate irreversibly a residue within the enzyme active site. In contrast to affinity labeling agents, suicide inhibitors are chemically unreactive. In addition, as the species that inactivates the enzyme is generated inside the active site, suicide inhibitors possess two tiers of selectivity: one derived from the binding and the other derived from the catalysis. These distinctive properties make suicide inhibitors particularly useful for *in vivo* applications. As a first approximation, the kinetic behavior of affinity labeling agents and suicide inhibitors is similar (Palfreyman *et al.*, 1987). An apparent dissociation constant K_i, which measures the affinity of the inhibitor for the enzyme, and a time of half-inactivation of the enzyme at infinite concentration of inhibitor τ_{50}, which is a measure of the rate of inactivation, can be derived from the kinetic analysis which is usually performed according to the method developed by Kitz and Wilson (1962) for affinity labeling agents.

A. Pyridoxal Phosphate-Dependent Enzymes

The pyridoxal phosphate (PLP) dependent enzymes are the first class of enzymes to have attracted a lot of attention as targets for design of mechanism-based inhibitors (Bey, 1981). As indicated in Scheme 1, PLP-dependent enzymes catalyze a variety of distinguishable reactions on physiologically important α-amino acid or amine substrates. The mechanism of action of these enzymes is well established. An obligatory intermediate for all PLP enzymes is the carbanionic quinonoid species **1** or **2**. This intermediate has different possibilities for subsequent reactions which are controlled by apoenzymes. A common approach to the design of mechanism-based inhibitors for these enzymes has been the incorporation of functionalities into their substrates or products that can alter the fate of the quinonoid intermediate to generate electrophilic species which eventually would alkylate a nucleophilic residue of the enzyme. The inhibition of ornithine decarboxylase (ODC) and γ-amino butyric acid (GABA) transaminase (GABA-T) by α-difluoromethylornithine **3** (DFMO; eflornithine) (Metcalf *et al.*, 1978) and γ-vinyl GABA **4** (GVG; vigabatrin) (Lippert *et al.*, 1977), respectively, will serve as generic illustrations of such mechanism-based inhibitors for

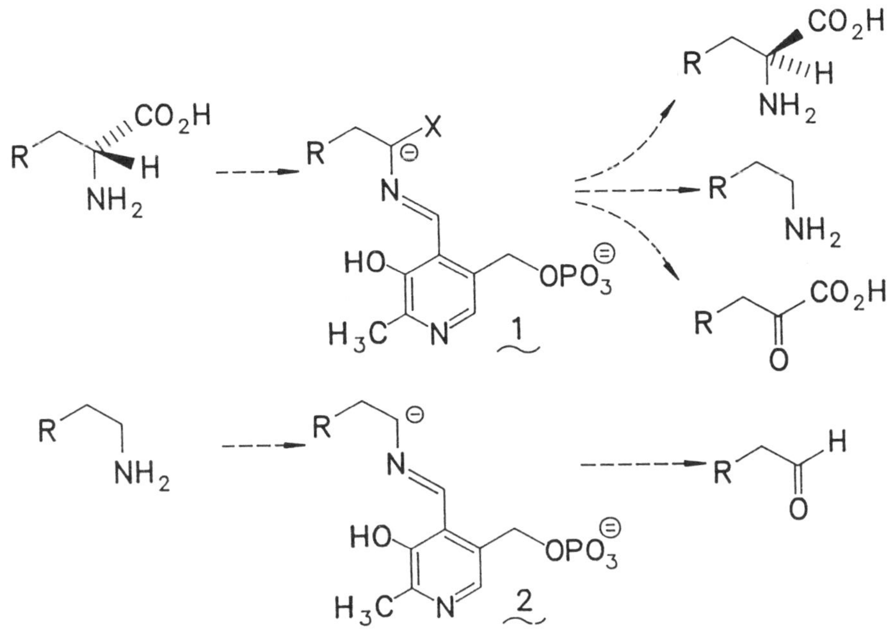

Scheme 1. *Reactions catalyzed by pyridoxal phosphate-dependent enzymes. The first line represents the racemization, the decarboxylation, and the transamination of an α-amino acid substrate whereas the second line corresponds to the transamination of an amine substrate. In all cases, a stabilized carbanionic intermediate (1 or 2) is formed. X can be either H or CO_2H. The charge is delocalized in the pyridine ring of the pyridoxal phosphate cofactor.*

α-amino acid decarboxylases and transaminases. The therapeutic relevance of ODC and GABA-T inhibitors is discussed in detail in a later chapter of this book (Schechter and Sjoerdsma, chapter 14, 1989).

DFMO, a close analog of the ODC substrate ornithine, was conceived on the premise that it would be accepted by the active site and decarboxylated by the catalytic machinery of ODC to form the carbanionic quinonoid intermediate **5** (Scheme 2). The normal electron flow to the PLP ring system could be expected to be redirected by the presence of the difluoromethyl group as one of the fluorine atoms is ideally positioned to undergo an elimination reaction. The resulting reactive electrophilic imine **6** then would eventually form a covalent bond with an adventitious nucleophilic residue (Nu) in the active site. Kinetic analysis of the inhibition of ODC by DFMO was in agreement with the mechanism of Scheme 2. Subsequent experiments with DFMO labeled with ^{14}C at the 1- or 5-position demonstrated that 3 moles of DFMO are decarboxylated for each mole of ODC inactivated (Pegg *et al.*, 1987); one mole of a DFMO metabolite is bound, most probably to the Lys-298 residue, per mole of inactivated ODC; and the covalent binding of the DFMO metabolite to ODC is so specific and so stable that

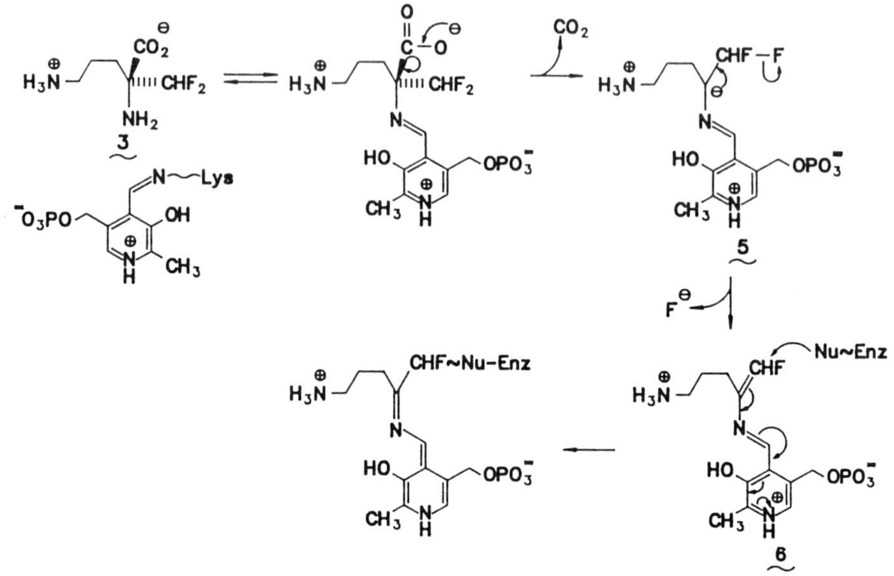

Scheme 2. *Inhibition of ornithine decarboxylase by α-difluoromethylornithine (DFMO). In the resting enzyme, the PLP cofactor is covalently bound to the apoenzyme through an internal aldimine with the terminal amino group of an active site lysine residue. See text for details.*

[^{3}H]DFMO can be used to titrate the number of active molecules of ODC in tissue extracts (Pegg, 1987). It is noteworthy that the structure of the DFMO-ODC adduct has not yet been established.

The proposed mechanism of inactivation of GABA-T by GVG is depicted in Scheme 3. As a close analog of GABA, GVG **4** is recognized and processed by GABA-T to generate inside the active site the electrophilic conjugated imine **7** which alkylates an adventitious active site nucleophilic residue (Nu). This mechanism is supported by the kinetics of the inactivation (Lippert *et al.*, 1977), the preferential inhibition by the (S) enantiomer of GVG (Danzin and Jung, 1984), and the primary isotope effect exhibited by γ-deuterio-GVG (Lippert *et al.*, 1980). The enzyme-inhibitor complex appears to be extremely stable, as dialysis of the inactivated enzyme for 5 days did not restore more than 5% of the activity.

Definitive proof for the mechanism of inhibition and the nature of the interaction between the enzyme and the mechanism-based inhibitor requires the determination of the structure of the adduct between the inhibitor and the inactivated enzyme. Very few structures of enzyme-inhibitor adducts are known. For GABA-T, only the structure of the adducts with the mechanism-based inhibitors γ-fluoromethyl GABA **8** and gabaculine **9** have been established. The structure of the γ-fluoromethyl GABA-GABA-T adduct (Silverman and Invergo, 1986) confirmed the formation of a covalent bond between the inhibitor and the protein. It also revealed that the inactivation mechanism

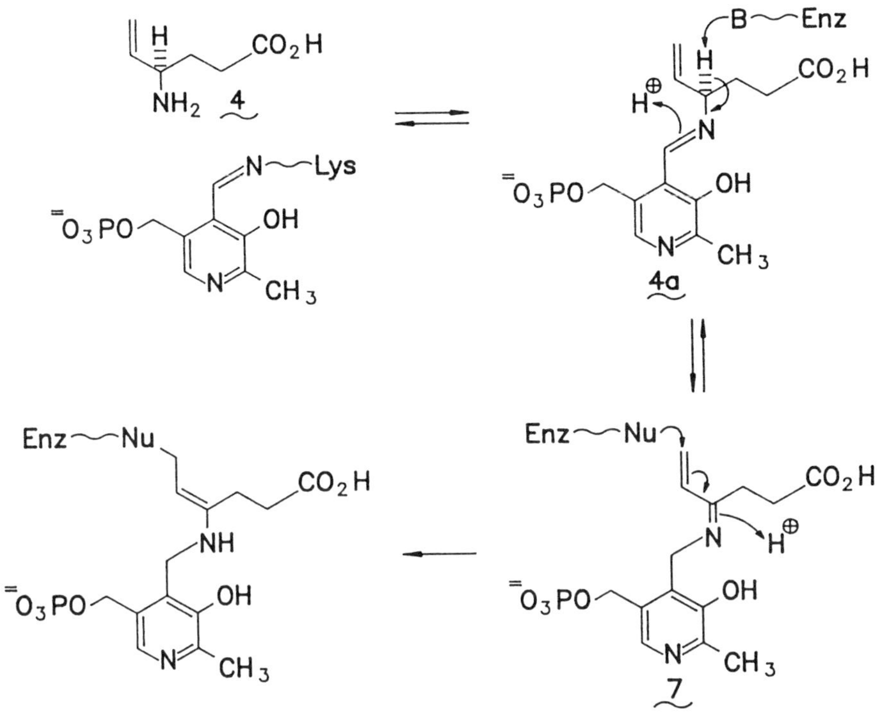

Scheme 3. *Postulated mechanism for the inactivation of GABA-T by γ-vinyl GABA. Enzymatic tautomerization of the imine 4a generates an electrophilic conjugated imine which alkylates a nucleophilic residue (Nu) from the active site.*

proceeds through the so-called Metzler enamine pathway (Likos *et al.*, 1982) (Scheme 4, path **b**) and not the conjugated imine pathway (Scheme 4, path **a**) as originally postulated. Gabaculine **9** (5-amino-1,3-cyclohexadienylcarboxylic acid), a natural dihydrophenyl analog of GVG and potent irreversible inhibitor of GABA-T, is processed by the enzyme into the reactive imine intermediate **10** which isomerized rapidly to form a stable *m*-anthranilic acid adduct with the cofactor **11** as illustrated in Scheme 5. The inhibitor-cofactor adduct remains tightly bound to the active site and is liberated only under conditions which denature the enzyme (Rando, 1977). The structure of the inhibitor-cofactor adduct clearly rules out the formation of a covalent bond between the protein and the inhibitor. This mechanism of inhibition is therefore different from that of GVG in the sense that, although a reactive species is generated in the active site, irreversible inactivation occurs without any covalent bond formation between the inhibitor and the enzyme. The two other isomers of gabaculine, i.e. 3-amino-1,5- and 3-amino-1,4-cyclohexadienyl carboxylic acids and the two dehydroaromatic amino acids, 4-amino-4,5-dihydro-2-thiophene carboxylic acid (**12**, X=S) and the 4-amino-4,5-dihydrofuran carboxylic acid (**12**, X=O), were also reported to irreversibly inhibit

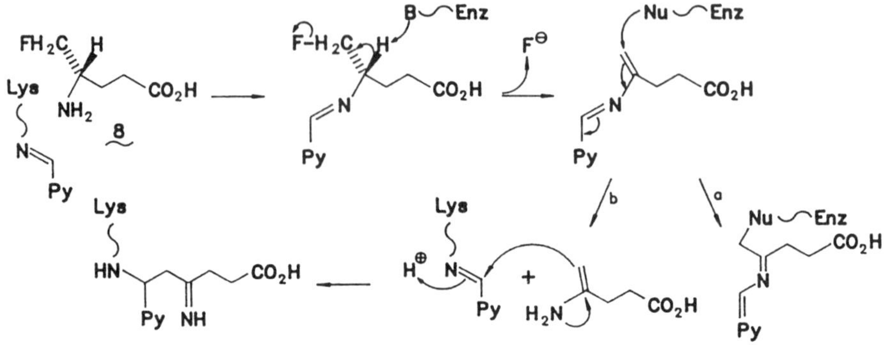

Scheme 4. *Mechanism of inhibition of GABA-T by γ-fluoromethyl GABA (8). The conjugated imine generated after elimination of fluoride does not alkylate a nucleophilic residue (Nu) of the enzyme as originally proposed (path **a**). Instead, it undergoes a transaldimination with the active site lysine residue to regenerate the internal PLP-ε-lysyl aldimine and a noncovalently bound enamine which attacks and covalently modifies the internal aldimine (path **b**).*

GABA-T (Lippert *et al.*, 1985), most probably by the same aromatization mechanism of gabaculine (Scheme 6).

Such mechanism where an enzyme catalyzes its own time-dependent inactivation by generating a nonreactive product that binds tightly to its active site is not unprece-

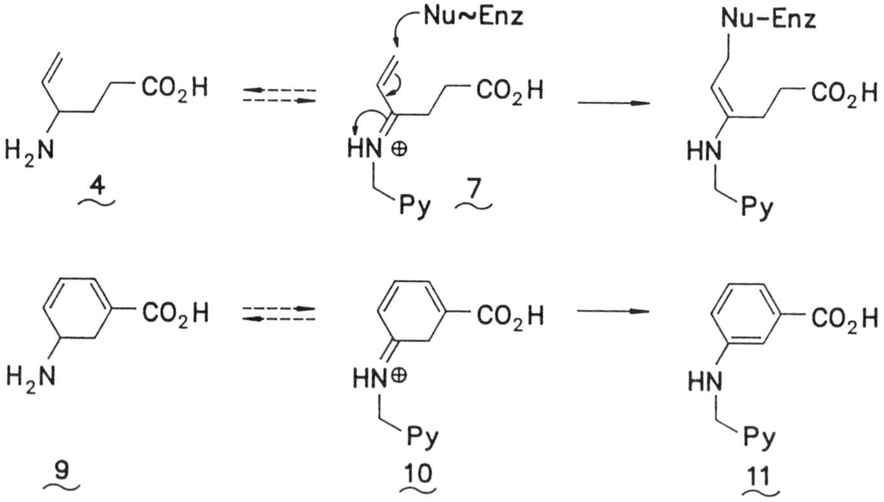

Scheme 5. *Mechanism of inhibition of GABA-T by gabaculine **9** (bottom line). The postulated mechanism for the inactivation of GABA-T by GVG **4** is indicated on the top line. Although similar intermediates **10** and **7** are generated, the mechanism of inhibition by gabaculine and GVG are clearly different. Py represents the pyridoxal phosphate ring and NuEnz, a nucleophilic residue of the active site.*

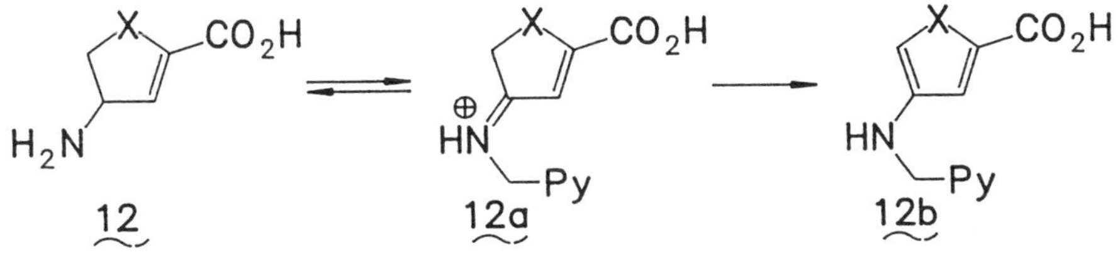

Scheme 6. *Postulated mechanism for the inhibition of GABA-transaminase by dihydro-thiophene (X=S) and furan (X=O) analogs. The imines **12a** resulting from the enzyme catalyzed tautomerization of the Schiff base between pyridoxal phosphate (Py=pyridoxal phosphate ring) and **12** isomerize to give the stable aromatic adducts **12b** which remain tightly bound to the enzyme active site.*

dented. An early example is the inhibition of xanthine oxidase by allopurinol (Cha *et al.*, 1975). Without formation of any reactive intermediates allopurinol is oxidized to oxipurinol which dissociates very slowly from the active site as long as the enzyme remains in the reduced state; the half-life of the reduced enzyme-oxipurinol complex is about 6 hours.

B. Inhibition of Flavin-Dependent Enzymes

Flavin-dependent enzymes constitute another class of enzymes well suited for design of mechanism-based inhibitors (Palfreyman *et al.*, 1987). In this class, monoamine oxidase (MAO) is of particular interest. MAO plays a key role in the metabolism of aminergic neurotransmitters as well as in the detoxification of exogenous amines. MAO inhibitors were introduced in the early fifties for the treatment of depression. However, the appearance of the so-called "cheese effect," i.e. a hypertensive crisis associated with the concomitant ingestion of tyramine-rich foodstuffs, severely restricted their use. Today, it is established that MAO exists as isoenzymes A and B which differ in their substrate specificity, their sensitivity to inhibitors, and their tissue distribution. It is believed that the beneficial effect in depression results from the inhibition of MAO A in the brain whereas the "cheese effect" is a consequence of the inhibition of MAO A in the periphery (Ilett *et al.*, 1980), most particularly the gut, liver and vascular neuronal enzyme pools. Thus, an MAO A inhibitor acting selectively in the brain can be expected to be a safe drug for depression. Furthermore, the dopa "sparing" effect of *L*-deprenyl (Presthus and Hajba, 1983), the only clinically used mechanism-based MAO B inhibitor, and the observation that *L*-deprenyl prevents the neurotoxicity of MPTP (Burks, 1985), a contaminant of a designer drug that causes clinical signs indistinguishable from those of Parkinson's disease, suggest that a specific MAO B inhibitor could be a valuable adjunct to *L*-dopa therapy of Parkinson's disease.

The best accepted mechanism of action of MAO is depicted in Scheme 7 with noradrenalin as substrate. Based on the reaction catalyzed by MAO, we designed and synthesized 2-substituted 3-haloallylamines as a new class of mechanism-based inhibitors of MAO (Bey *et al.*, 1984). Structure activity relationship studies establish that maximum inhibitory activity resides with the fluoro analogs where the fluorine atom is *trans* to the methylenamino group (McDonald *et al.*, 1985). For reasons which are not yet understood, most 2-substituted 3-fluoroallylamines preferentially inhibit the B isoenzyme. (*E*)-2-(3,4-Dimethoxyphenyl)-3-fluoroallylamine **13** (MDL 72,145) was initially selected for clinical evaluation. Because of untoward toxicity not related to MAO inhibition, MDL 72,145 was later replaced by (E)-2-(4-fluorophenylethyl)-3-allylamine **14** (MDL 72,974) which *in vivo* is 25 times more potent and 5 times more selective for the B isoenzyme than *L*-deprenyl **15** (Figure 1) (Zreika *et al.*, 1989). *In vitro* inhibition of MAO by these fluoroallylamines is time-dependent and irreversible, as indicated by the absence of recovery of enzyme activity after extensive dialysis of the inactivated MAO. Pseudo-first-order kinetics are usually observed over several half-lives. The mechanism of inactivation of MAO by these fluoroallylamines has been rationalized as indicated in Scheme 7. Preliminary experiments with ^{14}C labeled MDL 72,145 and purified beef liver MAO B indicate a stoichiometric binding of the inhibitor to the protein associated with a quench of the flavin UV absorption, suggesting a reduction of the cofactor (Singer and Salach). However, all attempts to further characterize the MAO-inhibitor adduct have been frustrated by its instability under

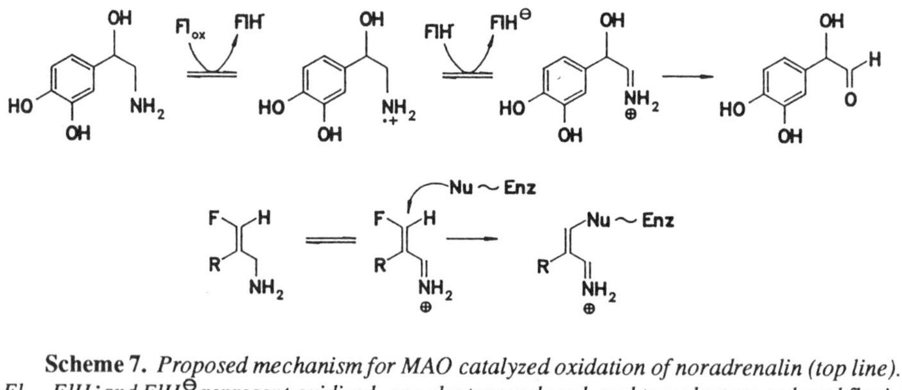

Scheme 7. *Proposed mechanism for MAO catalyzed oxidation of noradrenalin (top line). Fl_{ox}, FlH· and $FlH^{\ominus}$ represent oxidized, one electron reduced, and two electron reduced flavin adenine dinucleotide cofactor, respectively. The last step in the sequence corresponding to the imine hydrolysis is believed not to be enzyme catalyzed. The bottom line is a representation of the postulated mechanism of MAO inactivation by fluoroallylamines. Upon turnover by MAO, the fluoroallylamines are expected to generate an electrophilic conjugated imine which would eventually alkylate a nucleophilic residue (NuEnz) from the active site through an addition-elimination Michael reaction.*

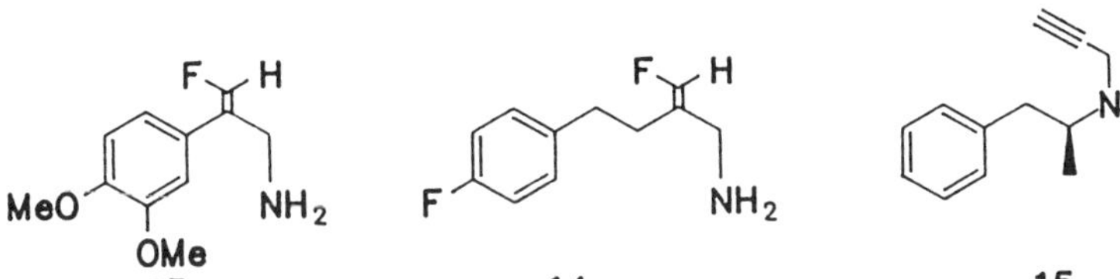

	13	14	15
MAO A	13 X 10^{-6}	0.7 X 10^{-6}	
MAO B	0.3 X 10^{-6}	0.003 X 10^{-6}	

Fig. 1. *Structures of the MAO B inhibitors: MDL 72,145 (13), MDL 72,974 (14), and L-deprenyl (15).*

conditions that denature the protein, raising doubt about the proposed covalent interaction between the enzyme and the inhibitor.

The design of an MAO A inhibitor acting selectively in the brain was a more challenging problem which began with the observation that ring hydroxylated 2-phenyl-3-fluoroallylamines were the only inhibitors in this series to show some selectivity for MAO A. In particular, (*E*)-2-(3-hydrophenyl)-3-fluoroallylamine **16** (MDL 72,392) (Figure 2), which displayed an IC_{50} of $5x10^{-8}$M, was about 10 times more selective for the A isoenzyme. Structurally, MDL 72,392 closely resembles the neurotransmitter noradrenalin. The synthesis of this neurotransmitter occurs predominantly in aminergic neurons and involves the obligatory decarboxylation of dopa, which is catalyzed by *L*-aromatic amino acid decarboxylase (AADC). AADC is not very selective. It can decarboxylate various phenylalanine analogs incorporating a hydroxyl substitutent in the *meta* or *ortho* position of the aromatic ring, and tolerate substituents on the β-carbon atom of the side chain (Figure 2). With this information in mind, it was postulated that the α-amino acid analogs of ring-hydroxylated 2-phenyl-3-fluoroallylamines could serve as substrates of AADC and selectively release the MAO inhibiting product in aminergic nerve endings where AADC is localized. To test this hypothesis, (*E*)-β-fluoromethylene-*m*-tyrosine **17** (MDL 72,394; FMMT), the α-amino acid corresponding to MDL 72,392 was synthesized (Figure 2) (McDonald *et al.*, 1984). As anticipated, *in vitro* FMMT was efficiently and stereoselectively decarboxylated by AADC to the fluoroallylamine MDL 72,392. Its K_m and V_{max} values are 600 μM and 41 nmol/min/mg of protein, respectively, compared to 55 μM and 99 nmol/min/mg of protein for dopa. No inhibition of AADC could be detected when FMMT was tested as a time-dependent inhibitor up to a concentration of 2 μM. *In vivo*, FMMT was demonstrated to enter the brain unchanged and be decarboxylated intraneuronally (Palfreyman *et al.*, 1985). FMMT is the prototype of a mechanism-based inhibitor being metabolically activated by the enzyme catalyzing the formation of the substrate for the target enzyme. The potential of this new approach to site selective enzyme inhibition is discussed in detail in later chapters (Jung *et al.*, chapter 10 and Palfreyman *et al.*, chapter 11, 1989).

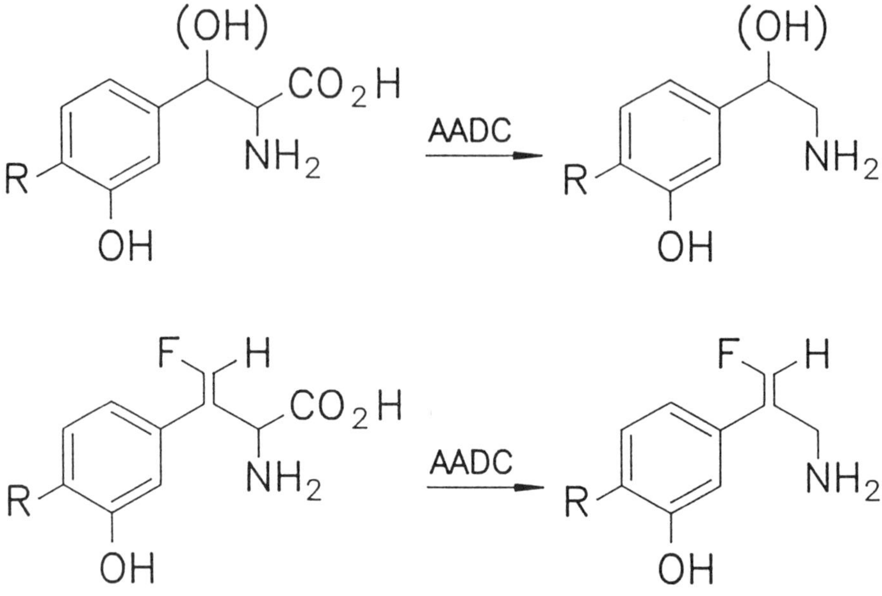

17 (R=H) 16 (R=H)

Fig. 2. *Substrates and corresponding products of AADC (top line). Structure of MAO inhibitor **16** MDL 72,392 and corresponding bioprecursor **17** MDL 72,394 (bottom line).*

C. Additional Examples

In addition to MAO, many mammalian tissues also contain a semicarbazide sensitive amine oxidase (SSAO) which belongs to the class of Cu^{++}-dependent amine oxidases. No physiological functions are know for this enzyme which is particularly abundant in vascular tissues. The preferred substrate of SSAO is benzylamine. SSAO uses a second organic cofactor which has long been thought to be pyridoxal phosphate, but is now believed to be methoxatin **18**, a pyrroloquinoline-quinone (PQQ) (Moog *et al.*, 1986). 2-Phenyl-3-haloallylamine derivatives are also potent inhibitors of SSAO (Palfreyman *et al.*, 1986). The inhibition is time-dependent and apparently irreversible as suggested by the absence of recovery of enzyme activity upon dialysis of the inhibited enzyme. A plausible, but still tentative mechanism for the inactivation, is suggested in Scheme 8. Interestingly, in contrast to MAO, the inhibition of SSAO does not appear to be sensitive to the nature of the halogen (Lyles *et al.*, 1987). For example, (*E*)-2-phenyl-3-chloroallylamine **19** and (*E*)-2-phenyl-3-fluoroallylamine **20** are almost equipotent inhibitors of SSAO with IC_{50} values of 8 nM and 2.5 nM, respectively, whereas against MAO the fluoro derivative is about 1000 times more active than the corresponding chloro analog. Thus, (*E*)-2-phenyl-3-chloroallylamine is a potent inhibitor of SSAO with sufficient specificity to explore the functional role of this enzyme.

The fluoromethylene moieties have also proven to be a useful latent reactive group to design mechanism-based inhibitors for *S*-adenosyl-*L*-homocysteine (SAH) hydrolase, the main enzyme involved in the catabolism of SAH. As SAH is a potent competitive inhibitor of most *S*-adenosylmethionine dependent methyl transferases, inactivation of SAH hydrolase could be expected to result in inhibition of biological methylation. SAH hydrolase is a NAD^+-containing enzyme which catalyzes the hydrolytic cleavage of SAH to adenosine and *L*-homocysteine. Its mechanism of action is well established and is represented in Scheme 9 (path **a**). The 3'-hydroxyl group of the substrate is first oxidized to the 3'-ketone while the NAD^+ is reduced to NADH. The 4'-hydrogen is then abstracted and *L*-homocysteine is eliminated to generate the 3'-keto-4'-methylene intermediate **21** (X=H). Michael addition of water and stereospecific reduction of the 3'-ketone by NADH finally gives adenosine and regenerates NAD^+. Knowing that 4',5'-didehydro-5'-deoxy-adenosine **22** is a substrate of SAH hydrolase, the vinyl fluoride **23** and its geometric isomer **24** were designed and synthesized (McCarthy *et al.*, 1989) on the premise they would be turned over by the enzyme and generate a β-fluoro-α,β-unsaturated ketone **21** (X=F) susceptible to alkylate a residue of the active site through an addition-elimination reaction as depicted in Scheme 9 (path **b**). As anticipated these two vinyl fluoride derivatives are potent time-dependent inhibitors of purified rat liver SAH hydrolase; the kinetic parameters K_i and τ_{50} for **23** and **24** determined according to Kitz and Wilson, are 0.55

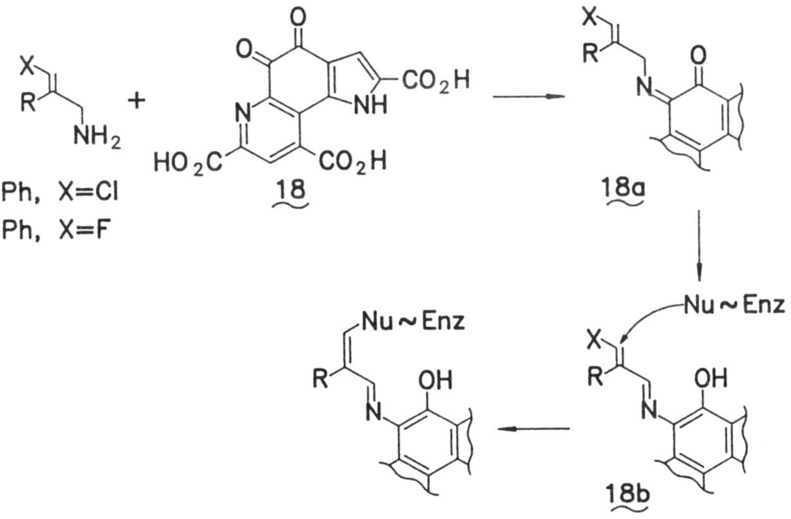

Scheme 8. *Proposed mechanism of inactivation of SSAO by 2-substituted-3-haloallylamine. Enzyme catalyzed tautomerization of the Schiff base **18a** formed between the inhibitor and the cofactor **18** generates the conjugated imine **18b** which can alkylate an active site nucleophilic residue (Nu) through an addition-elimination Michael reaction.*

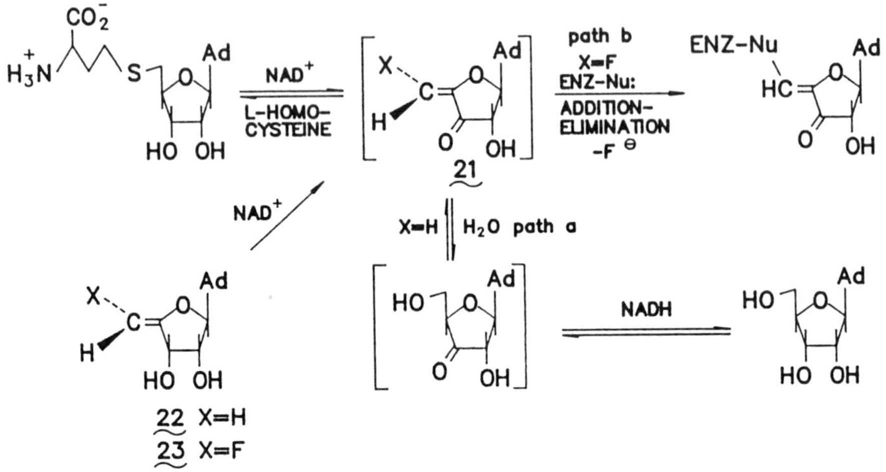

Scheme 9. *Mechanism of action of SAH hydrolase and proposed mechanism of inactivation by* ***23****. See text for details.*

μM and 2.5 min, and 1.04 μM and 2.7 min, respectively. The proposed mechanism of inactivation is supported by the demonstration that the inhibition of one mole of SAH hydrolase is accompanied by a) release of one mole of fluoride, b) progressive and stoichiometric formation of NADH, and c) stoichiometric binding of labeled inhibitor to the protein. However, although the inhibition is not reversed by dialysis, gel filtration, or addition of NAD^+, the rapid dissociation of radioactive material from the protein upon mild denaturation of the enzyme inhibited by labeled inhibitor **23** raises doubt about the proposed covalent interaction between the enzyme and the inhibitor (Mehdi *et al.*, 1989).

III. DESIGN OF MECHANISM-BASED INHIBITORS BASED ON THE CONCEPT OF TRANSITION-STATE ANALOGS

Enzymes catalyze reactions by lowering their free energy of activation, i.e. the energy required to reach the transition state. A thermodynamic analysis of this proposal for a single substrate, single-step reaction reveals that the ratio between the equilibrium constants for the dissociation from the enzyme of the substrate in the Michaelis complex and the altered substrate in the transition state is equal to the ratio of the rate constants for the enzyme-catalyzed and the nonenzymatic reactions (Wolfenden and Frick, 1987). The latter ratio for enzymes which do not proceed through covalently bound intermediates can be determined experimentally; it always exceeds 10^7 and can be as large as 10^{15}. This analysis shows that a stable analog of the altered substrate in the transition state would be expected to bind the enzyme 10^7 to 10^{15} times more tightly than the substrate in the ground state. If the reaction involves several substrates or proceeds through formation of covalent intermediates, the situation becomes more complicated and offers other possibilities for design of high affinity inhibitors. For multisubstrate or multiproduct reactions, analogs may be designed that incorporate electronic and binding determinants from each substrate or product critical for activation of the overall reaction; these analogs, which exhibit greater affinity for the enzyme than individual substrates or products, are often referred to as "multisubstrate" or "multiproduct" analogs. Also, analogs may be designed to be transformed by the catalytic capabilities of the target enzymes into species resembling the transition state. It must be emphasized that transition states are metastable species, i.e. chemically unstable; their structures can only be approximated by mimicking either their geometry or their charge distribution. In practice, transition-state inhibitors can be stable analogs of any intermediates along the reaction pathway; the higher the energy of the intermediate, the closer the resemblance to the actual transition state.

A. Inhibition of Glycoside Hydrolases

Glycosidases constitute a class of hydrolases that attack glycosidic bonds in carbohydrates, glycoproteins and glycolipids. The reactions catalyzed by these enzymes are formally nucleophilic substitutions at saturated carbon centers which, as schematically represented in Scheme 10, can occur with retention or inversion of configuration at the anomeric center (Sinnott, 1987). The mechanism of the "retaining" enzymes has been well studied; there is good evidence that a covalent glycosylated enzyme intermediate is involved. For the "inverting" enzymes, the sparse mechanistic data suggest that a direct displacement takes place. In both cases, various studies suggest that the high energy oxocarbonium ions **25** are intermediates along the reaction pathway.

Sucrase is an intestinal α-glucohydrolase which catalyzes the hydrolysis of sucrose into glucose and fructose (Scheme 11). Inhibition of intestinal α-glucohydrolases is expected to attenuate the blood glucose elevation from ingested carbohydrate and be an effective palliative therapy for diabetes mellitus and possibly obesity (Truscheit *et al.*, 1981; Dimitriadis *et al.*, 1985). Lysosomal α-glucosidase is an exoglucosidase which hydrolyzes the terminal nonreducing glucosyl unit of glycogen (Scheme 11).

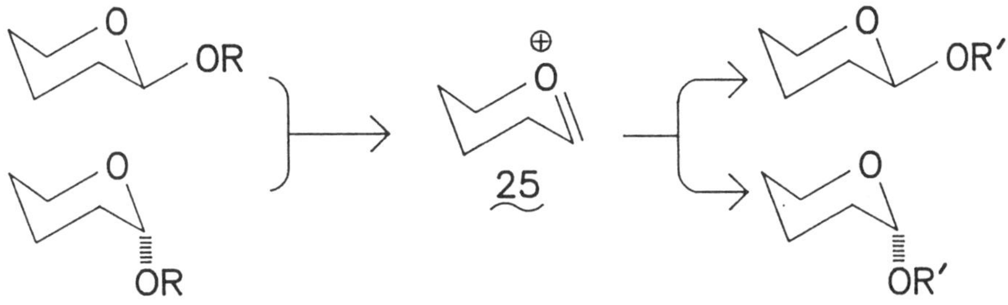

Scheme 10. *Schematic representation of reactions catalyzed by glycosyl transferases. The pyran ring is meant to represent any carbohydrate moiety. R can be any mono, poly, or oligosaccharide. In the case of glycosyl hydrolases, R' is hydrogen.*

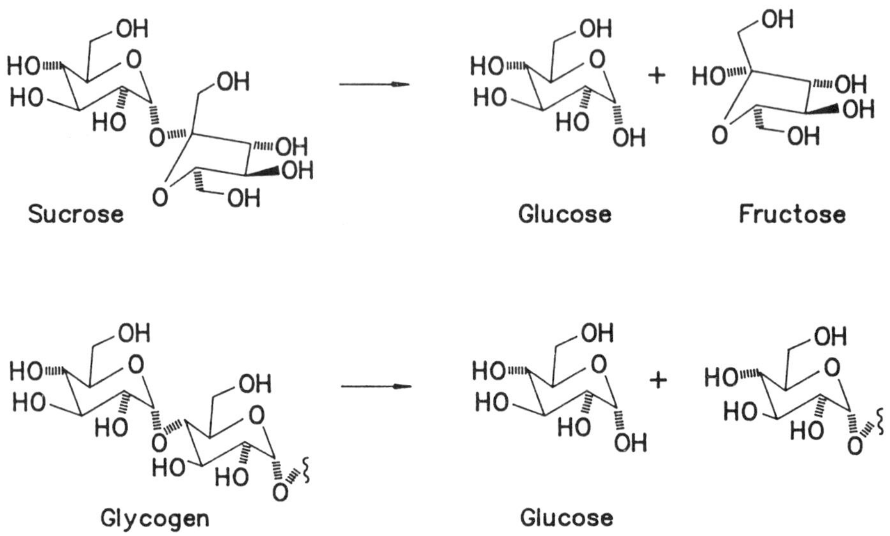

Scheme 11. *Reactions catalyzed by sucrose-α-D-glucohydrolase (top line) and α-D-glucosidase (bottom line).*

Its inhibition is probably not desirable since its absence in Pompe's disease results in lysosomal glycogen accumulation (Williams and Howell, 1983).

Deoxynojirimycin (Inouye *et al.*, 1968) **26** and castanospermine (Hohenschutz *et al.*, 1981) **27** are two alkaloids isolated from plants, which in their protonated forms are isosteric and isoelectronic with the glucosyl oxocarbonium ion (Figure 3). They were found to be potent inhibitors of a number of α- and β-glucosidases (Sinnott, 1987). Following these discoveries, many nitrogen containing analogs of carbohydrates were synthesized as potential transition-state analog inhibitors of glucohydrolases. Those depicted in Figure 3 were synthesized[1] (Liu, 1987), and/or evaluated in our Institute as inhibitors of sucrase and lysosomal α-glucosidase.[2] As is often observed with analogs of intermediates of the enzymic reaction pathway, the equilibrium between the inhibitors, the enzyme-inhibitor complexes, and sucrase or lysosomal α-glucosidase was found to occur slowly, i.e. the inhibition was time-dependent. Kinetic analysis

[1]*MDL 73,945 was synthesized by Ducep, J.B., Merrell Dow Research Institute, Strasbourg Center.*

[2]*Most of the kinetic constants in Tables I and II were determined by Danzin, C. and Ehrhard, A., Merrell Dow Research Institute, Strasbourg Center.*

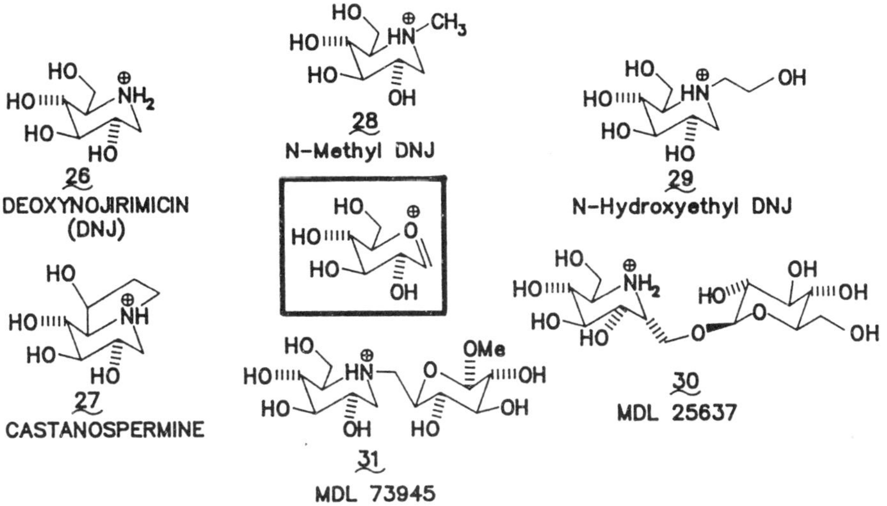

Fig. 3. *Amino sugar inhibitors of α-D-glucosidase and sucrose-α-glucohydrolase from natural and synthetic origin. Inserted in the box is the structure of the glucosyl oxocarbonium ion intermediate that the inhibitors are believed to mimic.*

using well-established methodologies developed by Morrison (1969), Duggleby *et al.*, (1982), and Cha (1975) revealed that they were either slow-binding or slow-tight binding inhibitors; their kinetic behavior can be accounted for by a single, slowly formed complex or rapid formation of an enzyme-inhibitor complex prior to slow formation of another enzyme-inhibitor complex. The inhibition of sucrase (Table I) fits the single enzyme-inhibitor complex model (Danzin and Ehrhard, 1987) whereas the inhibition of lysosomal α-glucosidase (Table II) conforms to the two enzyme-inhibitor complex model (Ellmers *et al.*, 1987).The half-lives of sucrase-inhibitor complexes calculated from their dissociation rate constants vary from 3.5 min for deoxynojirimycin to 75 hr for MDL 73,945, indicating that within this series of inhibitors, the inhibition changes gradually from a reversible to an irreversible pattern. In the case of lysosomal α-glucosidase, the rate of release of the inhibitor from the noncovalent enzyme-inhibitor complexes is so slow (less than 5% of enzyme activity is recovered after dialysis of the enzyme inhibited by castanospermine for 5 days) that, for all practical purposes, the enzyme is inactivated irreversibly. At that kinetic limit, these slow-tight binding inhibitors behave exactly as covalent inhibitors.

Interestingly, although not entirely conclusive, a study of the pH dependence of sucrase inhibition by deoxynojirimycin reported by Hanozet *et al.* (1981) indicates that the amino sugar could bind the enzyme in the unprotonated form and then get protonated by one of the two catalytic carboxylic acid residues in the active site, suggesting that the transition-state analog inhibitors could be generated by the catalytic capability of the target enzyme.

B. Inhibition of Proteases

The inhibition of proteases by peptidyl substrate analogs containing atom assemblies with electrophilic carbonyl groups in place of the scissile amide unit is a more clear-cut case of transition-state analog inhibitors generated by the target enzymes. Proteases are enzymes involved in almost every aspect of life. Their roles include basic functions in digestion as well as key regulatory mechanisms in peptide hormone or transmitter actions. Since excess protease activity is suspected or established in disease states such as emphysema, inflammation, tumor metastasis, muscular dystrophy, and hypertension, strong cases can be made for potential therapeutic usefulness of selective inhibitors of proteases (Schnebli and Braun, 1986).

Proteases constitute a large family of enzymes which catalyze the hydrolytic cleavage of amide linkages in proteins. From a mechanistic standpoint, proteases can be classified on the basis of their cofactor or essential catalytic residue as metallo, serine, cysteine or aspartic proteases (Polgar, 1989). Their mechanism of action is represented schematically in Scheme 12. A key feature is the formation of an unstable tetrahedral intermediate (**32** or **33**) as an early event along the catalytic pathway of all proteases. In the case of serine and cysteine proteases, the tetrahedral intermediates **33** are covalently bound to the enzyme as they arise from the addition of the catalytic serine hydroxyl or cysteine thiol group to the carbonyl of the scissile amide unit of the substrates. Collapse of the tetrahedral intermediates leads to an acyl-enzyme adduct **34** and an amine product. Deacylation involves the microscopic reverse of acylation where the attacking molecule of water forms a second unstable tetrahedral intermediate

Table I. Kinetic constants for the inhibition of sucrose-D-α-glucohydrolase from rat small intestine by amino sugar derivatives.

INHIBITORS	k_1 ($\mu M^{-1}s^{-1}$)	k_2 (s^{-1})	$t_{1/2}$	K_I (μM)
Deoxynojirimicin (DNJ)	6.3×10^{-2}	3.3×10^{-3}	3.5 min	0.05
MDL 25637	5.3×10^{-4}	2.6×10^{-3}	4.4 min	5.0
N-Hydroxyethyl DNJ	2.8×10^{-3}	1.9×10^{-3}	6 min	0.7
N-Methyl DNJ	6×10^{-3}	1.3×10^{-4}	90 min	0.02
Castanospermine	6.5×10^{-3}	3.6×10^{-6}	54 hr	0.0005
MDL 73945	10^{-4}	$<0.9 \times 10^{-6}$	>200 hr	<0.009

K_M for sucrose $\simeq$ 20mM

$$E + I^{\ddagger} \underset{k_2}{\overset{k_1}{\rightleftharpoons}} E \cdots I^{\ddagger *} \qquad K_I^{*} = \frac{k_2}{k_1} \qquad t_{1/2} = \frac{0.693}{k_2}$$

Table II. Kinetic constants for the inhibition of lysosomal α-D-glucosidase from rat liver by amino sugar derivatives.

	K_I (μM)	$T_{1/2}$ (min)
MDL 73945	460	0.5
CASTANOSPERMINE	21	0.4
MDL 25637	76	0.7

$$E + I^{\ddagger} \underset{k_2}{\overset{k_1}{\rightleftharpoons}} E \cdots I^{\ddagger} \underset{k_4}{\overset{k_3}{\rightleftharpoons}} E \cdots I^{\ddagger *} \qquad \mathcal{T}_{50} = \frac{0.693}{k_3}$$

$$\text{(fast)} \qquad\qquad \text{(slow)} \qquad\qquad K_I = \frac{k_2}{k_1}$$

35 that collapses to liberate the acid product and regenerates the enzyme catalytic residue. A reasonable corollary to this mechanism is that substrate-like molecules which could form stable hemiketal adducts with the catalytic serine or cysteine residues should be potent inhibitors of these proteases.

For years it has been known that peptidyl microorganism metabolites with C-terminal aldehyde functions, discovered as part of an extensive screening program for protease inhibitors, are potent inhibitors of serine and cysteine proteases (Umezawa, 1976). It was suggested, and confirmed later by NMR and crystallographic studies, that the inhibition was the result of the formation of a covalent hemiacetal bond between the electrophilic aldehyde carbonyl of the inhibitor and the catalytic serine hydroxyl or cysteine thiol group of the protease. We and others have extended this concept of inhibition of proteases to novel analogs of protease substrates in which the carbonyl group corresponding to that of the scissile amide unit is made electron deficient with adjacent electron withdrawing groups such as fluorine (Imperiali and Abeles, 1986; Stein *et al.*, 1987; Peet *et al.*, 1989) or alkyloxycarbonyl (Peet *et al.*, 1989; Burkhart *et al.*, 1988; Hori *et al.*, 1985). The peptidyl fluorinated ketones and α-keto esters listed in Table III will serve as illustrative examples of such inhibitors designed for human neutrophil elastase (HNE), a serine protease implicated in a number of disease states involving tissue destruction, such as pulmonary emphysema and rheumatoid arthritis. HNE has an extended substrate binding site. The peptide portions of the inhibitors were selected on the basis of the known binding preferences of HNE derived from synthetic substrates and affinity labeling agents. With the exception of AcProAlaPro-

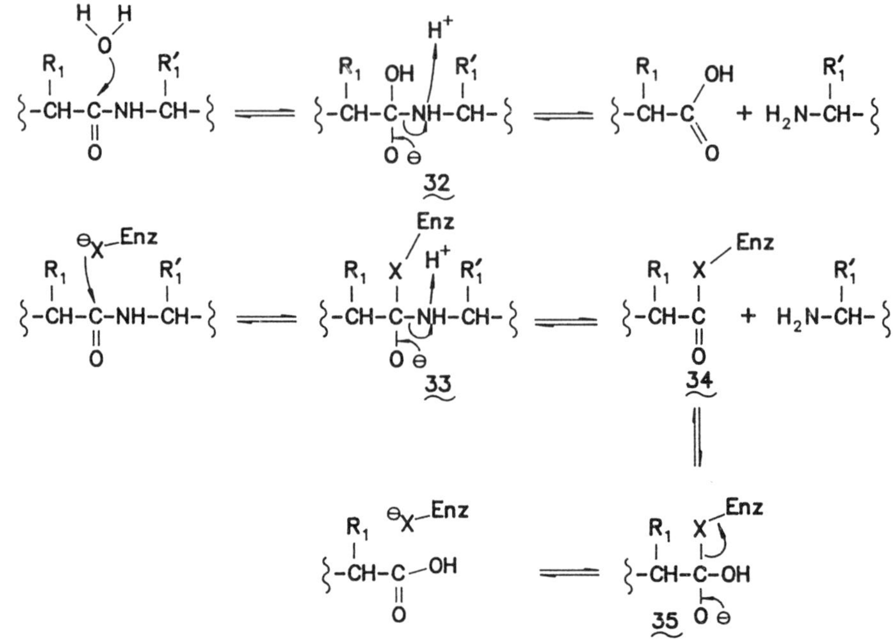

Scheme 12. *Tetrahedral intermediates in the hydrolysis of amide bonds catalyzed by metallo- and aspartic acid- (top line), and serine- and cysteine- (2nd and 3rd lines) proteases. X represents either the oxygen atom of the hydroxyl group of serine or the sulfur atom of the sulfhydryl group of cysteine.*

AlaCF3 and AcProAlaProAlaCO2CH3 which are straight competitive inhibitors, all the peptidyl fluorinated ketones and α-keto esters of Table III exhibit slow binding or slow-tight binding kinetics which have been found to conform to either the single complex (SC) or two complex (TC) models when analyzed according to the method of Morrison (Morrison, 1969). Similar to the case of glycosidase inhibitors, the nature of the inhibition gradually shifts from reversible towards irreversible as the half-life of the inhibitor-complex increases from 1 to more than 1200 min. Formation of a covalent tetrahedral hemiketal adduct between the carbonyl group of the peptidyl trifluoromethyl ketones and the catalytic serine hydroxyl group was clearly established by ^{13}C and ^{19}F spectroscopy and/or X-ray diffraction in the cases of the inhibition of chymotrypsin by AcLeuPheCOCF3 (Lang and Abeles, 1987; Abeles, 1988) and of porcine pancreatic elastase by AcAlaProValCF3 (Takahashi *et al.*, 1988). The mechanism of inhibition of serine proteases by peptidyl α-keto ester could, however, be different. According to preliminary X-ray diffraction studies on the complex between chymotrypsin and MeoSucAlaAlaProPheCO2CH3 (Ringe), it would appear that the catalytic serine hydroxyl does not form a hemiketal by adding to the α-ketone, but most probably has engaged in a transesterification reaction to displace the ester methoxy group of the inhibitor to form a stable α-keto acyl enzyme.

Table III. Kinetic constants for the inhibition of human leukocyte elastase by peptidyl fluoromethyl ketones and α-keto esters.

	K_I (μM)	$t_{1/2}$ (min)	Inhibition pattern
Ac Pro Ala Pro Ala CF_3	41	-	Comp.
Ac Ala Ala Pro Ala CF_3	24	1	T.C.
BOCAla Ala Pro Val CF_3	0.044	12	S.C.
MeOSuc Ala Ala Pro Val CF_3	0.014	55	T.C.
MeOSuc Lys(AdaSO$_2$) Pro Val CF_3	0.00058	125	S.C.
Z Val Pro Val CF_3	0.0016	325	S.C.
Z Lys(Z) Val Pro Val CF_3	< 0.0001	> 1200	S.C.
Ac Pro Ala Pro Ala CO_2CH_3	0.85	-	Comp
MeO Suc Ala Ala Pro Val CO_2CH_3	0.20	3	S.C.

The rate enhancement of enzymic reactions can be attributed to many factors whose relative contribution to overall catalysis is often difficult to assess. For multi-substrate reactions one of these factors is bringing together the substrates in a spatial arrangement that is conducive to reaction. As previously mentioned, combining the binding determinants of two substrates or products in one molecule is an approach that has been used very successfully to design competitive enzyme inhibitors. A classic example of this approach is the inhibition by captopril (Cushman *et al.*, 1977) and enalaprilat (Patchett *et al.*, 1980) of angiotensin-converting enzyme (ACE), a metalloprotease that cleaves the penultimate amide bond of the decapeptide angiotensin I to give the potent vasoconstrictor octapeptide angiotensin II and the dipeptide HisLeu, as indicated in Scheme 13. The antihypertensive drugs, captopril and enalapril, the bioavailable pro-drug of enalaprilat, incorporate in their structures a fragment mimicking the dipeptide HisLeu and a fragment oriented to bind the zinc ion in the active site. Based on a similar rationale, a rigid aromatic tricyclic surrogate of HisLeu was synthesized and coupled to the zinc-liganding 4-phenylbutanoic acid of enalaprilat to obtain **36** (MDL 27,088), the most potent competitive inhibitor of ACE ever reported (Giroux *et al.*, 1989). Kinetic analysis reveals that MDL 27,088 is a slow-tight binding inhibitor of ACE and supports a mechanism of inhibition involving a rapid formation of an initial complex which isomerizes slowly to a second complex. The overall apparent dissociation constant K_I and the half-life of the MDL 27,088-ACE complex were determined to be 3 pM and 436 min, respectively. As for captopril and enalaprilat, it can be argued that the extraordinary affinity of MDL 27,088 is not solely due to entropic factors, but also to a resemblance to an activated intermediate along the reaction pathway.

IV. CONCLUSION

In the last 15 years the concepts of "suicide" inhibition and transition-state analogs have been the main theoretical rationale behind the considerable research effort aimed

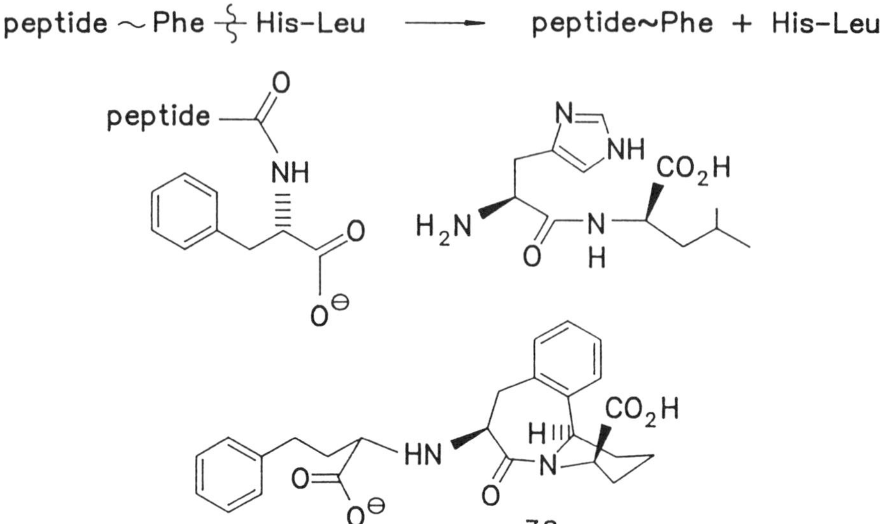

36

Scheme 13. *Reactions catalyzed by angiotensin-converting enzyme (top line) and structures of reaction products (2nd line) and of the multi-product analog inhibitor, MDL 27,088 (bottom line).*

at the design of new specific enzyme inhibitors. Today "suicide" inhibitors as originally defined from the work of Bloch and transition-state analogs as defined by Wolfenden constitute the two extremes of a large and unique class of inhibitors referred to as mechanism-based inhibitors. No single experimental or theoretical criterion can be used to define these inhibitors. They either mimic one of the intermediates along the catalytic pathway and/or require for activity some participation of the catalytic capability of the target enzymes. As illustrated by the few examples discussed in this review, a functional distinction between the various types of mechanism-based inhibitors is often a matter of degree, which explains the nomenclature and definition difficulties. The reversibility or irreversibility of the inhibition can no longer be defined in terms of time-dependency or covalent modification of the enzyme, but has to be related to the rate of breakdown of the enzyme-inhibitor complex whether or not the inhibitor covalently binds the enzyme. K_I, the apparent dissociation constant, τ_{50}, the half-life of the enzyme in presence of saturating concentration of inhibitor, and/or $t_{1/2}$, the half-life of the enzyme inhibitor complex, are the most relevant kinetic constants to predict the overall potency of the inhibitor *in vivo*. Notwithstanding the nomenclature and definition difficulties, there is a double bonus in working with mechanism-based inhibitors: first, the intellectual satisfaction of being able to design **rationally** potent inhibitors for enzymes whose mechanism of action is established or could be postulated; and second, but not least, the ability to produce highly specific, biologically active molecules which, at worst, would be useful tools to study the physiological role of enzymes in higher organisms and, at best, could become drugs if

Table IV. Mechanism-based enzyme inhibitors of potential therapeutic relevance developed at the Merrell Dow Research Institute.

Target Enzymes	Inhibitors	References
Ornithine decarboxylase	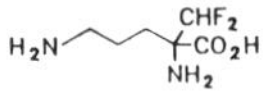	(Metcalf *et al.*, 1978)
	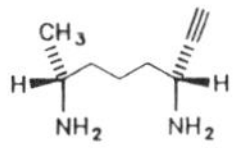	(Danzin *et al.*, 1983)
	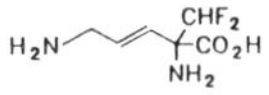	(Bey *et al.*, 1983)
GABA Transaminase	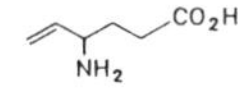	(Lippert *et al.*, 1977)
Arginine decarboxylase	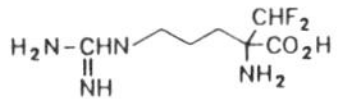	(Kallio *et al.*, 1981)
Polyamine oxidase	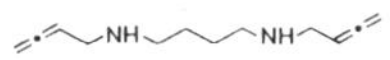	(Bey *et al.*, 1985)
Monoamine oxidase	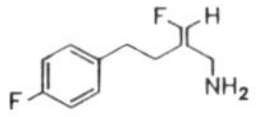	(Zreika *et al.*, 1989)
	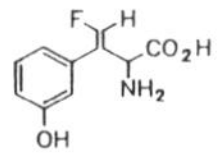	(McDonald *et al.*, 1984)
Dopamine β-hydroxylase	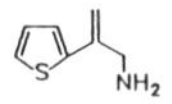	(Bargar *et al.*, 1986)
	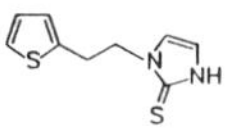	(McCarthy *et al.*, 1989)
Aromatase	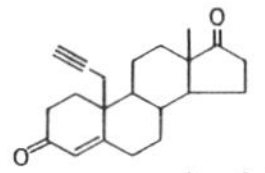	(Metcalf *et al.*, 1981)
5α-Reductase	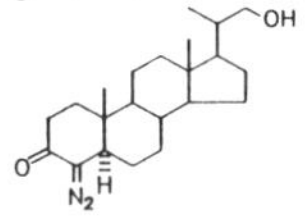	(Blohm *et al.*, 1980)
S-Adenosylhomocysteine hydroxylase	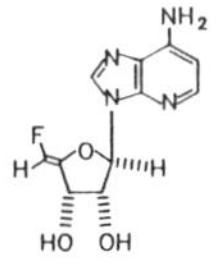	(McCarthy *et al.*, 1989)
Sucrase	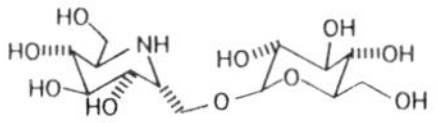	(Liu, 1986)
Glucohydrolase	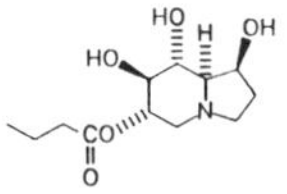	(Sunkara *et al.*, 1989)
Angiotensin Converting Enzyme	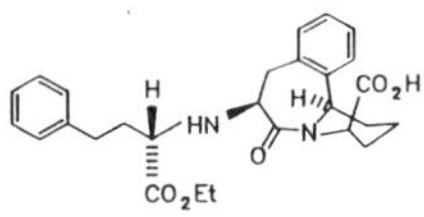	(Giroux *et al.*, 1989)

their target enzymes are of therapeutic relevance. Table IV lists a selection of mechanism-based inhibitors for enzymes of potential therapeutic relevance that have been discovered at the Merrell Dow Research Institute under the tenure of Al Sjoerdsma. Some of these inhibitors are on the verge of becoming drugs.

ACKNOWLEDGMENT

I am indebted to all the scientists from the Merrell Dow Research Institute who over the years have contributed to the design and synthesis of the mechanism-based inhibitors discussed in this paper. The names of most of them appear in the references cited. Special recognition, however, goes to Dr. M.J. Jung and Dr. B.W. Metcalf for their key roles in the inception of this research at the Merrell Dow Research Institute. I wish to express my gratitude to Dr. A. Sjoerdsma for initially stimulating my interest in enzyme inhibition as an approach to novel drug design and also for his continuous support, encouragement, and guidance over the sixteen years of our collaboration.

REFERENCES

Abeles, R.H. (1988). Abstr. Papers Am. Chem. Soc., 195th Meet. (Abstr. Biol. 12).

Abeles, R.H. and Maycock, A.L. (1976). *Acc. Chem. Res.* **9**, 313-319.

Bargar, T.M., Broersma, R.J., Creemer, L.C., McCarthy, J.R., Hornsperger, J.M., Palfreyman, M.G., Wagner, J., and Jung, M.J. (1986). *J. Med. Chem.* **29**, 315-317.

Bey, P. (1981). *Chem. Ind. (London)*, 139-144.

Bey, P., Gerhart, F., Van Dorsselaer, V., and Danzin, C. (1983). *J. Med. Chem.* **26**, 1551-1556.

Bey, P., Fozard, J., Lacoste, J.M., McDonald, I.A., Zreika, M., and Palfreyman, M.G. (1984). *J. Med. Chem.* **27**, 9-10.

Bey, P., Bolkenius, F.N., Seiler, N., and Casara, P. (1985). *J. Med. Chem.* **28**, 1-2.

Blohm, T.R., Metcalf, B.W., Laughlin, M.E., Sjoerdsma, A., and Schatzman, G.L. (1980). *Biochem. Biophys. Res. Commun.* **95**, 273-280.

Burkhart, J.P., Peet, N.P., and Bey, P. (1988). *Tetrahedron Lett.* **291**, 3433-3436.

Burks, T.F. (ed.) (1985). *Life Sci.* **36**, 199-254.

Cha, S. (1975). *Biochem. Pharmacol.* **24**, 2177-2185.

Cha, S., Agarwell, R.P., and Parks, R.E., Jr. (1975). *Biochem. Pharmacol.* **24**, 2187-2197.

Cushman, D.W., Cheung, H.S., Sabo, E.F., and Ondetti, M.A. (1977). *Biochemistry* **16**, 5484-5491.

Danzin, C. and Ehrhard, A. (1987). *Arch. Biochem. Biophys.* **257**, 472-475.

Danzin, C. and Jung, M.J. (1984). *Prog. Clin. Biol. Res.* **144**, 377-385.

Danzin, C., Casara, P., Claverie, N., Metcalf, B.W., and Jung, M.J. (1983). *Biochem. Biophys. Res. Commun.* **116**, 237-243.

Dimitriadis, G.D., Tessari, P., Go, V.L.W., and Gerich, J.E. (1985). *Metabolism* **34**, 261-265.

Duggelby, R.C., Attwood, P.V., Wallace, J.C., and Keech, D.B. (1982). *Biochemistry* **21**, 3364-3370.

Ellmers, B.R., Rhinehart, B.L., and Robinson, K.M. (1987). *Biochem. Pharmacol.* **36**, 2381-2385.
Endo, K., Helmkamp, M., Jr., and Bloch, K. (1970). *J. Biol. Chem.* **245**, 4293-4296.
Giroux, E., Beight, D.W., Dage, R.C., and Flynn, G.A. (1989). *J. Enzym. Inhib.* **2**, 269-277.
Hanozet, G., Pircher, H-P., Paolo, V., Oesch, B., and Semenza, G. (1981). *J. Biol. Chem.* **256**, 3703-3711.
Hohenschutz, L.D., Bell, E.A., Jewess, P.J., Leworthy, D.P., Pryce, R.J., Arnold, E., and Clardy, J. (1981). *Phytochemistry* **20**, 811-814.
Hori, H., Yasutake, A., Minematsu, Y., and Powers, J.C. (1985). In "Peptides, structure and function" (C.M. Deber, V.J. Hruby, and K.D. Kopple, eds.), pp. 819-822. Rockford, IL, Pierce Chemical Co.
Ilett, K.F., George, C.F., and Davies, D.S. (1980). *Biochem. Pharmacol.* **29**, 2551-2556.
Imperiali, B. and Abeles, R.H. (1986). *Biochemistry* **25**, 3760-3767.
Inouye, S., Tsuruoka, T., Ito, T., and Niida, T. (1968). *Tetrahedron* **24**, 2125-2144.
Jencks, W.P. (1966). In "Current aspects of biochemical energetics" (N.O. Kaplan and E.P. Kennedy, eds.), pp. 273-298. New York, Academic Press.
Jung, M.J. (1989). This book, pp. 121-137.
Kallio, A., McCann, P.P., and Bey, P. (1981). *Biochemistry* **20**, 3163-3166.
Kitz, R. and Wilson, I.B. (1962). *J. Biol. Chem.* **237**, 3245-3249.
Lang, T.C. and Abeles, R.H. (1987). *Biochemistry* **26**, 7603-7608.
Likos, J.J., Ueno, H., Feldhaus, R.W., and Metzler, D.E. (1982). *Biochemistry* **21**, 4377-4386.
Lippert, B., Metcalf, B.W., Jung, M.J., and Casara, P. (1977). *Eur. J. Biochem.* **74**, 441-445.
Lippert, B., Jung, M.J., and Metcalf, B.W. (1980). *Brain Res. Bull.* **5** (Suppl. 2), 375-379.
Lippert, B.J., Resvick, R.J., Burkhart, J.P., Holbert, G.W., Adams, J.A., and Metcalf, B.W. (1985). *Fed. Proc.* **44**, 1399.
Liu, P.A. (1987). *J. Org. Chem.* **52**, 4717-4721.
Lyles, G.A., Marshall, S.C.M., McDonald, I.A., Bey, P., and Palfreyman, M.G. (1987). *Biochem. Pharmacol.* **36**, 2847-2853.
McCarthy, J.R., Jarvi, E.T., Matthews, D.P., Edwards, M.L., Prakash, N.J., Bowlin, T.L., Mehdi, S., Sunkara, P.S., and Bey, P. (1989). *J. Am. Chem. Soc.* **111**, 1127-1128.
McCarthy, J.R., Matthews, D.P., Broersma, R.J., McDermott, R.D., Kastner, P.R., Hornsperger, J.M., Demeter, D.A., Weintraub, H.J.R., and Whitten, J.P. (1989). *J. Med. Chem.* Submitted for publication.
McDonald, I.A., Lacoste, J.M., Bey, P., Wagner, J., Zreika, M., and Palfreyman, M.G. (1984). *J. Am. Chem. Soc.* **106**, 3354-3356.
McDonald, I.A., Lacoste, J.M., Bey, P., Palfreyman, M.G., and Zreika, M. (1985). *J. Med. Chem.* **28**, 186-193.

Mehdi, S., Jarvi, E.T., Koehl, J.R., McCarthy, J.R., and Bey, P. (1989). *J. Enzym. Inhib.*, Submitted for publication.
Metcalf, B.W., Bey, P., Danzin, C., Jung, M.J., Casara, P., and Vevert, J.P. (1978). *J. Am. Chem. Soc.* **100**, 2551-2553.
Metcalf, B.W., Wright, C.L., Burkhart, J.P., and Johnston, J.O. (1981). *J. Am. Chem. Soc.* **103**, 3221-3222.
Moog, R.S., McGuirl, M.A., Cote, C.E., and Dooley, D.M. (1986). *Proc. Natl. Acad. Sci. U.S.A.* **83**, 8435-8439.
Morrison, J.F. (1969). *Biochem. Biophys. Acta* **185**, 269-286.
Palfreyman, M.G., McDonald, I.A., Fozard, J.R., Mely, Y., Sleight, A.J., Zreika, M., Wagner, J., Bey, P., and Lewis, P.J. (1985). *J. Neurochem.* **45**, 1850-1860.
Palfreyman, M.G., McDonald, I.A., Bey, P., Danzin, C., Zreika, M., Lyles, G.A., and Fozard, J.R. (1986). *Biochem. Soc. Trans.* **14**, 410-413.
Palfreyman, M.G., Bey, P., and Sjoerdsma, A. (1987). In "Essays in biochemistry" (R.D. Marshall and K.F. Tipton, eds.), pp. 28-81. London, Academic Press.
Palmer, J.L. and Abeles, R.H. (1979). *J. Biol. Chem.* **254**, 1217-1226.
Patchett, A.A., Harris, E., Tristram, E.W., Wyvratt, M.J., Wu, M.T., Taub, D., Peterson, E.R., Ikeler, T.J., Ten Broeke, J., Payne, L.G., Ondeyka, D.L., Thorsett, E.D., Greenlee, W.J., Lohr, N.S., Hoffsommer, R.D., Joshua, H., Ruyle, W.V., Rothrock, J.W., Aster, S.D., Maycock, A.L., Robinson, F.M., and Hirschmann, R. (1980). *Nature* **288**, 280-283.
Pauling, L. (1946). *Chem. Eng. News* **24**, 1375-1377.
Peet, N.P., Burkhart, J.P., Angelastro, M.R., Giroux, E.L., Mehdi, S., Kolb, M., Neises, B., Schirlin, D., and Bey, P. (1989). *J. Med. Chem.* In press.
Pegg, A.E., McGovern, K.A., and Wiest, L. (1987). *Biochem. J.* **241**, 305-307.
Pegg, A.E. (1987). In "Inhibition of polyamine metabolism" (P.P. McCann, A. Pegg, and A. Sjoerdsma, eds.), pp. 107-116. Orlando, Academic Press.
Polgar, L. (1989). Mechanisms of Protease Action. Boca Raton, CRC Press Inc.
Presthus, J. and Hajba, A. (1983). *Acta Neurol. Scand.* **95** (Suppl.), 127-133.
Rando, R.R. (1977). *Biochemistry* **16**, 4604-4610.
Ringe, D. Private communication.
Schechter, P. (1989). This book, pp. 201-210.
Schnebli, H.P. and Braun, N.J. (1986). *Res. Mongr. Cell Tissue Physiol.* **12**, 613-627.
Silverman, R.B. (1988). In "Mechanism-based enzyme inactivation: chemistry and enzymology", Vols. I and II. Boca Raton, CRC Press.
Silverman, R.B. and Invergo, B.J. (1986). *Biochemistry* **25**, 6817-6820.
Singer, T. and Salach, J.I. Private communication.
Sinnott, M.L. (1987). In "Enzyme mechanisms" (M.I. Page, and A. Williams, eds.), pp. 259-297. London, The Royal Society of Chemistry.
Stein, R.L., Strimpler, A.M., Edwards, P.D., Lewis, J.J., Mauger, R.C., Schwartz, J.A., Stein, M.M., Trainor, D.A., Wildonger, R.A., and Zottola, M.A. (1987). *Biochemistry* **26**, 2682-2689.
Sunkara, P.S., Taylor, D.L., Kang, M.S., Bowlin, T.L., Liu, P.S., Tyms, A.S., and Sjoerdsma, A. (1989). *Lancet*, 1206.

Takahashi, L.H., Radhakrishnan, R., Rosenfield, R.E., Jr., Meyer, E.F., Jr., Trainor, D.A., and Stein, M. (1988). *J. Mol. Biol.* **201**, 423-428.

Truscheit, E., Frommor, W., Junge, B., Muller, L., Schmidt, D.D., and Wingender, W. (1981). *Angew. Chem. Int. Ed. Eng.* **20**, 744-761.

Umezawa, H. (1976). *Methods Enzymol.* **45**, 678-695.

Williams, J.C. and Howell, R.R. (1983). In "Metabolic basis of inherited disease" (J.B. Sturbury, J.B. Wyngaarden, D.S. Fredrickson, J.L. Goldstein, and M.S. Brown, eds.), pp. 141-166. New York, McGraw-Hill.

Wolfenden, R. (1969). *Nature* **223**, 704-705.

Wolfenden, R. and Frick, L. (1987). In "Enzyme mechanisms" (M.I. Page and A. Williams, eds.), pp. 97-122. London, The Royal Society of Chemistry.

Zreika, M., Fozard, J.R., Dudley, M.W., Bey, P., McDonald, I.A., and Palfreyman, M.G. (1989). *J. Neural Transm.* In press.

8

COFACTOR DEPENDENCY IN THE INHIBITION OF STEROID 5 ALPHA-REDUCTASE

Brian W. Metcalf
Dennis A. Holt
Mark A. Levy

Department of Medicinal Chemistry
Smith Kline & French Laboratories
King of Prussia, PA 19406

I. INTRODUCTION

It is now generally accepted that the prostate gland is subject to stimulation by 5α-dihydrotestosterone (DHT) rather than by the classic testicular hormone, testosterone (T). This notion is a consequence of accumulating observations. For example, the presence of a pyridine nucleotide dependent enzyme in the prostate capable of converting T to DHT was demonstrated in 1968 by Bruchovsky and Wilson. Moreover, the binding of DHT to prostatic nuclei markedly exceeded that of T (Bruchovsky and Wilson, 1968). In the same year Anderson and Liao showed that the nuclear chromatin from prostate, but not from other tissue insensitive to androgens, contains an androgen receptor which selectively retains DHT (Anderson and Liao, 1968). Also in 1968 Baulieu suggested that the stimulatory action of T on the prostate was mediated by local metabolism and that DHT was implicated in the control of cell division (Baulieu *et al.*, 1968).

The importance of steroid 5α-reductase, the NADPH-dependent enzyme which converts T to DHT, was later dramatically underlined by the description of the occurrence of a genetic steroid 5α-reductase deficiency in male pseudohermaphrodites (Imperato-McGinley *et al.*, 1979). This study of patients with the genetic steroid 5α-reductase deficiency has allowed the categorization of androgen activity which is T-mediated compared with that which is DHT-mediated (Figure 1).

In steroid 5α-reductase deficient man the prostate is often undetectable, but can be demonstrated to enlarge on DHT administration (Peterson *et al.*, 1977). Furthermore, DHT levels in the prostates of patients with benign prostatic hyperplasia have been reported to be elevated (Habib, 1980). While this conclusion, first arrived at by Siiteri and Wilson (1970) has been confirmed independently by several workers, it has become controversial owing to recent studies by Walsh, Hutchins and Ewing (1983). These workers found that there was no significant difference between the DHT content

Enzymes as Targets for Drug Design

T-mediated	DHT-mediated
Anabolic actions	Increased facial, body hair
muscle mass increase	Acne
penis enlargement	Scalp hair recession
scrotum enlargement	Prostate enlargement
vocal cords enlargement	
Spermatogenesis	
Male sex drive, performance	

Fig. 1.

in normal and hyperplastic prostatic tissue. The difference in the results is ascribed to the use of cadaverous tissue for "normal" levels of DHT in the earlier studies, compared to the use of tissue obtained from open surgical procedures where normal and hyperplastic tissue was harvested simultaneously in the later studies. On the other hand, increased activity of steroid 5α-reductase has been observed in hyperplastic tissue compared to normal. Using the activity of 3β-hydroxysteroid dehydrogenase, which does not vary from hyperplastic prostate to normal, as a reference, it has been found that the ratio of steroid 5α-reductase to 3α(β)-hydroxysteroid dehydrogenase, is dramatically increased in hyperplastic tissue compared to the normal prostate (Bruchovsky and Lieskovsky, 1979). Thus it has been proposed that the inhibition of steroid 5α-reductase would allow for the treatment of some major androgen-mediated diseases without interference to the desirable masculine characteristics mediated by T. Our initial target was benign prostatic hyperplasia.

II. BIOSYNTHETIC PATHWAYS FOR STEROID HORMONES

The biosynthesis of all the steroid hormones from cholesterol is shown in Figure 2. Clearly, inhibition of steroid 5α-reductase should leave the other pathways intact. Androstanediol (ADIOL), the product of further metabolism of DHT should be depleted. This is considered inconsequential as the biological activity attributed to ADIOL is thought to result from its oxidation back to DHT (Jacobi and Wilson, 1976).

III. TISSUE AND SPECIES DIFFERENCES IN STEROID 5α-REDUCTASE

Steroid 5α-reductase has been solubilized from rat liver (Liang *et al.*, 1983a), from rat epididymus (Scheer and Robaire, 1983), and from human prostate (Houston *et al.*, 1985). All attempts to purify the enzyme from any source have met with little success owing to the instability of the enzyme and low recoveries. This has led to the general use of microsomal preparations for mechanistic and inhibitor studies.

Steroid 5α-reductase from rat liver appears to be different from that of rat prostate. For example, 11-hydroxysteroids such as cortisol are reduced by the liver enzyme, but not by the prostatic enzyme. The K_m for T has been reported to be 1×10^{-4} M for the liver enzyme but 1×10^{-6} M for the prostatic enzyme (Frederiksen and Wilson, 1971).

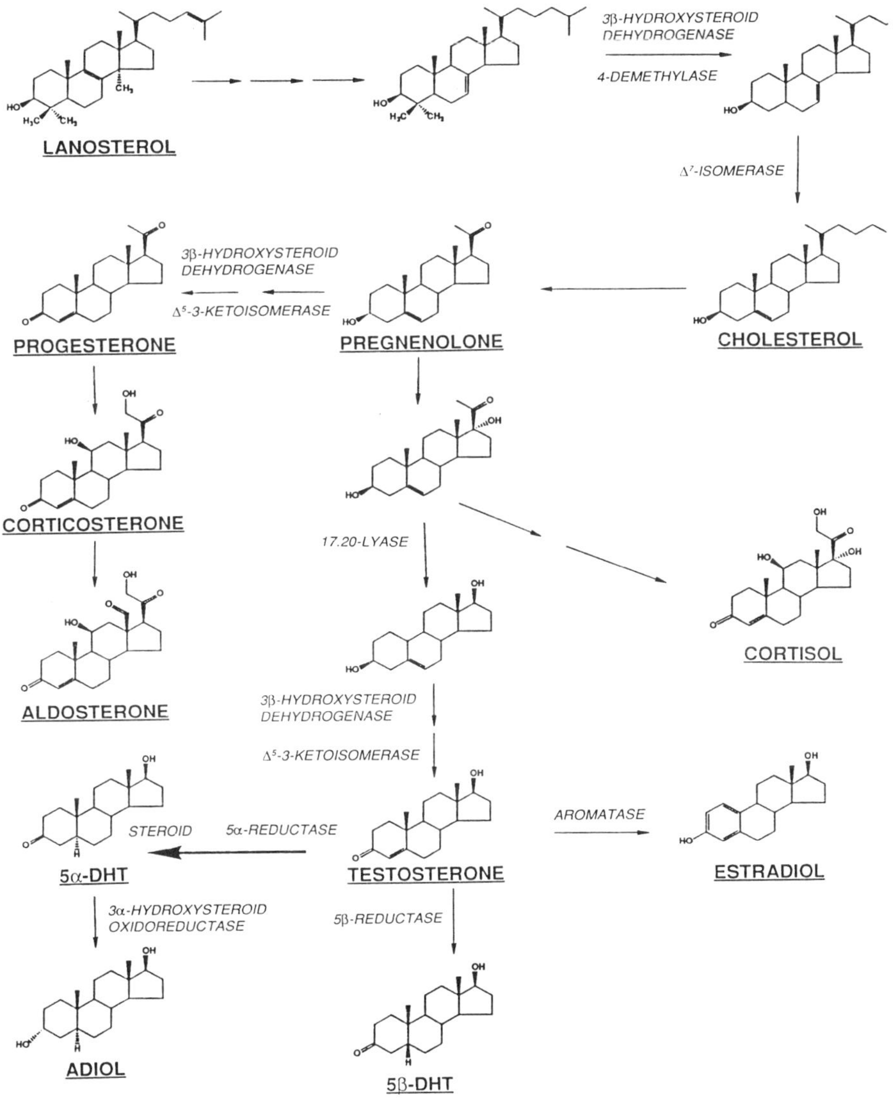
LANOSTEROL
3β-HYDROXYSTEROID
DEHYDROGENASE
4-DEMETHYLASE
Δ7-ISOMERASE
CHOLESTEROL
PREGNENOLONE
3β-HYDROXYSTEROID
DEHYDROGENASE
Δ5-3-KETOISOMERASE
PROGESTERONE
CORTICOSTERONE
ALDOSTERONE
17.20-LYASE
CORTISOL
3β-HYDROXYSTEROID
DEHYDROGENASE
Δ5-3-KETOISOMERASE
STEROID
5α-REDUCTASE
TESTOSTERONE
AROMATASE
ESTRADIOL
5α-DHT
3α-HYDROXYSTEROID
OXIDOREDUCTASE
5β-REDUCTASE
ADIOL
5β-DHT

Fig. 2.

Furthermore, the pH optimum of the enzyme activity varies in the human (pH 4.8 to 5.2), dog (pH 6.2 to 6.8), and rat (pH 6.5 to 7.0). On the other hand the rank order of inhibition of a number of competitive inhibitors was recently found to be the same for rat ventral prostatic and liver 5α-reductases, although the inhibitory potencies were quite distinct (Liang *et al.*, 1983a).

Species differences between the prostatic steroid 5α-reductases of rat, dog and human have been demonstrated with a series of steroidal inhibitors containing varying substituents at position 17 (see Figure 3 for the conventional steroid numbering system). Those compounds with a 17β-dialkylcarbamoyl group are approximately equipotent inhibitors of the rat and human enzymes, but are only 0.1-15% as potent on the dog enzyme. Inhibitors with a 17β-spiroether substituent are most potent with the rat enzyme, being only 15-50% and 0.2-0.4% as effective against the activity derived from dog and human respectively. Appropriate functionalization at C-17 is therefore of particular importance in inhibitor design (Liang *et al.*, 1985).

IV. CHEMICAL MECHANISM OF STEROID 5α-REDUCTASE

In experiments with rat liver steroid 5α-reductase, the use of $NADP^3H$, in place of NADPH, leads to incorporation of tritium into DHT, which can not be exchanged out. The complementary experiment of reduction of T in tritiated water afford tritiated DHT, where the tritium can be exchanged out by gentle treatment with aqueous base. These experiments establish that hydride transfer from NADPH occurs to the 5α-position of T with protonation from solvent at the 4-position (Wilton and Ringold, 1968). It has been established that it is the 4-pro-S-hydrogen of the pyridine nucleotide cofactor that is transferred to steroid in the case of the rat liver enzyme (Bjorkhem, 1969), and for the enzyme from rat seminal vesicles (Suzuki and Tomaoki, 1974). A mechanism of action consistent with these results involves direct hydride donation from NADPH to the 5 position of T leading to enolate formation at C-3, C-4. The enolate presumably would be stabilized by some electrophilic residue (E^+) in the active site. This process may be viewed alternatively as activation of the enone by E^+ leading to a positively-polarized species which accepts a hydride from NADPH at C-5. Enzyme-mediated tautomerism then leads to the product DHT with release of $NADP^+$ (Figure 4).

It appears that the enol-keto tautomerism is enzyme-mediated since a report has described the incorporation of exchangeable tritium into DHT on incubation in ^3H_2O in the presence, but not absence, of enzyme. Thus enolization, but not reduction, is reversible, at least for the rat liver enzyme (Wilton and Ringold, 1968). The stereo-

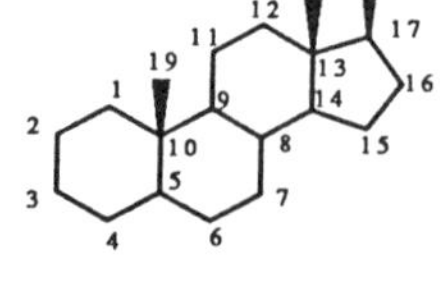

Fig. 3.

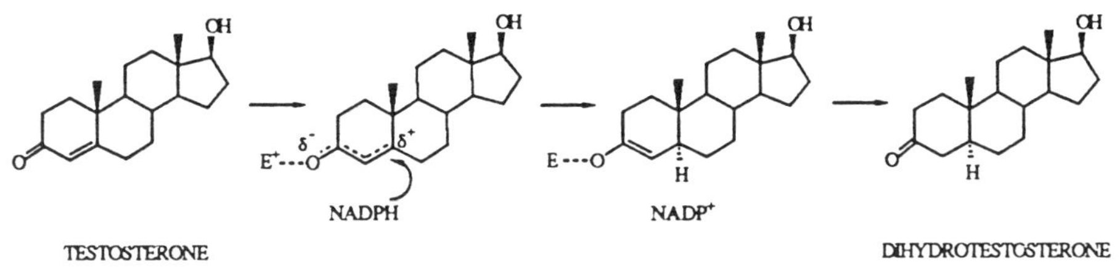

Fig. 4.

chemistry of the protonation step is unknown. The effects of electron-withdrawing substituents at any of positions 2, 4 and 6 on the regiochemistry of reduction also require clarification. It has been reported that rat liver microsomes can effect an abnormal enzymatic reduction with compounds of this type leading to 3-ol-4, 5-ene formation (Ringold *et al.*, 1964; Starka *et al.*, 1969). A more recent report which proposes that a steroidal 3-keto-4-cyano-4, 5-ene undergoes reduction in the normal 4, 5-mode with the rat prostatic enzyme offers no experimental evidence to support this regiochemistry of reduction (Rasmusson *et al.*, 1983), and would seem to be at variance with the earlier work. In contrast, a recent report describes the lack of substrate activity of 4-chloroprogesterone for the rat prostatic enzyme (Batzold, 1981).

It would be of interest to confirm the change in regiochemistry of reduction elicited by an electron-withdrawing group at C-4 and to clarify whether the responsible enzyme is actually steroid 5α-reductase or the 3β-hydroxysteroid dehydrogenase. The former enzyme usually reduces 3-keto-4, 5-enes to saturated ketones, while the latter normally only reduces saturated 3-ketones to alcohols.

V. KNOWN INHIBITORS OF STEROID 5α-REDUCTASE

The first inhibitor of steroid 5α-reductase to be described was the 17β-carboxylic acid 1 by Hsia and Voigt (1974). Compound 1 was reported to have a K_i of 1.1×10^{-6} M which is similar to the K_m found for T in these microsomal preparations of steroid 5α-reductase from human foreskin. Hsia and Voigt also demonstrated that 1 could block the androgenic effects of T but not those of DHT on the hamster sebaceous gland when administered topically.

The secosteroid 2, reported by Covey and Robinson (1976) to be an inactivator of Δ-ketosteroid isomerase, is also an affinity label for steroid 5α-reductase with a reported K_i of 0.98×10^{-6} M (Robaire *et al.*, 1977).

The prostate-reducing effect of the secosteroid 2 on i.p. administration to rats has been described. It is not clear whether this effect results from steroid 5α-reductase inhibition, or from inactivation of the isomerase. The latter would be an undesirable mechanism because all steroid hormone biosynthesis would be affected as this isomerization is common to the biosynthesis of all the steroid hormones.

The diazoketone 3 (Metcalf *et al.*, 1980) has been reported to be a potent, time-dependent inhibitor of steroid 5α-reductase from rat prostate with a K_i of 3.5×10^{-8}

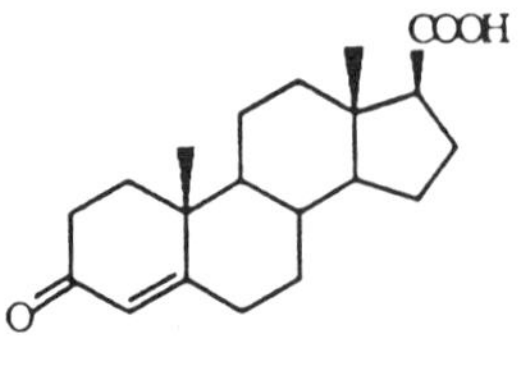

1

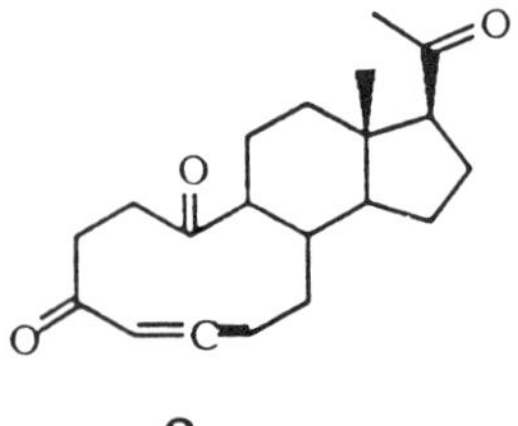

2

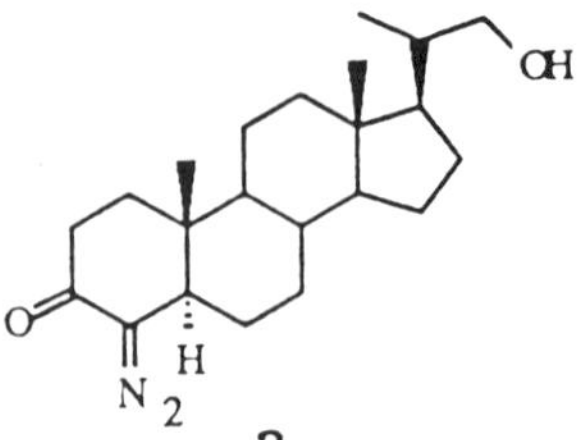

3

M (Blohm *et al.*, 1980). Time-dependency was considered to be indicative of irreversibility. In this case it was proposed that the protonation steps implicit in the normal enzymatic transformation described above activate the diazoketone functionality to a diazonium ion which alkylates some nucleophilic residue at the active site. This report was the first to point out the dramatic increase in affinity associated with an inhibitor bearing a 5α-ring junction and having sp^2 hybridization at C-3 and C-4. Compound 3 appears to be a specific inhibitor of steroid 5α-reductase in that aromatase, Δ^3-ketoisomerase, 3β-hydroxysteroid dehydrogenase and steroid 5β-reductase, all of which process T or DHT (Figure 2) are unaffected *in vitro* at concentrations up to 10^{-6} M. When administered intragastrically to rats at 100 mg/kg, the diazoketone 3 produced a reduction of prostatic steroid 5α-reductase activity of 60%, which was maintained for at least 12 hours. Coincident with inhibition of steroid 5α-reductase, was an attenuation of T-induced, but not DHT-induced prostatic growth in the castrate rat (Blohm *et al.*, 1986).

In keeping with the high affinity associated with a 5α-ring junction and sp^2 hybridization at C-3 and C-4 is the high affinity (K_i = 5 nM) of the lactam 4 (4-MA) to the enzyme from rat prostate (Liang *et al.*, 1983b). The inhibition by 4-MA is competitive with T, but uncompetitive with NADPH. These inhibition patterns are consistent with binding of NADPH prior to binding of either substrate of inhibitor (Bertics *et al.*, 1984). In studies using radiolabelled inhibitor, it was demonstrated that

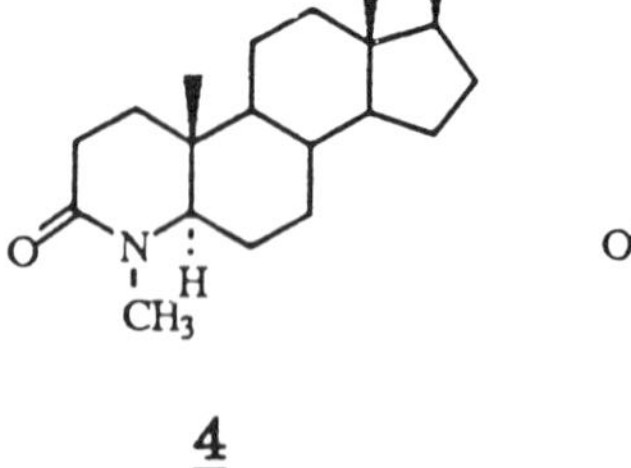

4

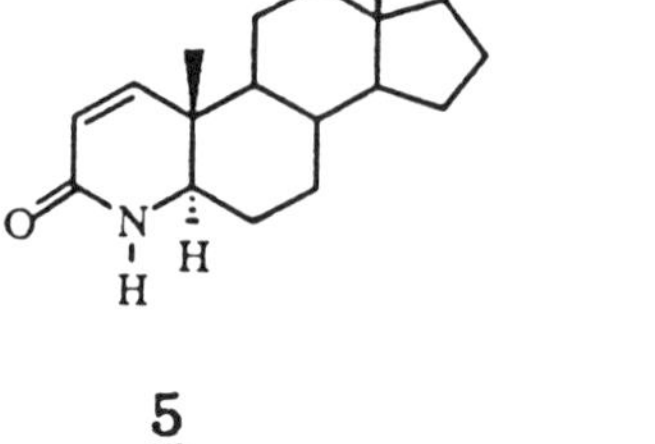

5

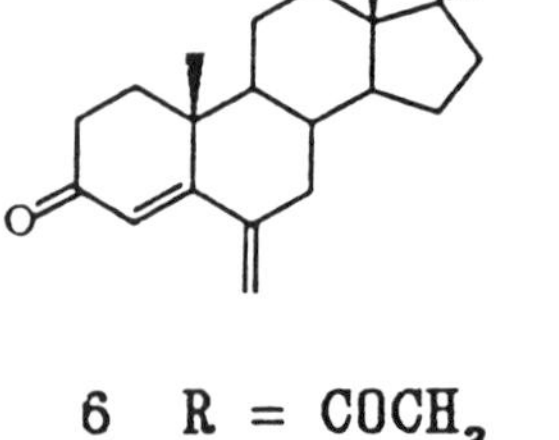

6 R = $COCH_3$
7 R = H

binding of tritiated 4 to rat liver microsomes was dependent on NADPH, and not on $NADP^+$ (Lang *et al.*, 1983a).

In intact male rats, the lactam 4 (1 mg/rat, s.c.) decreased prostatic DHT concentration 35% and increased T concentration 380%. The total DHT plus T levels remained unchanged (Brooks *et al.*, 1982). The lactam 4 attenuates the growth of the prostate and seminal vesicles but has little effect on body weight, testes, kidneys and adrenals. In addition there is no effect on fertility. Such activity for a steroid 5α-reductase inhibitor is consistent with the effects anticipated by the clinical reports of humans lacking the enzyme (Imperato-McGinley *et al.*, 1979).

4-MA was replaced in development by the $\Delta^{1,2}$-lactam, 5 (MK-906), apparently owing to antagonism elicited at the androgen receptor *in vivo* by 4-MA. MK-906 has been reported to have a K_i of 26 nM with the human prostatic enzyme and 6 nM with the rat enzyme. It is a competitive inhibitor versus T (Liang *et al*, 1985). In early clinical reports, MK-906 has been reported to suppress plasma DHT levels 50% at 24 hours after a single administration of doses as low as 0.5 mg (approximately 0.01 mg/kg) with DHT levels returning to normal only after 5-7 days (Giaguli *et al.*, 1988). Major metabolites of MK-906 result from oxidative metabolism of the t-butyl side chain, and reduction of the Δ^1 double bond (Carlin *et al.*, 1987).

The 6-methylene steroid 6 has been shown to be a time-dependent inactivator of steroid 5α-reductase from rat prostatic microsomes (Petrow *et al.*, 1981). The proposed mechanism for inactivation of steroid 5α-reductase by compound 6 is priming of its dieneone functionality by elctrophilic activation (see Figure 4) towards nucleophilic attack at the 6-methylene group. It seems likely, however, that this could be a nonspecific mechanism as protein cysteine groups could probably add to compound 6 in this fashion without priming by steroid 5α-reductase. That the inhibition is irreversible as implied by the time-dependent kinetics, is further supported by results in a system using explants from rat ventral prostate. Inhibition of steroid 5α-reductase in these cultured explants by the desacetyl analog of 6, 6-methylene-17α-hydroxyprogesterone (7), could not be reversed on dilution (Kadohama *et al.*, 1983). Administration of 7 to immature rats results in a marked inhibition of prostatic growth. Inhibition of overall body weight increase as well as effects on kidney and adrenal growth were also apparent, suggesting a lack of biological specificity (Petrow *et al.*, 1982). There have been no reports relating to the relative enzyme specificity of 7. It is conceivable that it also inhibits the Δ^5-3-ketoisomerase which may lead to the evident lack of organ specificity.

VI. KINETIC MECHANISM OF STEROID 5α-REDUCTASE

The kinetic mechanism for steroid 5α-reductase has been studied for the 1, 4-reduction of progesterone from rat anterior pituitary (Campbell *et al.*, 1986), and for the human prostatic enzyme (Houston *et al.*, 1987). For the progesterone 5α-reductase, initial velocity data suggested a sequential mechanism. $NADP^+$ was found to be competitive versus NADPH, but noncompetitive versus progesterone. Dead end inhibition patterns using a lactam analog related to 4-MA (4), were found to be competitive versus progesterone, and uncompetitive versus NADPH, consistent with an ordered kinetic mechanism where NADPH binds before progesterone. Catalysis occurs with the initial release of dihydroprogesterone, and finally $NADP^+$. A similar conclusion was reached for the kinetic mechanism of human prostatic steroid 5α-reductase, with NADPH being the first substrate to bind, and $NADP^+$ being the last product to be released (Houston *et al.*, 1987). This kinetic mechanism is illustrated in Figure 5.

VII. THE DESIGN OF UNCOMPETITIVE INHIBITORS OF STERIOD 5α- REDUCTASE

This ordered binding of substrates and release of products demands the existence of discrete binary complexes of enzyme with NADPH or $NADP^+$ and suggests that the design of inhibitors which selectively bind to either of these forms should be possible. Inhibitors which are competitive versus testosterone, such as the 4-azasteroids exemplified by 4, would be expected to require NADPH for association to steroid 5α-reductase. The requirement of NADPH for the binding of one such 4-aza inhibitor to the enzyme in rat prostatic microsomes supports this interpretation (Liang *et al.*, 1983a). On the other hand, an inhibitor which binds exclusively to the $NADP^+$ form would be uncompetitive versus testosterone in a double reciprocal plot (1/V versus 1/S), while noncompetitive patterns would result if complexes with each oxidation state were involved. In such cases of uncompetitive or noncompetitive inhibition, complexation of the inhibitor with E-$NADP^+$ would preclude relief of inhibition by high concentrations of testosterone. Such high concentrations of testosterone could result *in vivo* if testosterone biosynthesis continues in the presence of an inhibitor of steroid 5α-reductase. Indeed, in the case of the inhibition of aspartate transcarbamylase by the experimental antineoplastic drug, PALA, such "metabolic resistance" was proposed to be responsible for the failure of PALA to block the pyrimidine biosynthetic pathway *in vivo* (Duggleby and Christopherson, 1984). We therefore hoped that an inhibitor of steroid 5α-reductase, which binds to an enzyme-$NADP^+$ complex could offer significant advantages over one which is competitive versus testosterone *in vivo*. Consequently, we elected to design inhibitors of steroid 5α-reductase which would be uncompetitive versus testosterone.

The tight binding of the diazoketone (3), and of the lactam (4-MA, 4) presumably is a consequence of the sp^2 hybridization at C-3 and C-4, and of the 5α-ring junction. These features clearly impose on these inhibitors the same conformation as the putative enzyme-generated enolate intermediate (Figure 4). At least in the case of 4-MA (4), they are, however, not sufficient to allow binding to the E-$NADP^+$ complex (the cofactor requirements for the binding of the diazoketone (3) have not been reported).

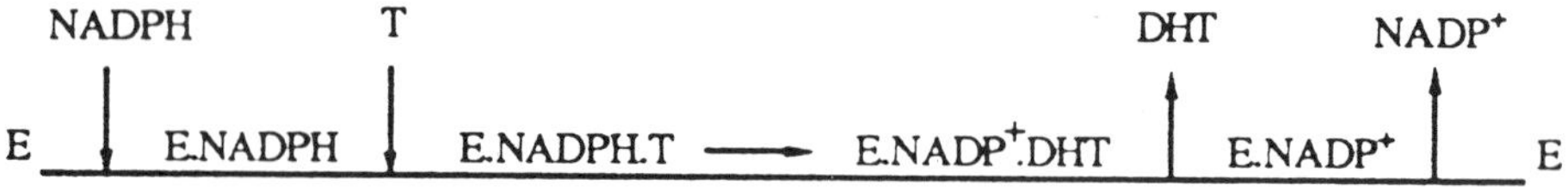

Fig. 5.

We concluded that for high affinity, inhibitors of steroid 5α-reductase should be mimics of the enolate of DHT, but that additional features which take into account the E-NADP$^+$ complex should be included. A survey of the cofactor dependency of the inhibition of other pyridine nucleotide dependent enzymes suggested some possibilities.

A. Pyrazoles

In 1963 Theorell and Yonetani reported that pyrazole forms a very tight ternary complex with liver alcohol dehydrogenase and NAD$^+$ and on the basis of spectrometric and titrimetric data, that the formation of this complex is associated with the release of one proton equivalent to solvent. They concluded that the pyrazole combines with the active site zinc ion. Release of one proton from the pyrazole nitrogen then generates the zinc-bound pyrazole anion which they surmised adds in covalent fashion to the NAD$^+$ at the positively-charged C-4 position of the nicotinamide cofactor (Theorell *et al.*, 1969).

A recent X-ray structure determination of alcohol dehydrogenase at 2.9 A-resolution confirms the binding of pyrazole nitrogen to the zinc ion with displacement of the usual water ligand. The other pyrazole nitrogen is a 2 A° from the C-4 atom of the NAD$^+$, which is consistent with either a covalent bond between the pyrazole and pyridine nucleotide cofactor or close association of pyrazole and NAD$^+$ as a consequence of electrostatic attraction. It was noted, however, that "the conformation of the oxidized factor is essentially the same as that of the reduced coenzyme NADH in the NADH-dimethyl sulfoxide complex" (Eklund *et al.*, 1982). Later, Becker and Roberts demonstrated, using ^{15}N-labelled NAD$^+$, that the ^{15}N chemical shift in the ternary complex formed from alcohol dehydrogenase, NAD$^+$ and pyrazole corresponded to that of 4-substituted dihydropyridines (1984). The ternary complex therefore results from covalent addition of the pyrazole to the 4-position of NAD$^+$. This is shown schematically in Figure 6. These observations, which have not previously been extrapolated to the inhibition of other pyridine nucleotide-dependent enzymes, suggested a generic approach to the design of inhibitors of such enzymes, including steroid 5α-reductase.

While there is no evidence that steroid 5α-reductase is zinc-dependent, it clearly must contain some functional equivalent of a zinc ion to stabilize the enzyme-generated

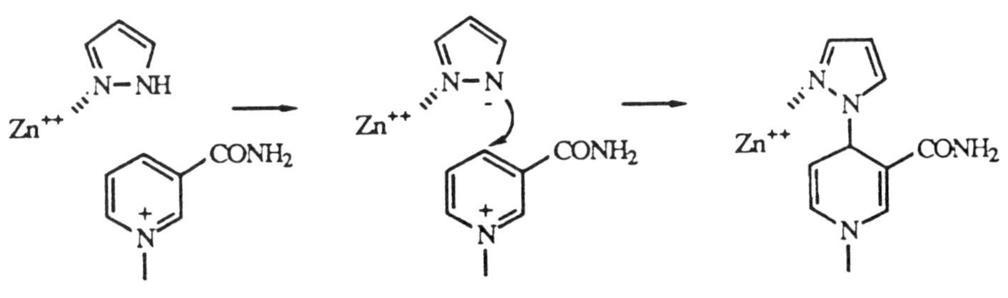

Fig. 6.

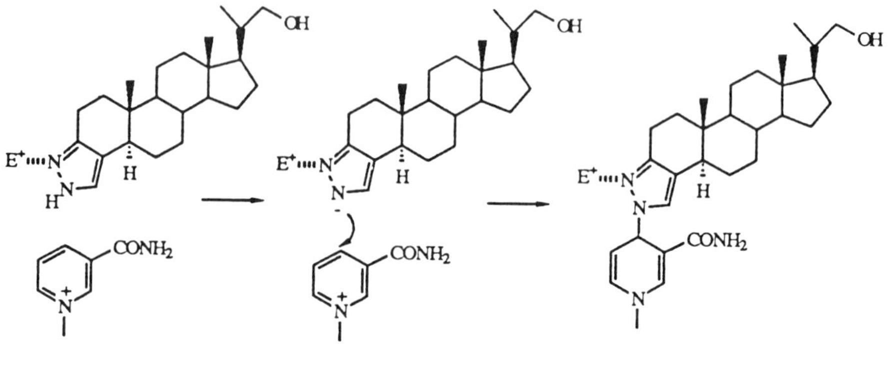

Fig. 7.

enolate (Figure 4). Other pyridine nucleotide-dependent dehydrogenases such as lactate dehydrogenase, for example, use a protonated imidazole rather than Zn^{++} as the active site electrophile (Branden and Eklund, 1980).

We proposed that a pyrazole-containing steroid such as 8 which possesses sp^2 hybridization at C-3 and C-4, and a 5α-ring junction, may prove to be an inhibitor of steroid 5α-reductase. These features, incorporated into the A ring, should elicit the conformation of the enolate of DHT. The rationale for inhibition, which is illustrated in Figure 7, demands that the cofactor be in the oxidized form. The 3,4-fused pyrazole (8), was prepared, but found to have no significant inhibitory activity towards either the human or rat prostatic steroid 5α-reductase, even in the presence of increasing amounts of $NADP^+$. On the assumption that this inactivity was a consequence of the inappropriate positioning of the pyrazole ring, the novel A-ring pyrazole 9 was

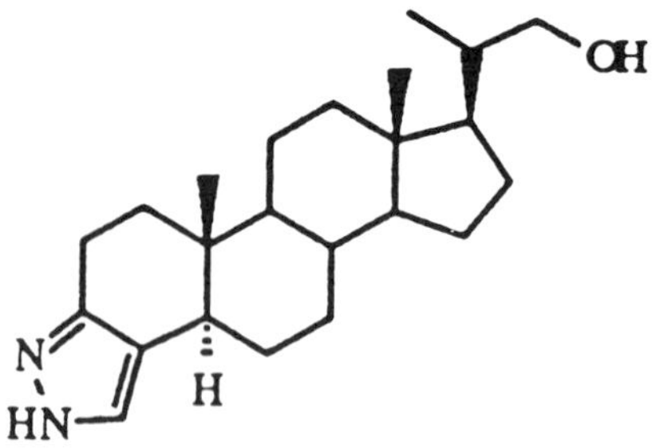

8

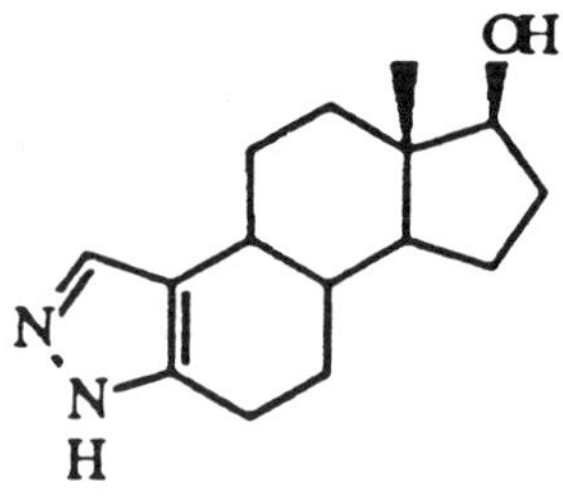

9

prepared, but it too was essentially inactive. Being unwilling to dismiss the design principles on the basis of these results, we turned to the study of the NAD^+ dependent 3(17)β-hydroxysteroid dehydrogenase from *Pseudomonas testosteroni*. This enzyme reversibly catalyzes the dehydrogenation at C-17 of testosterone to afford androstenedione.

In comparison to the K_m for testosterone of 700 nM for 3(17)β-hydroxysteroid dehydrogenase, compound 8 was found to have a K_i of 5 nM, when measured in the testosterone to androstenedione direction (Levy *et al.*, 1987). The A-ring pyrazole 9 was also found to be a potent inhibitor of this transformation with a K_i of 90 nM (Holt *et al.*, 1986). For the case of the fused pyrazole 8, which was studied in greater detail, it was found that inhibition occurred via an exclusive ternary complex involving NAD^+, but that there was no deprotonation event in the formation of this complex. In addition, the purified enzyme contained no zinc. Inactivation of the enzyme by the histidine-selective reagent, diethylpyrocarbonate, and the detection of an enzyme-associated ionizable residue with a pK_a near 6.5 suggest that the active site electrophile in 3(17)β-hydroxysteroid dehydrogenase is a histidine residue (Levy *et al.*, 1987). This mode of inhibition then clearly differs from the alcohol dehydrogenase case. It is proposed that the lone pair of electrons of the imidazole nitrogen forms a hydrogen bond to the hydrogen of the pyrazole N-H. The second pyrazole nitrogen could coordinate to the electropositive nicotinamide. The proposal is represented in Figure 8.

We have thus uncovered a novel mode of inhibition of pyridine nucleotide-dependent enzymes, but unfortunately, this mode of inhibition does not apply to steroid 5α-reductase. The next section describes an alternative structural motif which has been found to lead to the inhibition of pyridine nucleotide-dependent enzymes via ternary complexes involving oxidized cofactor. Application to the inhibition of steroid 5α-reductase follows.

B. Carboxylates

1. Lactate Dehydrogenase

Lactate dehydrogenase, which catalyzes the freely reversible interconversion of lactate and pyruvate, follows a kinetic mechanism analogous to that of steroid 5α-re-

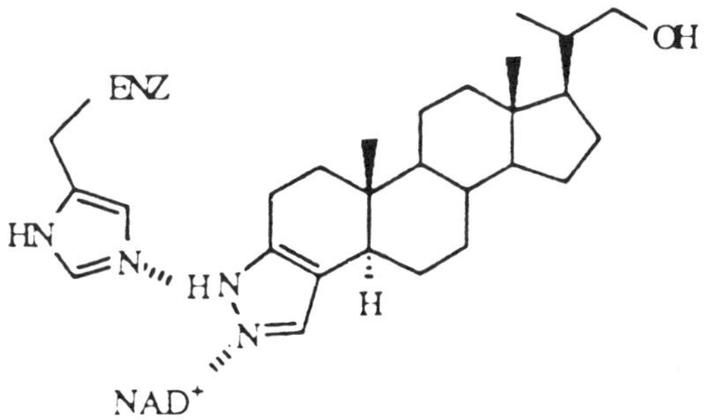

Fig. 8.

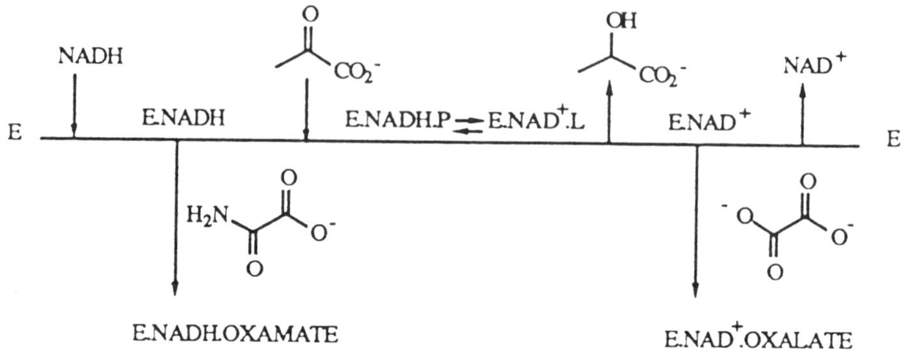

Fig. 9.

ductase with the pyridine nucleotides binding before the substrates, lactate or pyruvate (Figure 9). When assayed in the lactate to pyruvate direction, oxalate is a competitive inhibitor versus lactate, while in the alternate direction, pyruvate to lactate, oxamate (the half amide of oxalate) is a competitive inhibitor versus pyruvate. Given the compulsory order of binding of pyridine nucleotide and lactate or pyruvate, oxalate must form a ternary complex with NAD^+, while oxamate must form a ternary complex with NADH (Novoa *et al.*, 1961). This conclusion has been supported by X-ray studies (White *et al.*, 1976).

2. Alcohol Dehydrogenase

In the oxidation of long chain alcohols to aldehydes by alcohol dehydrogenase, fatty acids compete with the alcohol substrate and form ternary complexes with NAD^+. In the reverse direction, reduction of aldehydes to alcohols, fatty acid amides are competitive with aldehydes and form ternary complexes with NADH. The formation of these complexes can be followed by changes in the fluorescence spectrum. No change in the spectrum is seen when fatty acids are added to the NADH complex, nor is there any effect when an amide is added to the NAD^+ complex (Winer and Theorell, 1960). Thus, for both lactate dehydrogenase and alcohol dehydrogenase, inhibitory acids bind to E-NAD^+, while the corresponding amides bind to E-NADH. It seemed that these conclusions, if combined with structural features which would mimic an enolate, could be extrapolated to the inhibition of steroid 5α-reductase.

C. Carboxylate as an Enolate Mimic

Carboxymethyl-CoA (10) is a powerful inhibitor of citrate synthase which catalyzes the condensation of oxaloacetate and acetyl-CoA (Bayer *et al.*, 1981). The normal catalytic mechanism is shown in Figure 10. In the obligatory presence of oxaloacetate, acetyl-CoA is deprotonated to afford the thioester enolate which then adds to oxaloacetate. Carboxymethyl-CoA (10) has a low affinity itself towards the synthase ($K_d = 230\ \mu M$), however in the presence of oxaloacetate the affinity improves

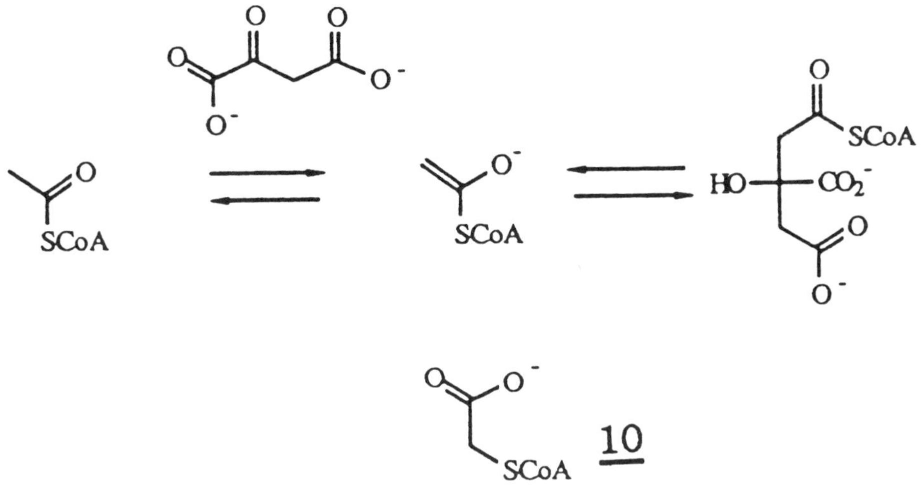

10

Fig. 10.

3000 fold ($K_i = 0.07$ μM). In the normal transformation the enolate of acetyl-CoA is generated only when oxaloacetate effects a conformational change of the protein. Carboxymethyl-CoA (10) thus mimics the enolate of acetyl-CoA and binds tightly only to the enzyme form which stabilizes that enolate.

The next section describes the application of the carboxylate motif in the successful design of inhibitors of steroid 5α-reductase which are uncompetitive versus testosterone.

D. Acrylates

Steroidal acrylates, such as compound 11, appeared particularly attractive. The carboxyl moiety should not only supplant the oxyanion of the enzyme-generated enolate it is designed to mimic, but as described above for lactate dehydrogenase and

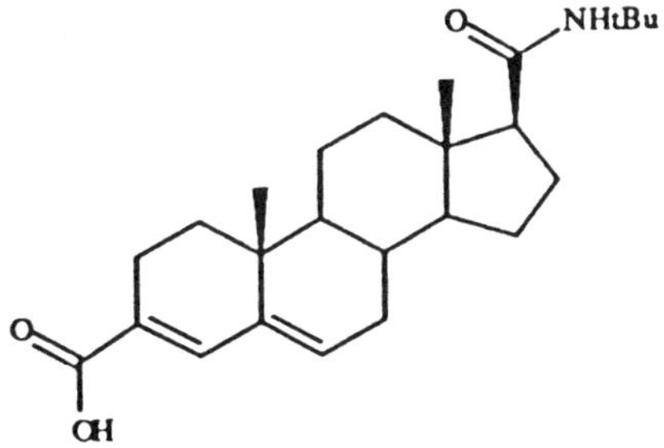

11

alcohol dehydrogenase, carboxylates prefer to form ternary complexes with oxidized cofactor. In addition, sp^2 hybridization at C-3, C-4, and at C-5 should mimic the intermediate delocalized enone shown in Figure 4. Compound 11 also incorporates a side chain at C-17 which has been demonstrated to be favorable for binding to the human prostatic enzyme (Liang *et al.*, 1985).

The apparent affinity constant ($K_{i,app}$) of acrylate 11 for the human prostatic enzyme was 30-36 nM. This apparent dissociation constant was derived by Dixon analysis (Dixon, 1953) in which the concentrations of testosterone and NADPH are held constant at 1.3 μM and 400 μM respectively. This analysis does not address cofactor dependency. Subsequently, double reciprocal analysis of the diene 11 with the human prostatic steroid 5α-reductase, demonstrated uncompetitive inhibition versus testosterone. These preliminary results are consistent with this steroidal carboxylate binding to the E-$NADP^+$ complex (Metcalf *et al.*, in press).

VIII. CONCLUSION

There is no pharmacological treatment at present for benign prostatic hypertrophy, a disease which afflicts a large percentage of men over 50 years of age. There is therefore a significant medical need for a successful approach. As DHT is a causative factor in the development of benign prostatic hyperplasia, the inhibition of steroid 5α-reductase offers potential treatment for this disease.

Uncompetitive inhibition is comparatively rare (Cornish-Brown, 1986) possibly owing to a lack of generic design principles. We have therefore been interested in developing generic concepts for the design of uncompetitive inhibitors. In the case of steroid 5α-reductase, the ordered kinetic mechanism with its interplay of two binary complexes involving reduced and oxidized cofactor, has allowed us to design such an inhibitor which is uncompetitive versus testosterone. Future clinical studies might demonstrate superiority of an uncompetitive inhibitor versus testosterone, compared to a competitive one for steroid 5α-reductase, owing to metabolic resistance *in vivo* to competitive inhibitors.

REFERENCES

Anderson, K.K. and Liao, S. (1968). *Nature* **219**, 277-279.

Batzold, F.H. (1981). *Prog. Clin. Biol. Res.* **75B**, 269-282.

Baulieu, E.E., Lasnitzki, I., and Robel, P. (1968). *Nature* **219**, 1155-1156.

Bayer, E., Bauer, B., and Eggerer, H. (1981). *Eur. J. Biochem.* **120**, 155-160.

Becker, N.N. and Roberts, J.D. (1984). *Biochemistry* **23**, 3336-3340.

Bertics, P.J., Edman, C.F., and Karavolas, H.J. (1984). *J. Biol. Chem.* **259**, 107-111.

Bjorkhem, I. (1969). *Eur. J. Biochem.* **8**, 345-351.

Blohm, T.R., Metcalf, B.W., Laughlin, M.E., Sjoerdsma, A., and Schatzman, G.L. (1980). *Biochem. Biophys. Res. Commun.* **95**, 273-280.

Blohm, T.R., Laughlin, M.E., Benson, H.D., Johnston, J.O., Wright, C.L., Schatzman, G.L., and Weintraub, P.M. (1986). *Endocrinology* **119**, 959-966.

Branden, C.I. and Eklund, H. (1980). In "Dehydrogenases requiring nicotinamide coenzymes" (J. Jeffrey, ed.), (Experientia 36) (Suppl), pp. 40-84. Basel, Birkhauser Verlag.

Brooks, J.R., Berman, C., Hichens, M., Primka, R.L., Reynolds, G.F., and Rasmusson, G.H. (1982). *Proc. Soc. Exp. Biol. Med.* **169**, 67-73.

Bruchovsky, N. and Lieskovsky, G. (1979). *J. Endocrinol.* **80**, 289-301.

Bruchovsky, N. and Wilson, J.D. (1968). *J. Biol. Chem.* **243**, 2012-2021.

Campbell, J.S., Bertics, P.J., and Karavolas, H.J. (1986). *J. Steroid Biochem.* **24**, 801-806.

Carlin, J., Christofalo, P., Arison, B., Berman, C., Brooks, J., Rasmusson, G., Rosegay, A., Stoner, E., and VandenHeuvel, W. (1987). *Pharmacologist* **29**, 149.

Cornish-Brown, A. (1986). *FEBS Lett.* **203**, 3-6.

Covey, D.F. and Robinson, C.H. (1976). *J. Am. Chem. Soc.* **98**, 5038-5040.

Dixon, M. (1953). *Biochem. J.* **55**, 170-171.

Duggleby, R.G. and Christopherson, R.I. (1984). *Eur. J. Biochem.* **143**, 221-226.

Eklund, H., Samama, J-P., and Wallen, L. (1982). *Biochemistry* **21**, 4858-4866.

Frederiksen, D.W. and Wilson, J.D. (1971). *J. Biol. Chem.* **246**, 2584-2593.

Giaguli, V.A., Vermeulen, A., DeSchepper, P., Buntix, A., and Stoner, E. (1988). *Endocrinology* **122**, 210s.

Habib, F.K. (1980). In "Steroid receptors, metabolism and prostatic cancer" (Intl. Cong. Ser. No. 494) (F.H. Schroder and H.J. de Voogt, eds.), pp. 157-164. Oxford, Exerpta Medica.

Holt, D.A., Levy, M.A., Brandt, M., and Metcalf, B.W. (1986). *Steroids* **48**, 213-222.

Houston, B., Chisholm, G.D., and Habib, F.K. (1985). *J. Steroid Biochem.* **22**, 461-467.

Houston, B., Chisholm, G.D., and Habib, F.K. (1987). *Steroids* **49**, 355-369.

Hsia, S.L. and Voigt, W. (1974). *J. Invest. Dermatol.* **62**, 224-227.

Imperato-McGinley, J., Peterson, R.E., Gautier, T., and Sturla, E. (1979). *J. Steroid Biochem.* **11**, 637-645.

Jacobi, G.H. and Wilson, J.D. (1976). *Endocrinology* **99**, 602-610.

Kadohama, N., Petrow, V., Lack, L., and Sandberg, A.A. (1983). *J. Steroid Biochem.* **18**, 551-558.

Levy, M.A., Holt, D.A., Brandt, M., and Metcalf, B.W. (1987). *Biochemistry* **26**, 2270-2279.

Liang, T., Heiss, C.E., Ostrove, S., Rasmusson, G.H., and Cheung, A. (1983a). *Endocrinology* **112**, 1460-1468.

Liang, T., Rasmusson, G.H., and Brooks, J.R. (1983b). *J. Steroid Biochem.* **19**, 385-390.

Liang, T., Cascieri, M.A., Cheung, A.H., Reynolds, G.F., and Rasmusson, G.H. (1985). *Endocrinology* **117**, 571-579.

Metcalf, B.W., Holt, D.A., Levy, M.A., Erb, J., Heaslip, J.I., Brandt, M., and Oh, H-J. (1989). *Bioorganic Chem.* In press.

Metcalf, B.W., Jund, K., and Burkhart, J.P. (1980). *Tetrahedron Lett.* 21, 15-18.

Novoa, W.B. and Schwert, G.W. (1961). *J. Biol. Chem.* **236**, 2150-2153.

Peterson, R.E., Imperato-McGinley, J., Gautier, T., and Sturla, E. (1977). *Am. J. Med.* **62**, 170-191.

Petrow, V., Wang, Y., and Lack, L. (1981). *Steroids* **38**, 121-140.

Petrow, V., Padilla, G.M., Kendle, K., and Tantawi, A. (1982). *J. Endocrinol.* **95**, 311-313.

Rasmusson, G.H., Liang, T., and Brooks, J.R. (1983). In "Gene regulation by steroid hormones II" (A.K. Roy and J.H. Clark, eds.), pp. 311-334. New York, Springer-Verlag.

Ringold, H.J., Ramachandran, S., and Forchielli, E. (1964). *Biochim. Biophys. Acta* **82**, 143-157.

Robaire, B., Covey, D.F., Robinson, C.H., and Ewing, L.L. (1977). *J. Steroid Biochem.* **8**, 307-310.

Scheer, H. and Robaire, B. (1983). *Biochem. J.* **211**, 65-74.

Siiteri, P.K. and Wilson, J.D. (1970). *J. Clin. Invest.* **49**, 1737-1745.

Starka, Von L., Siekmann, L., Hoippen, H.O., and Breuer, H. (1969). *Arzneimittelforsch.* **19**, 2022-2025.

Suzuki, K. and Tomaoki, B. (1974). *J. Steroid Biochem.* **5**, 249-256.

Theorell, H., Yonetani, T., and Sjoberg, B. (1969). *Acta Chem. Scand.* **23**, 255-260.

Walsh, P.C., Hutchins, G.M., and Ewing, L.L. (1983). *J. Clin. Invest.* **72**, 1772-1777.

White, J.L., Hackert, M.L., Buehner, M., Adams, M.J., Ford, G.C., Lentz, P.J., Jr., Smiley, I.E., Steindel, S.J., and Rossman, M.G. (1976). *J. Mol. Biol.* **102**, 759-779.

Wilton, D.C. and Ringold, H.J. (1968). In "Abstracts of brief communications. Third International Congress of Endocrinology, Mexico, 1968" (Intl. Congr. Ser. No. 157) (C. Gual, ed.), p. 105. Amsterdam, Excerpta Medica.

Winer, A.D. and Theorell, H. (1960). *Acta Chem. Scand.* **14**, 1729-1742.

9

HYDROXYLASES AS TARGETS FOR DRUG DESIGN

Walter Lovenberg

Merrell Dow Research Institute
Strasbourg Research Center
67084 Strasbourg Cedex, France

I. INTRODUCTION

Hydroxylations are utilized by mammalian organisms for a variety of functions. These include both anabolic and catabolic processes. Most often the hydroxylation reactions are carried out by enzymes known as monooxygenases or mixed function oxidases. This term implies that in addition to an electron acceptor, usually oxygen, an electron donor is also required as a cosubstrate. In such reactions one atom from the oxygen molecule is incorporated into the hydroxyl group, while the other atom appears in the water molecule that is formed. The general reaction for monooxygenases is as follows:

$$R\text{-}H + O_2 + ED\text{-}H_2 \rightarrow ROH + H_2O + ED$$

On the metabolic side, many reactions that are used to catabolize toxic or foreign substances and drugs are carried out by a family of enzymes known as the microsomal hydroxylases. Teleologically it is thought that these hydroxylating enzymes have evolved by organisms to make organic molecules more water soluble and thus more easily excreted. These drug metabolizing enzymes have been studied in detail, however, no attempt will be made to review this important family of enzymes in this chapter. Three types of hydroxylases will be discussed as examples of potential drug targets. These are the aromatic amino acid hydroxylases, dopamine β-hydroxylase and the hydroxylating reaction in the steroid biosynthetic transformations. With regard to the latter group, particular focus will be directed toward the aromatase reaction.

II. STEROID BIOSYNTHESIS

A number of very important humoral agents are synthesized in mammalian organisms with cholesterol being the precursor molecule. Many of the transformations involve specific monooxygenases and in themselves, provide potential therapeutic targets. A summary of the biosynthetic pathway for these steroid hormones is shown in Figure 1.

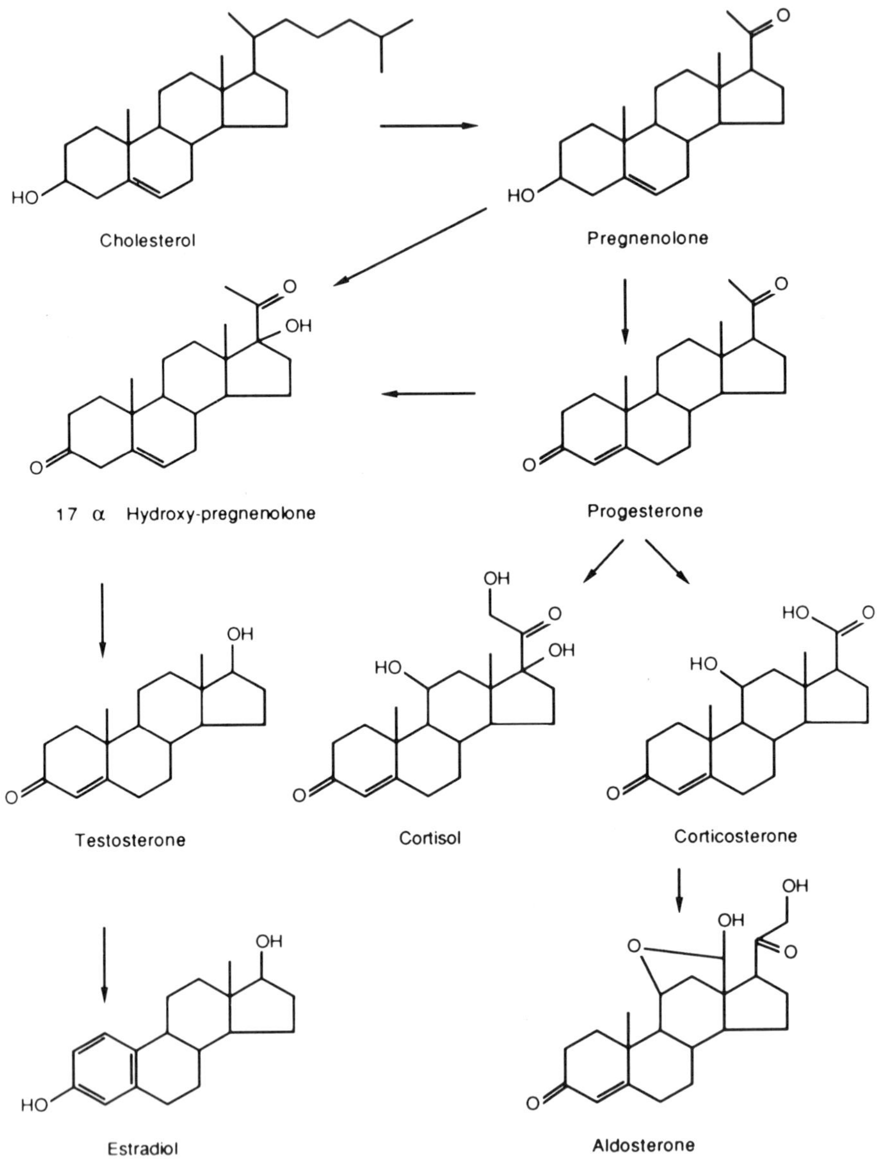

Fig. 1. *Steroid hormone biosynthesis.*

In this pathway alone there are multiple specific hydroxylation steps. These monooxygenases are located in either the mitochondria or microsomes of the adrenal cortical and other hormone producing cells. The enzymes have properties which are fairly similar to those found in the liver for drug metabolism. They are of the cytochrome P-450 type of monooxygenase and appear to have similar reaction mechanisms. In general the P-450 hydroxylating systems found in the endoplasmic reticulum receive electrons from a NADPH reductase whereas those in mitochondria utilize an electron carrier known as adrenodoxin. A general reaction scheme for the microsomal type is shown in Figure 2. One of the interesting enzymes from a therapeutic point of view is the enzyme aromatase which converts androstenediones to estrogen-like compounds. This enzyme is present in many tissues, most notably placenta and is associated with the endoplasmic reticulum.

A. Properties of the Aromatase Enzyme

The aromatization of androstenedione was first reported to occur in microsomes from placenta (Ryan, 1959). This reaction required both NADPH and molecular oxygen. The enzyme was assumed to be a classical mixed function oxidase. Although early studies (Meigs and Ryan, 1971) questioned whether a cytochrome P-450 containing enzyme was involved in the aromatase reaction, it was subsequently concluded (Thompson and Siiteri, 1974a) that the aromatase reaction was carried out by one or more P-450 components in the placental microsomes. These studies demonstrated the known inhibitors of cytochrome P-450 catalyzed reactions were inhibitors of aromatase. These workers also showed (Thompson and Siiteri, 1974b) that in the

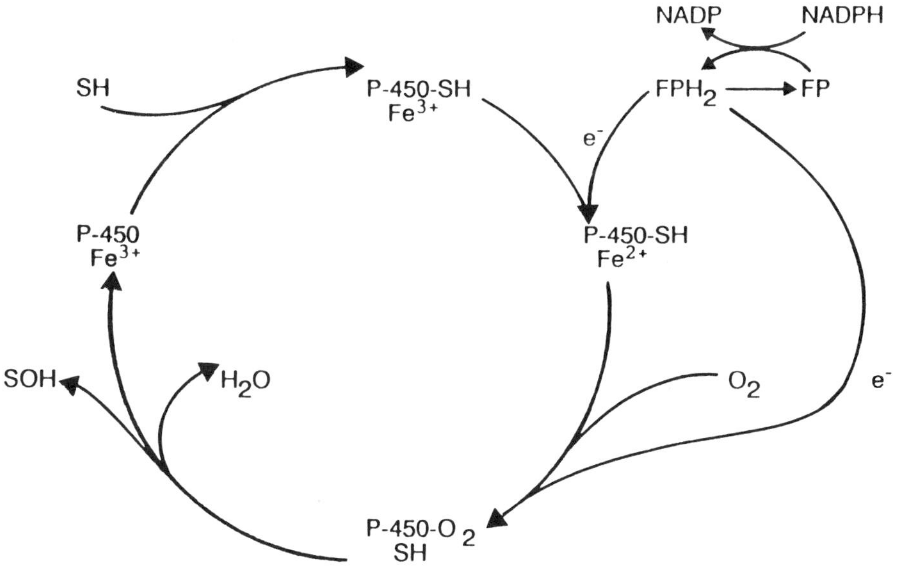

Fig. 2. *General reaction for microsomal sterol hydroxylases.*

presence of sufficient KCN to prevent spurious oxidation of NADPH exactly 3 moles of oxygen and 3 moles of NADPH were utilized per mole of estrone formed. This would be consistent with the execution of 3 successive hydroxylation reactions by the enzyme(s) involved in the aromatization process. Indeed it was found that 19-hydroxy- or 19-oxoandrostenedione could also be substrates for the reaction. Thus it was conceived that the aromatization reaction consisted of 3 successive hydroxylation reactions. The first two of these hydroxylations appear to occur on the 19-methyl group to yield the 19-oxo compound. It was thought that the 2 carbon of ring A was the likely site for the third hydroxylation. It was then shown (Hosoda and Fishman, 1974) that the triply hydroxylated steroid, 2-β-hydroxy-19-oxoandrostenedione, underwent very rapid spontaneous aromatization to estrone in the presence of aqueous conditions.

The overall sequence of the reactions leading to aromatization can be visualized as shown in Figure 3. In all these early studies crude microsomal preparations were used as the enzyme source. More recently the enzyme has been purified to apparent homogeneity by different techniques (Nakajin *et al.*, 1986; Hagerman, 1987; Kellis and Vickery, 1987a). The purified enzyme exhibits a cytochrome P-450 heme spectrum and has a molecular weight of about 55,000 daltons. There is an absolute requirement for the NADPH-cytochrome P-450 reductase, but it appears that a single hydroxylating enzyme carries out all three of the proposed hydroxylation steps. Of interest is the finding that the reductase isolated from placenta seems to be the most effective electron donating system. However, substantially higher concentrations of this reductase are necessary for optimal activity of the purified enzyme (Hagerman, 1987). The substrate specificity with regard to the steroid is very high with only androstenedione and testosterone of the compounds examined being converted to the

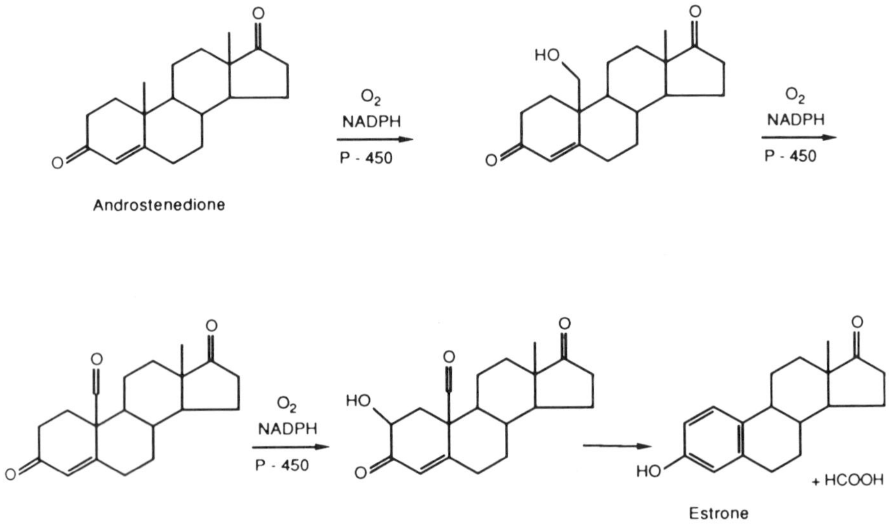

Fig. 3. *Aromatase reaction.*

estrogens (Kellis and Vickery, 1987a). A partial amino acid sequence has been attempted (Chen *et al.*, 1986) and some homology to other P-450 enzyme is suggested. Likewise fragmentary information on a cDNA specific for human aromatase has been reported (Evans *et al.*, 1986). More recently McPhaul *et al.* (1988) have isolated a full length cDNA clone for the chicken ovary aromatase. The deduced amino acid sequence for chicken aromatase is highly homologous with the human placental enzyme but only limited homology is seen with other microsomal P-450 enzymes. Of particular interest in this work was the fact that when the cDNA was transfected into COS-1 cells, these transfected cells expressed high levels of aromatase with subcellular distribution and kinetic properties that are very similar to the chicken ovary enzyme. The expression of this enzyme in the COS-1 cells will provide an excellent tool for inhibitory and regulation studies.

While considerably more must be learned about the enzyme, it is clear that aromatase is one of the family of P-450 enzymes and that this one heme protein carries out all 3 monooxygenase reactions associated with the aromatization of steroids. Now that pure enzyme is available it will be necessary to perform detailed kinetic and mechanistic studies to determine the exact mechanism of this enzyme. The scheme presented in Figure 2 could be modified to accommodate three successive hydroxylation reactions following initial binding of the androstenedione to the enzyme.

B. Inhibition of Aromatase

There is ample precedent for the therapeutic effectiveness of limiting estrogen production. This is particularly true for estrogen dependent malignancies of the breast. A number of different compounds are currently being studied as aromatase inhibitors. Since aromatase is a heme protein it was initially thought that both CN and CO should inhibit the activity. The fact that CO was a relatively poor inhibitor (Thompson and Siiteri, 1974a) and CN was primarily active in inhibiting the conversion of 19-keto-androstenediones caused considerable confusion in the early studies. It was thought that perhaps multiple enzymes were involved. The isolation of pure homogeneous hemoprotein catalyzing the entire aromatization reaction has resolved this question. Recent spectral studies (Kellis and Vickery, 1987b) using a pure preparation of aromatase from human placenta have shown that CN which is a good inhibitor of the aromatization of 19-norandrostenediones but not of the methylated analogue, probably does not bind in the presence of the natural substrate. Cyanide, therefore, is not a good inhibitor of the reaction. Other compounds include aminoglutethimide and SKF 525 which are known to be relatively non-selective inhibitors of microsomal cytochrome P-450 reactions have also been shown to be potent inhibitors of aromatase (Thompson and Siiteri, 1974a).

To design a compound that could be used selectively, a much more specific inhibitor of the enzyme is required. The recognition that inhibition of aromatase would be an attractive therapeutic target generated considerable research. In 1981, it was reported from two laboratories (Covey *et al.*, 1981; Metcalf *et al.*, 1981) that the addition of an acetylenic group to the C_{19} position of the normal substrate produced a powerful time-dependent inhibition of aromatase. Although several slightly more potent inhibitors have been discovered, this compound still remains an attractive candidate for clinical use. The mechanism of the interaction of the acetylenic substrate with the enzyme remains to be fully resolved. However, it is clear from the experiments of Metcalf *et al.* (1981) that it is a mechanism-based inhibition requiring both oxygen and NADPH. Furthermore, isotope studies and inhibitor intermediate compounds indicated that the reactive intermediate formed may be a highly reactive oxirene species. However shortly thereafter other studies (Ortiz de Montellano and Kunze, 1981; Ortiz de Montellano *et al.*, 1982) on the inhibition of cytochrome P-450 by propyne suggested that a vinyl radical intermediate is more likely (Figure 4). Other mechanisms of inhibition are not excluded. The specificity of this inhibitor appears to be quite good showing essentially no inhibitory activity toward general drug metabolizing cytochrome P-450 or the 11-β-hydroxylase system for the formation of corticosterone.

In a somewhat similar approach to inhibiting this enzyme system, Berdarski *et al.*, (1985) synthesized a 19-thio derivative of the normal substrate and found it to be a suicide inhibitor of similar potency. More recently Kellis *et al.* (1987) have synthesized 10-oxirane and 10-thiirane derivatives of the substrate. These compounds were simple competitive inhibitors, but were very potent with K_i's in the 2-10 nM range.

Brueggemeier *et al.* (1978; Snider and Brueggemeier, 1987) have found that androstenediones substituted in the 7-α position with a parathiol aniline are potent enzyme activated inhibitors. One of the most potent of these compounds is 7-α–(4'-amino)phenyl-thio-1,4-androstadione-3,17-dione. (Figure 5).

No mechanism has yet been proposed for this compound, but it is clear that a reactive intermediate is formed.

C. Biological Effects of Aromatase Inhibition.

It is now widely accepted that limitation of the action of estrogen is of substantial benefit in the treatment of estrogen dependent malignancies of the breast. Blockade of estrogen receptors is a commonly used treatment (Pearson *et al.*, 1982). Alternatively, specific inhibition of estrogen synthesis appears also to be a viable approach. This concept was reviewed in detail by Sjoerdsma (1981) particularly with regard to the use of suicide inhibitors of aromatase. Following the discovery of MDL 18,962 (Metcalf *et al.*, 1981) the biochemical properties of this compound were studied

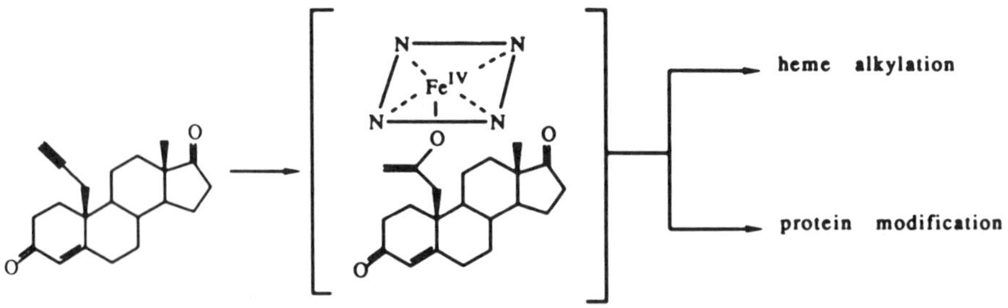

Fig. 4.

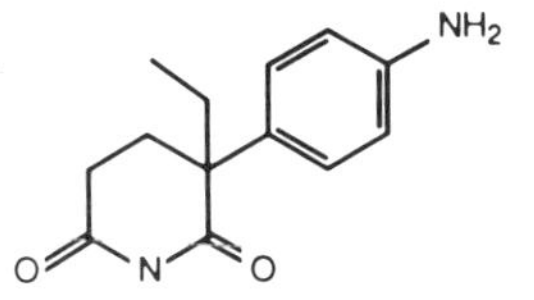

Amino Glutethimide

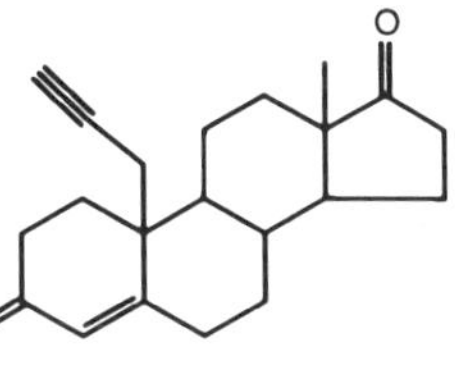

10 Propargylic Steroid
(MDL 18962)

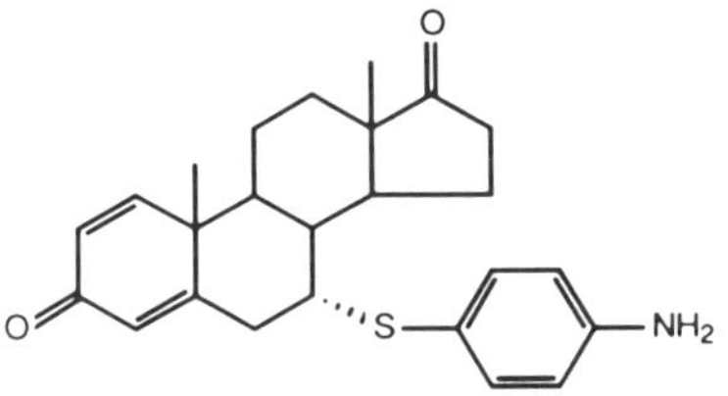

7 α (4' Amino) Phenylthio - 1,4 - Androstadiene -3,17- Dione
(7 α - A P T A DD)

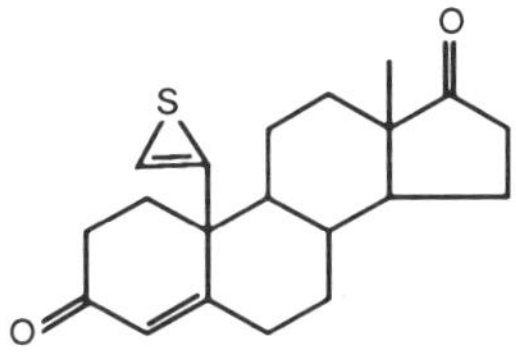

10 β Thiirane 4 - Androstene - 3,17 - dione

Fig. 5. *Some inhibitors of aromatase.*

(Johnston *et al.*, 1984). The compound was found to be highly specific for aromatase and had essentially no activity against other mitochondrial or microsomal steroid hydroxylating enzymes. The inhibition of aromatase *in vitro* was time-dependent and irreversible consistent with the originally proposed (Metcalf *et al.*, 1981) mechanism of action. This inhibitor did not bind to estrogen receptors and had no estrogen-like activity when administered *in vivo*, although inhibition of aromatase activity could be demonstrated *in vivo*. Subsequently MDL 18,962 has been shown to inhibit estrogen synthesis *in vivo* in baboons (Longcope *et al.*, 1988). This species was chosen because, compared to humans, it has similar rates of peripheral aromatization and an aromatase with similar sensitivity to the inhibitor. In these studies the conversion of either labelled androstenedione or testosterone to the corresponding estrogen was measured during the course of a 3.5 hour infusion. MDL 18,962 caused substantial inhibition of conversion of androstenedione in doses as low as 0.1 mg/kg when given I.V. Somewhat larger doses were needed when the drug was given orally. It is clear, however, that this compound should be very effective in limiting estrogen synthesis in humans and clinical trials are currently being initiated. Recently clinical studies have been done with 4-hydroxy-androstenedione in 58 post-menopausal breast cancer patients (Brodie *et al.*, 1987). This compound is a strong competitive inhibitor of aromatase and at doses of 250-500 mg/day orally reduced circulating estradiol levels from a mean of 7.2 pg/ml to 2.8 pg/ml. Partial or complete remissions were seen in 27% with another 10% showing stabilization of the disease. This study tends to substantiate the hypothesis that appropriate aromatase inhibitors may be effective therapeutic tools.

The role of aromatase in non-reproductive tissues is not well understood although significant amounts of this enzyme are found in brain. It is, however, clear that estrogen levels in neonates play a role in determining sexual behavior in adulthood (Erskine *et al.*, 1988). The activity of aromatase in brain appears to be critically regulated during pre- and neonatal development and may play an important role in the sexual differentiation of the central nervous system (Schumacher *et al.*, 1988). It would appear that inhibitors of this enzyme would be useful tools in probing the role of aromatase in brain development. Another interesting physiological system where aromatase may play a role is in the formation of the mineralocorticoid, 19-nor-deoxycorticosterone (19-nor-DOC). This compound has been suggested to have an important role in regulating blood pressure. In a genetically hypertensive strain of rat (SHR), there appears to be an excessive formation of 19-nor-DOC (Melby *et al.*, 1987). It is of interest that the aromatase inhibitor, MDL 18,962, was effective in reducing blood pressure in young 7.5 week old SHR. Daily administration of the inhibitor 10 mg/kg for the succeeding 8 weeks significantly retarded the development of hypertension. In this study it was also shown that the excretion of 19-nor-DOC was decreased by about 90% at the end of the first week of treatment. These results provide evidence for the role of aromatase in the *in vivo* formation of this powerful mineralocorticoid.

III. NEUROTRANSMITTER FORMATION

The synthesis of the catechol- and the hydroxyindoleamine neurotransmitters are dependent upon several key hydroxylation reactions catalyzed by monooxygenases. Modulation of these hydroxylase reactions through activation or inhibition offers an approach to modifying the activity of certain neuronal systems having implications for both blood pressure control and CNS activity. The two systems we will consider in the current review are the catecholamine system and the serotonin system. The biosynthetic pathway for these neurotransmitters (Figure 6) contains three important steps that are catalyzed by monooxygenases. The various types of catecholaminergic neurons can be characterized by the complement of biosynthetic enzymes which they express. Dopaminergic cells contain only tyrosine hydroxylase and aromatic *L*-amino-acid decarboxylase, whereas noradrenergic cells contain dopamine-β-hydroxylase in addition to the above enzymes. Serotoninergic cells, however, appear to express only tryptophan hydroxylase and aromatic *L*-amino-acid decarboxylase.

A. Dopamine-β-Hydroxylase

1. Biochemical Properties

This enzyme which catalyzes the conversion of dopamine to noradrenaline was the first studied by Kirshner in 1957 although it was first characterized by Levin and Kaufman (Levin *et al.*, 1960; Levin and Kaufman, 1961). The enzyme is heavily concentrated in the adrenal medullary cells and in sympathetic nerve terminals. The enzyme exists in the noradrenaline storage vesicles where it is thought to catalyze the formation by noradrenaline. However, significant amounts are also found in the cytosol. This enzyme is a typical mixed function monooxygenase and ascorbic acid

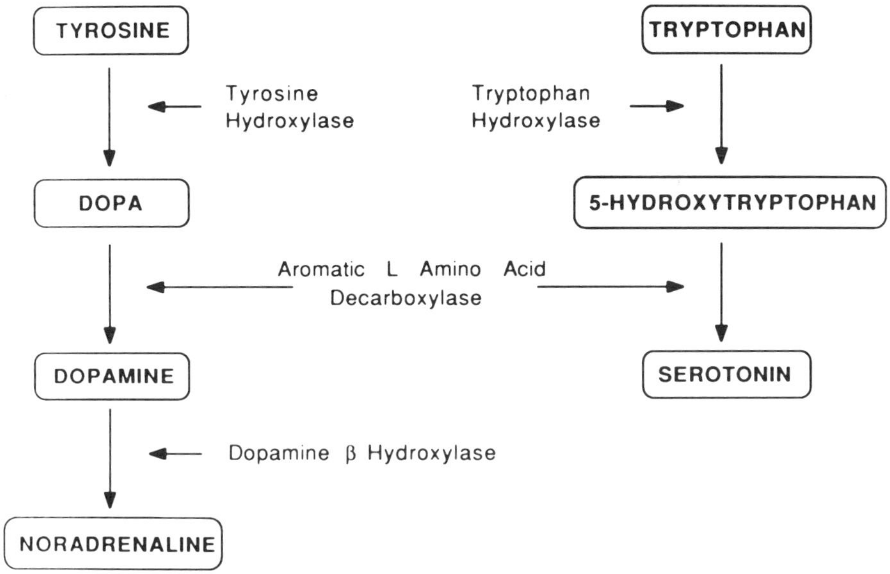

Fig. 6. *Biogenic amine synthesis.*

appears to be the endogenous electron donor. In the succeeding years much has been learned about this enzyme. It is a copper-containing enzyme with a molecular weight of about 300,000 daltons. In 1973, we reported that the enzyme was a tetrameric glycoprotein and that each of the subunits had molecular weight of about 75,000 daltons (Wallace *et al.*, 1973). The sugar residues appeared not to play any role in the catalytic activity. It also appears that two subunits are associated through disulfide interaction and that each dimer is associated with another by hydrogen bonds (Craine *et al.*, 1973). No activity has been observed in units smaller than dimers and this has only been seen in the enzyme obtained from human plasma (Rosenberg and Lovenberg, 1977). It is clear that the enzyme-bound copper plays an important role in the catalytic function of the enzyme but the exact nature of the active site remains to be determined. When DBH is isolated under conditions designed to yield optimal activity, the tetrameric enzyme contains about four copper atoms per molecule (Colombo *et al.*, 1987; Walker *et al.*, 1977). The exact relationship between the individual copper atoms remains to be determined. They are not in close proximity to one another since the EPR spectroscopy shows no spin coupling (Walker *et al.*, 1977; Blackburn *et al.*, 1988). While many of the earlier proposed mechanisms featured two copper atoms receiving two electrons from ascorbate, we now know that two sequential electrons additions occur (Rosenberg *et al.*, 1980; Diliberto and Allen, 1981). Therefore a single copper atom per active site could be sufficient for total activity. However there is evidence to suggest that fully active enzyme contains two copper atoms per subunit. Knowledge of the primary structure of the enzyme and the ligands that hold the copper will be of great benefit in understanding the active site and in the design of inhibitors. A recent study by Scott *et al.* (1988) using X-ray absorption techniques shows that each copper most likely has four nitrogen or oxygen ligands in the oxidized enzyme. Kinetic studies (Stewart and Klinman, 1987) with various electron donors suggest that the copper atom is first reduced by two electrons in the reaction sequence. The two substrates (O_2 and dopamine) bind in either sequence to form a one electron-reduced ternary complex which converts to an enzyme product complex which appears to remain until the complex is reduced again by an additional electron.

DBH shows very broad substrate specificity. It is interesting that its primary location of physiologic function is in the vesicles of nerve terminals and adrenal medullary cells. The specificity for the enzyme is thereby probably provided by the uptake system for dopamine. It is known that when sympathetic nerves release transmitter, a small amount of DBH is also released. In man quite high levels of enzyme accumulate in serum but it has been shown that this circulating enzyme is not active *in vivo*. In part, because of this broad substrate specificity, a large number of molecules have been found to be good inhibitors. The inhibition of DBH was reviewed several years ago by Rosenberg and Lovenberg (1980).

2. Inhibitors of DBH

In recent years a number of specific mechanism-based inhibitors of this enzyme have been discovered. A variety of approaches have been taken in the use of inhibitors to study the enzyme. Since DBH is a copper-containing enzyme and the copper is thought to participate in the active mechanism, a variety of metal chelating substances have been examined. These have proved more useful from a mechanistic point of view than from a potential therapeutic viewpoint. Substrate competitive inhibitors can be designed to be competitive with either ascorbate or dopamine. Most mechanism-based inhibitors, however, are based on the reactive intermediate formed when an abnormal substrate is hydroxylated. It is here that the broad substrate specificity of this enzyme has provided a wealth of approaches.

a. Mechanism-Based Inhibitors. The first such compound to be characterized as a "suicide" inhibitor was *p*-hydroxybenzylcyanide (Baldoni and Villafranca, 1980). This was an obvious compound to characterize since it is known that DBH actively hydroxylates a wide range of phenylethyl amine derivatives. This compound is an analogue of tyramine which is perhaps the best substrate for the enzyme. Analysis of the inactivation of the enzyme was difficult because potentially CN can be released from the product of hydroxylation (*p*-hydroxymandelonitrile). The CN ion is known to inhibit DBH presumably by interaction with the copper, although such interaction is reversible by dialysis. Clearly the product is interacting with enzyme in another manner and this has been studied in depth but not completely resolved (Colombo *et al.*, 1984). The next important finding in the search for suicide inhibitors of this enzyme was the observation of May *et al* (1983; Padgette *et al.*, 1985) that 1-phenyl-1-aminomethylethylene (PAME) appeared to be a mechanism-based inhibitor of DBH. In this case both an electron donor and oxygen were required for the time dependent inhibition. The enzyme appeared to catalyze a β-hydroxylation of this compound, possibly resulting in the formation of a highly reactive aziridine. Alternatively, one could visualize one electron oxidation of the olefin to form a radical cation which could partition between the formation of an epoxide or a nucleophilic attack by an active site residue of the enzyme. It was found (Padgette *et al.*, 1985) that labelled PAME was incorporated into inactivated DBH at a ratio of about 5 moles per mole of tetramer suggesting that there may be an active site in each of the subunits. Another compound that was examined as an inhibitor of DBH was cinnamylamine (May *et al.*, 1983). This compound was a very poor substrate but appeared to be a time dependent inhibitor of the enzyme. More recently, Bargar *et al* (1986) have found that analogues of PAME in which the phenyl group is replaced by either a thiophene or a furan result in even more potent inhibitors. An increase in inhibitory potency of about 1000 fold was demonstrated for the thiophene analogue (2-(2-thienyl)-allylamine). Hypothetically the increased activity of this compound may result from a stabilization of the radical cation intermediate. Equally interesting, both these heterocyclic analogues cause a significant and long lasting hypotension when administered to spontaneously hypertensive rats (Bargar *et al.*, 1986). These workers (Bargar *et al.*, 1988) then hypothesized that the radical cation formed during the attempted oxidation of the substrate analogue would be further stabilized in the case of 2-(1,3-dithiolan-2-ylidene) ethanamine. This compound proved to be about 10 times more active than the thiophene analogue described above. Furthermore, this is the most potent and first non-aromatic mechanism-based inhibitor for dopamine β-hydroxylase.

b. Competitive Inhibitors. In addition to these mechanism-based inhibitors, a number of potent competitive inhibitors have been designed. Several of 1-benz-imidazole-2-thiols have been found to be good inhibitors of DBH (Kruse *et al.*, 1986) the idea being to produce a compound which would recognize the substrate binding site and have a functionality to bind to the enzyme copper. The compound which showed the most promise 1-(3,5-difluorobenzyl)imidazole-2-thiol (SKF 102698) had a K_i value of 41 nM and like other strong DBH inhibitors was quite effective in lowering blood pressure. Of particular interest was the observation (Berkowitz *et al.*, 1988) that the changes in the dopamine/norepinephrine ration in mesenteric arteries resulting from inhibition of the enzyme correlated very well with the change in blood pressure. Based on their previous success in substituting thiophene for benzene in mechanism-based inhibitors, McCarthy *et al.* (1989) have prepared a series of 1-(thienylalkyl)imidazole-2-thiones. Of this series 1-[2-(thienyl)ethyl]-imidazole-2-thiol (MDL 43,925) was the most potent (K_i=16 nM). This compound was shown to be effective when given *in vivo* as indicated by a reduction of norepinephrine content and an increase in dopamine content in various rat tissues. MDL 43,925 was also quite effective in the reduction of blood pressure in experimental animals.

c. Turnover Dependent Cofactor Depletion. Recently May *et al.* (1988) have suggested a novel approach to manipulating the adrenergic system by reduction of dopamine β-hydroxylation. These workers found that the artificial substrate for DBH, phenyl-2-aminoethyl selenide, is converted to the corresponding selenoxide and that this product spontaneously undergoes a reduction with a second mole of ascorbate. Repetition of this reaction results in a catalytic depletion of ascorbate within chromaffin granules. While the effects of this selenium-containing substrate analogue on tissue catecholamines remain to be determined, it is clear that phenyl-2-aminoethyl selenide is a potent antihypertensive agent in spontaneously hypertensive rats (May *et al.*, 1988).

3. DBH Inhibitors as Therapeutic Agents

Reduction of norepinephrine synthesis and content in various organs has been a long sought goal in therapeutics. It is clear that we now have at our disposal the appropriate compounds to selectively inhibit the conversion of dopamine to norepinephrine. The full consequence of this inhibition in man has yet to be evaluated. Based on studies with experimental animals, one effect of such inhibition is likely to be the reduction of blood pressure in hypertensive subjects: the long-term effects of perturbing the relationship between norepinephrine and dopamine in brain must be examined carefully both from a neurological and psychiatric point of view. Of the compounds that have been studied in the hypertensive rat, it is not known whether they affect blood pressure primarily by a peripheral or a central mechanism. The recent description of an extremely potent mechanism-based inhibitor of DBH that exhibited rather small blood pressure responses (Kruse *et al.*, 1988) bring into question whether inhibiting peripheral DBH is sufficient to bring about a major reduction in blood pressure. Clearly while inhibition of this monooxygenase has important implications as a therapeutic approach to hypertension, much remains to be learned.

B. Aromatic Amino Acid Hydroxylases

The metabolism of phenylalanine and the formation of serotonin and the catecholamine neurotransmitters are dependent upon several key monooxygenases. The 4-hydroxylation of phenylalanine, the 3-hydroxylation of tyrosine and the 5-hydroxylation of tryptophan is each carried out by a specific monooxygenase that appear to belong to family of related enzymes. Although such a family of related enzymes had long been postulated, based upon biochemical properties, Grenett *et al.* (1987) were recently able to perform a primary structure analysis and to show that extensive sequence homology exists between enzymes which hydroxylate each of these three amino acids. It is now well established that each of these monooxygenases catalyze the reaction of three cosubstrates (oxygen, tetrahydrobiopterin and amino acid) to yield the hydroxylated amino acid, quinoid dihydrobiopterin and water. Each of these hydroxylases also appear to be regulated *in vivo* by protein phosphorylation reactions. However, the nature of these phosphorylations and the trigger mechanisms appear to be quite different for each enzyme. Because of the complexity of each of these reactions and their regulation, there are multiple points for the design of possible inhibitors.

1. Molecular and Regulatory Properties of Tyrosine and Tryptophan Hydroxylase

Tyrosine hydroxylase appears to have a molecular weight of about 240,000 and to probably consist of four identical subunits (Rosenberg and Lovenberg, 1983). The subunit molecular weight deduced from the sequence of the messenger RNA for tyrosine hydroxylase is about 56,000 (Grima *et al.*, 1985). It is thought that the molecular structure of tryptophan hydroxylase is similar although the only messenger

RNA analysis completed is for the rabbit pineal enzyme. This enzyme appears to have a subunit molecular weight of 51,000 (Grenett *et al.*, 1987). However, caution should be exercised since a variety of hydroxylating enzymes with various substrate specificities have been termed tryptophan hydroxylase. The pineal enzyme which is highly homologous to the rat liver phenylalanine hydroxylase actually catalyzes the hydroxylation of either tryptophan or phenylalanine at relatively similar rates. It should be noted that tryptophan can also serve as substrate for liver phenylalanine hydroxylase. Likewise an enzyme found in mast cells also actively hydroxylates both tryptophan and phenylalanine (Levine *et al.*, 1964). Perhaps the enzyme of greatest interest from a therapeutic point of view, however, is that which is localized in the serotonergic neurons of the CNS. This enzyme is highly specific for tryptophan although it also hydroxylates phenylalanine, but at a very slow rate.

The activity of tyrosine hydroxylase appears to be regulated in a complex manner with multiple factors participating. The basis for the regulatory events, however, appear to revolve around the binding of the election donor, tetrahydrobiopterin, to the enzyme. In the original work on the characterization of the enzyme, it was shown that the end-product of the reaction, *L*-DOPA, was a competitive inhibitor with regard to the reduced pterin (Nagatsu *et al.*, 1964). We later reported that the affinity of the enzyme for the pterin cofactor was controlled by a protein phosphorylation reaction (Loven-berg *et al.*, 1975). Cyclic AMP-dependent phosphorylation resulted in a marked decrease in the K_m value for tetrahydrobiopterin and consequently a loss of feedback inhibition by the end product (Ames *et al.*, 1978). Following a series of experiments (Levine *et al.*, 1981) designed to establish conditions within the terminals of dopamine neurons in the corpus striatum, it was concluded that:

1. The phosphorylated enzyme was fully active whereas the nonphosphorylated enzyme would probably exhibit little activity *in vivo*
2. Under resting conditions about 20% of the enzyme is in the active state
3. Pharmacological or physiological activation could create a condition where the uptake of the substrate tyrosine into the nerve terminal might become a limiting factor
4. A rather large increase in the intraneuronal content of tetrahydrobiopterin would be required to result in significant synthetic activity of the nonphosphorylated enzyme.

It is now known that all three major hydroxylases are regulated by protein phosphorylations. In the case of tryptophan hydroxylase, it appears that a Ca^{++}-calmodulin dependent protein kinase is responsible for the activation (Kuhn *et al.*, 1980). With serotonin synthesis, however, there appears to be little end-product inhibition and activation only takes place through a reduction in the K_m for tetrahydrobiopterin. Recently the deduced sequences for the three hydroxylases have been compared (Grenett *et al.*, 1987). The highest amount of homology appears to be in the central region of these molecules with the greatest variation occurring in the amino terminal domain. It has been postulated that the active sites for all these amino acid hydroxylases are similar and occur in the central portion of the molecule and that since each has quite different regulatory properties that the region responsible for regulation is in the amino terminal domain. It is known that partial proteolysis of tyrosine hydroxylase results in the formation of an active fragment that no longer is subject to the complex regulatory phenomena that apply to the parent molecule. It is thought serine 40 in the tyrosine hydroxylase molecule is at least one of the sites of phosphorylation. Using a specific kinase that copurifies with tyrosine hydroxylase from rat pheochromocytoma, it has been reported that limited trypsin proteolysis removes the phosphorylation site (Bonnefoy *et al.*, 1988). This enzyme which is reduced in size by about 6kDa is fully activated and appears to lack residue 40. In earlier work (Campbell *et al.*, 1986), however, it was suggested that the endogenous kinase phosphorylates serine 8 whereas cAMP-dependent kinase would be responsible for the phosphorylation at residue 40. One can conclude, however, that at least three serine residues in the N-terminal region of the protein can be phosphorylated by various protein kinases and that it is this domain of the protein that regulates the catalytic activity of the enzyme. What role the individual protein kinases play *in vivo* remains to be determined. In adrenal chromaffin cells cholinergic stimulation results in increased catecholamine synthesis, activation of tyrosine hydroxylase and rapid incorporation of phosphate into several sites on the enzyme; at least three of those sites appear to be subject to cAMP-dependent protein phosphorylation (Waymire *et al.*, 1988).

Although the deduced sequences of pineal tryptophan hydroxylase is known, the sites of phosphorylation and whether *in vivo* phosphorylation plays a regulatory role remain to be determined. It has recently been reported (Ichimura *et al.*, 1988) that a small brain specific protein mediates the activation of tryptophan hydroxylase by the calcium-calmodulin dependent protein kinase. It has also been shown (Boadle-Biber and Phan, 1987) that brain slices treated to increase intracellular calcium levels result in enhanced activity of tryptophan hydroxylase. This activation is blocked by specific antagonist of calmodulin suggesting the activation is mediated by a calcium-calmodulin kinase.

2. Pharmacological Regulation of Catecholamine and Serotonin Biosynthesis.

The biogenic amine neurotransmitters appear to participate in a wide range of CNS functions, thus gross inhibition or stimulation of their synthesis might have an impact on multiple functions. Therefore with the exception of unique tumors that generate excess of one or another of these neurohumoral substances, modulating overall biogenic amine synthesis is of unknown therapeutic potential. The availability of specific agents to modify biogenic amine synthesis would, however, be of substantial experimental value. It is well established that the initial aromatic amino acid hydroxylases are rate limiting steps in the reaction sequence and therefore the appropriate target for intervention. Since each of these enzymes has a complex reaction mechanism with multiple substrates and regulatory events, a variety of approaches to affecting the synthesis rate of biogenic amines is possible. The reader is referred to a recent review (Lovenberg, 1986) for references for the following discussion.

There are several disease states where it might be beneficial to increase synthesis of neurotransmitters, or the activity of the hydroxylases. These include certain types of dystonia, Parkinson's disease, depression, and atypical phenylketonuria which results from a defect in tetrahydrobiopterin synthesis. The simplest approach to this would be to increase the availability of precursor molecules. In the case of tryptophan hydroxylase the normal intracellular level of tryptophan appears to be near or below the K_m value and enhancing tryptophan levels leads to increased serotonin synthesis. This process is, however, more complex, because multiple factors control the uptake of amino acids by brain. It would appear, however, that tyrosine is normally near saturating concentrations for the enzyme and that catecholamine synthesis cannot be increased be raising the intracellular content of this amino acid. However, analysis of conditions within a dopaminergic nerve terminal would suggest that if tyrosine hydroxylase was fully activated by a phosphorylation process, the uptake of tyrosine into the nerve terminal might become the rate limiting step. Evidence for this has been shown experimentally. Another approach to increasing the activity of either tryptophan or tyrosine hydroxylase, is to enhance the levels of tetrahydrobiopterin. While increasing brain levels of the cofactor have been shown to increase biogenic amine synthesis, the peripheral administration of this compound leads to relatively small increases in brain cofactor levels. Since a number of related compounds can also serve as electron donors for these enzymes it may be possible to design a compound that will more readily cross the blood-brain barrier. Recently (Bigham *et al.*, 1987) a series of analogues have been synthesized and it was found that 6 ethyl tetrahydropterin had kinetic characteristics which are superior to the natural cofactor. It will be of interest to know whether this compound more readily penetrates the brain and if so, does it enhance serotonin and catecholamine synthesis.

On the inhibitor side few compounds have been described in recent years as new inhibitors of these enzymes. Once again inhibition could take place either through amino acid substrate analogues or through cofactor analogues. A recent study (Fitzpatrick, 1988) has described the inhibitory kinetics of 6-methyl-5-deazo tetrahydrobiopterin for tyrosine hydroxylase. Although this is a modestly good inhibitor, it is unlikely to be of therapeutic value since it probably does not readily penetrate the brain. Substrate analogue inhibitors for both these enzymes were described over 20 years ago (Sato *et al.*, 1967; Udenfriend *et al.*, 1965). With regard to tyrosine hydroxylase 3-iodotyrosine and α-methyl tyrosine remain today the most useful substrate analogue inhibitors of tyrosine hydroxylase. Substrate analogues of tryptophan hydroxylase proved less interesting. However, p-chlorophenylalanine proved to be a very interesting compound *in vivo*. Administration of a large dose of this amino acid to rats resulted in a profound and long-lasting loss of both serotonin (Koe and Weissman, 1966) and tryptophan hydroxylase in brain (Jequier *et al.*, 1967). It would appear that perhaps this amino acid is concentrated within serotonergic neurons causing a suicide type inhibition of the enzyme. The latter point has never been resolved.

3. Therapeutic Roles for Hydroxylase Regulation.

Perhaps the most obvious role for replacement therapy with hydroxylase cofactor molecules is in patients with specific genetic lesions in the cofactor synthesizing pathways. Indeed, in some cases of atypical phenylketonuria, this has proved to be quite but not uniformly effective (Kaufman, 1987; Niederwieser, 1987). Likewise, attempts to treat neurological diseases where significant loss of cofactor occurs has met with occasional success (LeWitt and Miller, 1987). The major problem with this approach is the uncertainty of getting a significant increase in brain levels of the cofactor. Potentially a cofactor molecule that readily penetrated the brain would circumvent these problems. Another disease where intriguing anecdotal reports have appeared is in depression. In 1983, we reported (Curtius *et al.*, 1983) a remarkable improvement in a patient with severe chronic depression following administration of tetrahydrobiopterin. Subsequent controlled studies (David Pickar, Robert Levine and Walter Lovenberg unpublished) failed to find any significant benefit of tetrahydrobiopterin in a group of depressed patients. Subsequent analysis of the CSF indicated only minimal increases in CSF cofactor levels. Recently, however, another anecdotal report appeared in which 6-R-tetrahydrobiopterin caused a dramatic recovery in a severely depressed patient (Ito *et al.*, 1988). This is clearly an approach to therapeutics which must be studied much more systematically in the future.

As indicated above the severe inhibition of biogenic amine synthesis in the CNS might have unwanted effects because these transmitters are utilized in many different and important neuronal mechanisms. It should be noted, however, that both α-methyltyrosine and p-chlorophenylalanine have found clinical use in tumors of the periphery. While these compounds had no effects on the progression of the tumor, they did limit the unwanted effects of overproduction of catecholamines and serotonin. α-Methyltyrosine was shown to be effective in limiting the overproduction of catecholamines in patients with pheochromocytoma (Engelman *et al.*, 1968). Likewise, p-chlorophenylalanine was effective in blocking excess serotonin production in patients with the carcinoid syndrome (Engelman *et al.*, 1967).

IV. CONCLUDING REMARKS

Monooxygenases participate in a wide range of important biochemical transformations and as such present valuable targets for drug design. In the preceding selective review, we have only addressed a very few of the possible points for inhibiting the formation of important metabolites. While much remains to be done, it is an area of considerable promise for future drug development.

REFERENCES

Ames, M.M., Lerner, P., and Lovenberg, W. (1978). *J. Biol. Chem.* **253**, 27-31.
Baldoni, J.M. and Villafranca, J.J. (1980). *J. Biol. Chem.* **255**, 8987-8990.
Bargar, T.M., Broersma, R.J., Creemer, L.C., McCarthy, J.R., Hornsperger, J.M., Palfreyman, M.G., Wagner, J., and Jung, M.J. (1986). *J. Med. Chem.* **29**, 315-317.

Bargar, T.M., Broersma, R.J., Creemer, L.C., McCarthy, J.R., Hornsperger, J.M., Atwood, P.V., and Jung, M.J. (1988) *J. Am. Chem. Soc.* **110**, 2975-2976.
Berdarski, P.J., Porubek, D.J., and Nelson, S.D. (1985). *J. Med. Chem.* **28**, 775-779.
Berkowitz, B.A., Arleth, A.J., Sung, C.P., Kruse, L.I., DeWolf, W.E., Jr., and Ohlstein, E.H. (1988). *J. Pharmacol. Exp. Ther.* **245**, 850-857.
Bigham, E.C., Smith, G.K., Reinhard, J.F., Jr., Mallory, W.R., Nichol, C.A., and Morrison, R.W., Jr. (1987). *J. Med. Chem.* **30**, 40-45.
Blackburn, N.J., Concannon, M., Shahiyan, S.K., Mabbs, F.E., and Collison, D. (1988). *Biochemistry* **27**, 6001-6008.
Boadle-Biber, M.C. and Phan, T.H. (1987). *Biochem. Pharmacol.* **36**, 1174-1176.
Bonnefoy, E., Ferrara, P., Rohrer, H., Gros, F., and Thibault, J. (1988). *Eur. J. Biochem.* **174**, 685-690.
Brodie, A.M., Coombes, R.C., and Dowsett, M. (1987). *J. Steroid Biochem.* **27**, 899-903.
Brueggemeier, R.W., Floyd, E.E., and Counsell, R.E. (1978). *J. Med. Chem.* **21**, 1007-1011.
Campbell, D.G., Hardie, D.G., and Vulliet, P.R. (1986). *J. Biol. Chem.* **261**, 10489-10492.
Chen, S., Shively, J.E., Nakajin, S., Shinoda, M., and Hall, P.F. (1986). *Biochem. Biophys. Res. Commun.* **135**, 713-719.
Colombo, G., Rajashekhar, B., Ash, D.E., and Villafranca, J.J. (1984). *J. Biol. Chem.* **259**, 1607-1615.
Colombo, G., Papadopoulos, N.J., Ash, D.E., and Villafranca, J.J. (1987). *Arch. Biochem. Biophys.* **252**, 71-80.
Covey, D.F., Hood, W.F., Parikh, V.D. (1981). *J. Biol. Chem.* **256**, 1076-1079.
Craine, J.E., Daniels, G.H., and Kaufman, S. (1973). *J. Biol. Chem.* **248**, 7838-7844.
Curtius, H.-Ch., Niederwieser, A., Levine, R.A., Lovenberg, W., Woggon, B., and Angst, J. (1983). *Lancet* **1**, 657-658.
Diliberto, E.J., Jr. and Allen, P.L. (1981). *J. Biol. Chem.* **256**, 3385-3393.
Engelman, K., Lovenberg, W., and Sjoerdsma, A. (1967). *N. Engl. J. Med.* **277**, 1103-1108.
Engelman, K., Horwitz, D., Jequier, E., and Sjoerdsma, A. (1968). *J. Clin. Invest.* **47**, 577-594.
Erskine, M.S., Tobet, S.A., and Baum, M.J. (1988). *Endocrinology* **122**, 524-530.
Evans, C.T., Ledesma, D.B., Schulz, T.Z., Simpson, E.R., and Mendelson, C.R. (1986). *Proc. Natl. Acad. Sci. U.S.A.* **83**, 6387-6391.
Fitzpatrick, P.F. (1988). *J. Biol. Chem.* **263**, 16058-16062.
Grenett, H.E., Ledley, F.D., Reed, L.L., and Woo, S.L.C. (1987). *Proc. Natl. Acad. Sci. U.S.A.* **84**, 5530-5534.
Grima, B., Lamouroux, A., Blanot, F., Faucon Biguet, N., and Mallet, J. (1985). *Proc. Natl. Acad. Sci. U.S.A.* **82**, 617-621.
Hagerman, D.D. (1987). *J. Biol. Chem.* **262**, 2398-2400.
Hosoda, H. and Fishman, J. (1974). *J. Am. Chem. Soc.* **96**, 7325-7329.

Ichimura, T., Isobe, T., Okuyama, T., Takahashi, N., Araki, K., Kuwano, R., and Takahashi, Y. (1988). *Proc. Natl. Acad. Sci. U.S.A.* **85**, 7084-7088.
Ito, T., Fujita, K., Matsuura, S., and Nagatsu, T. (1988). *Biog. Amines* **5**, 489-493.
Jequier, E., Lovenberg, W., and Sjoerdsma, A. (1967). *Mol. Pharmacol.* **3**, 274-278.
Johnston, J.O., Wright, C.L., and Metcalf, B.W. (1984). *Endocrinology* **115**, 776-785.
Kaufman, S. (1987). *Top. Neurochem. Neuropharmacol.* **1**, 1-28.
Kellis, J.T., Jr. and Vickery, L.E. (1987a). *J. Biol. Chem.* **262**, 4413-4420.
Kellis, J.T., Jr. and Vickery, L.E. (1987b). *J. Biol. Chem.* **262**, 8840-8844.
Kellis, J.T., Jr., Childers, W.E., Robinson, C.H., and Vickery, L.E. (1987). *J. Biol. Chem.* **262**, 4421-4426.
Kirshner, N. (1957). *J. Biol. Chem.* **226**, 821-825.
Koe, B.K. and Weissman, A. (1966). *J. Pharmacol. Exp. Ther.* **154**, 499-516.
Kruse, L.I., Kaiser, C., DeWolf, W.E., Jr., Frazee, J.S., Erickson, R.W., Ezekiel, M., Ohlstein, E.H., Ruffolo, R.R., Jr., and Berkowitz, B.A. (1986). *J. Med. Chem.* **29**, 887-889.
Kruse, L.I., Kaiser, C., DeWolf, W.E., Jr., Chambers, P.A., Goodhart, P.J., Ezekiel, M., and Ohlstein, E.H. (1988). *J. Med. Chem.* **31**, 704-706.
Kuhn, D.M., O'Callaghan, J.P., Juskevich, J., and Lovenberg, W. (1980). *Proc. Natl. Acad. Sci. U.S.A.* **77**, 4688-4691.
Levin, E.Y. and Kaufman, S. (1961). *J. Biol. Chem.* **236**, 2043-2049.
Levin, E.Y., Levenberg, B., and Kaufman, S. (1960). *J. Biol. Chem.* **235**, 2080-2086.
Levine, R.J., Lovenberg, W., and Sjoerdsma, A. (1964). *Biochem. Pharmacol.* **13**, 1283-1290.
Levine, R.A., Miller, L.P., and Lovenberg, W. (1981). *Science* **214**, 919-921.
LeWitt, P.A. and Miller, L.P. (1987). *Top. Neurochem. Neuropharmacol.* **1**, 157-172.
Longcope, C., Femino, A., and Johnston, J.O. (1988). *Endocrinology* **122**, 2007-2011.
Lovenberg, W. (1986). *Nutr. Rev.* **44** (Suppl.), 6-11.
Lovenberg, W., Bruckwick, E.A., and Hanbauer, I. (1975). *Proc. Natl. Acad. Sci. U.S.A.* **72**, 2955-2958.
McCarthy, J.R., Matthews, D.P., Broersma, R.J., McDermott, R., Kastner, P.R., and Hornsperger, J.M. (1989). *J. Med. Chem.* In press.
McPhaul, M.J., Noble, J.F., Simpson, E.R., Mendelson, C.R., and Wilson, J.D. (1988). *J. Biol. Chem.* **263**, 16358-16363.
May, S.W., Mueller, P.W., Padgette, S.R., Herman, H.H., and Phillips, R.S. (1983). *Biochem. Biophys. Res. Commun.* **110**, 161-168.
May, S.W., Wimalasena, K., Herman, H.H., Fowler, L.C., Ciccarello, M.C., and Pollock, S.H. (1988). *J. Med. Chem.* **31**, 1066-1068.
Meigs, R.A. and Ryan, K.J. (1971). *J. Biol. Chem.* **246**, 83-87.
Melby, J.C., Holbrook, M., Griffing, G.T., and Johnston, J.O. (1987). *Hypertension* **10**, 484-487.
Metcalf, B.W., Wright, C.L., Burkhart, J.P., and Johnston, J.O. (1981). *J. Am. Chem. Soc.* **103**, 3221-3222.
Nagatsu, T., Levitt, M., and Udenfriend, S. (1964). *J. Biol. Chem.* **239**, 2910-2917.

Nakajin, S., Shinoda, M., and Hall, P.F. (1986). *Biochem. Biophys. Res. Commun.* **134**, 704-710.
Niederwieser, A. (1987). *Top. Neurochem. Neuropharmacol.* **1**, 107-118.
Ortiz de Montellano, P.R. and Kunze, L.K. (1981). *Biochemistry* **20**, 7266-7271.
Ortiz de Montellano, P.R., Kunze, L.K., Beilan, H.S., and Wheeler, C. (1982). *Biochemistry* **21**, 1331-1339.
Padgette, S.R., Wimalasena, K., Herman, H.H., Sirimanne, S.R., and May, S.W. (1985). *Biochemistry* **24**, 5826-5839.
Pearson, O.H., Manni, A., and Arafah, B.M. (1982). *Cancer Research* **42** (Suppl.), 3424s-3429s.
Rosenberg, R.C. and Lovenberg, W. (1977). *Mol. Pharm.* **13**, 652-661.
Rosenberg, R.C. and Lovenberg, W. (1980). *Essays Neurochem. Neuropharmacol.* **4**, 163-209.
Rosenberg, R.C., and Lovenberg, W. (1983). *J. Neurochem.* **40**, 1529-1533.
Rosenberg, R.C., Gimble, J.M., and Lovenberg, W. (1980). *Biochim. Biophys. Acta* **613**, 62-72.
Ryan, K.J. (1959). *J. Biol. Chem.* **234**, 268-272.
Sato, T.L., Jequier, E., Lovenberg, W., and Sjoerdsma, A. (1967). *Eur. J. Pharmacol.* **1**, 18-25.
Schumacher, M., Hutchison, R.E., and Hutchison, J.B. (1988). *Brain Res.* **441**, 98-110.
Scott, R.A., Sullivan, R.J., DeWolf, W.E., Jr., Dolle, R.E., and Kruse, L.I. (1988). *Biochemistry* **27**, 5411-5417.
Sjoerdsma, A. (1981). *Clin. Pharmacol. Ther.* **30**, 3-22.
Snider, C.E. and Brueggemeier, R.W. (1987). *J. Biol. Chem.* **262**, 8685-8689.
Stewart, L.C. and Klinman, J.P. (1987). *Biochemistry.* **26**, 5302-5309.
Thompson, E.A., Jr. and Siiteri, P.K. (1974a). *J. Biol. Chem.* **249**, 5373-5378.
Thompson, E.A., Jr. and Siiteri, P.K. (1974b). *J. Biol. Chem.* **249**, 5364-5372.
Udenfriend, S., Zaltzman-Nirenberg, P., and Nagatsu, T. (1965). *Biochem. Pharmacol.* **14**, 837.
Walker, G.A., Kon, H., and Lovenberg, W. (1977). *Biochim. Biophys. Acta* **482**, 309-322.
Wallace, E.F., Krantz, M.J., and Lovenberg, W. (1973). *Proc. Natl. Acad. Sci. U.S.A.* **70**, 2253-2255.
Waymire, J.C., Johnston, J.P., Hummer-Lickteig, K., Lloyd, A., Vigny, A., and Craviso, G.L. (1988). *J. Biol. Chem.* **263**, 12439-12447.

10

ALPHA-FLUOROMETHYLTYROSINE, ALPHA-FLUOROMETHYLTRYPTOPHAN AND BIOGENIC AMINE SYNTHESIS

Michel J. Jung [1]
Jean-Marie Hornsperger
Daniel Schirlin
Fritz Gerhart

Merrell Dow Research Institute
Strasbourg Research Center
67084 Strasbourg, France

Modern drug discovery is based upon the steadily improving understanding of cause and effect in disease conditions. Our knowledge or educated guesses of enzyme mechanisms, receptor topology, and signal transport systems (inter- and intra-cellularly) give the innovative, adventurous chemist the possibility of focusing his efforts (see other chapters of this book). This approach will lead to the discovery of a rationally designed molecule. However, such a specialized, sophisticated compound will have many hurdles to clear before reaching its target site when administered to a patient. Numerous investigators, in academia and much more so in private research institutions, have been searching for physical or chemical means to achieve and eventually maintain the right concentration of the drug at the right place. To cite just a few of these devices: osmotic pumps, liposomes, live or dead cells, nanoparticles, antibodies, synthetic polymers (see for instance, Buri and Gumma, 1985, and references cited); all these means have been in the limelight for a time but were only successful in isolated cases.

In parallel with the search of physical means of selective delivery, the chemist can possibly conceive inactive precursor forms which may distribute evenly through the body but become activated only at the desired site of action by the help of an enzyme specifically expressed at or near the site of action. Ideally it could be an enzyme of the same metabolic route. The names pro-drug (Gardner and Alexander, 1985) bio-precursor (Jung *et al.*, 1985) or dual enzyme-activated inhibitor (McDonald *et al.*,

[1] *Present address: Merrell Dow Research Institute, Lepetit Research Center, 21040 Gerenzano (Varese), Italy.*

Enzymes as Targets for Drug Design

1986) have been proposed for such a two-stage inhibitor. We suggest calling pro-drugs those compounds which are activated by a serendipitous enzyme (i.e. a hydrolase), to reserve the name bio-precursor to those compounds needing an activation which is not necessarily part of the metabolic cascade (oxidation or hydroxylation for instance) and to reserve the term dual enzyme-activated inhibitors to cases where the activating enzyme is by design part of the same metabolic route. Thus, for instance, the methyl ester of α-methyl-*p*-tyrosine (α-MT) would be a pro-drug for the insoluble tyrosine hydroxylase inhibitor; some antibiotic quinones are administered in their reduced bio-precursor form (see for instance, Gottlieb and Shaw, 1967). We wish to describe in this chapter two examples of dual enzyme-activated inhibitors dealing with our attempts to regulate biogenic amine synthesis in selected neurons by using the site and substrates specific, finely regulated aromatic amino acid hydroxylases. This work has appeared in full length in a Ph.D. Thesis (Hornsperger, 1984) and in three brief publications (Jung *et al.*, 1985; Jung *et al.*, 1984; Schirlin *et al.*, 1988). For the sake of clarity we shall review separately the studies of α-monofluoromethyl-*p*-tyrosine (α-MFMT) and α-fluoromethyl-tryptophan; in the conclusion we will describe briefly a brilliant and successful attempt to limit monoamine oxidase inhibition to nerve terminals in the CNS based on similar principles (McDonald *et al.*, 1984; McDonald *et al.*, 1986).

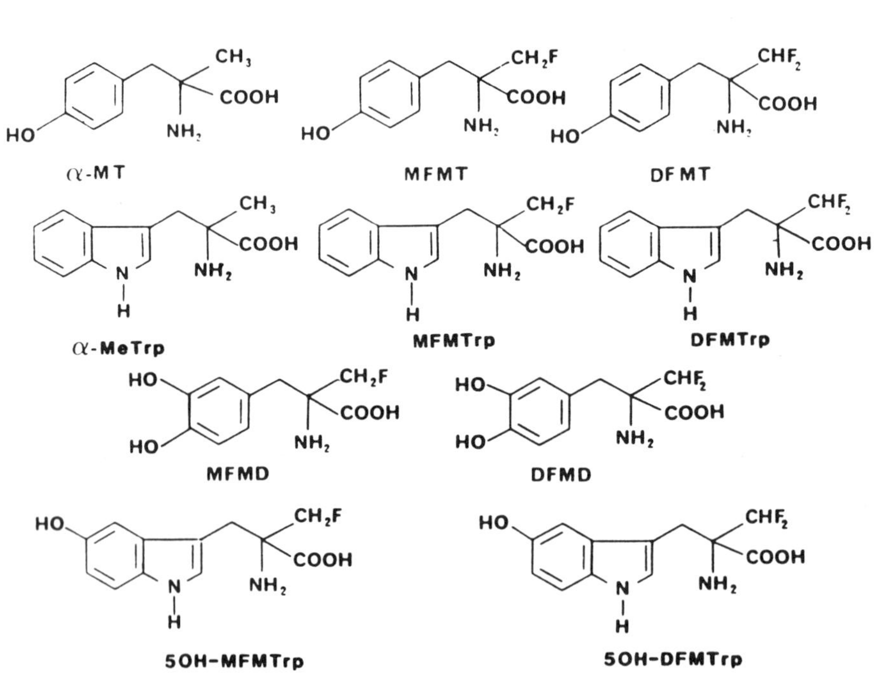

Chemical structure of α-methyl and α-fluoromethyl analogues of tyrosine, tryptophan and their corresponding hydroxyfluoromethyl derivatives.

I. α-MONOFLUOROMETHYL-*p*-TYROSINE (α-MFMT)

α-Monofluoromethyldopa (MFMD) is a very potent irreversible inhibitor of aromatic aminoacid decarboxylase (AADC) (Kollonitsch *et al.*, 1978). This compound has a profound effect on biogenic amine synthesis both in the peripheral sympathetic and in the central nervous system (Fozard *et al.*, 1981; Bey *et al.*, 1980). Despite its catechol structure, MFMD inhibits both catechol and indoleamine synthesis to similar extents. This is due to the similarity if not absolute identity of AADC in dopaminergic, noradrenergic and serotoninergic neurons (Borri-Voltattorni *et al.*, 1979). Unlike AADC, the enzymes preceding the decarboxylation step i.e., tyrosine

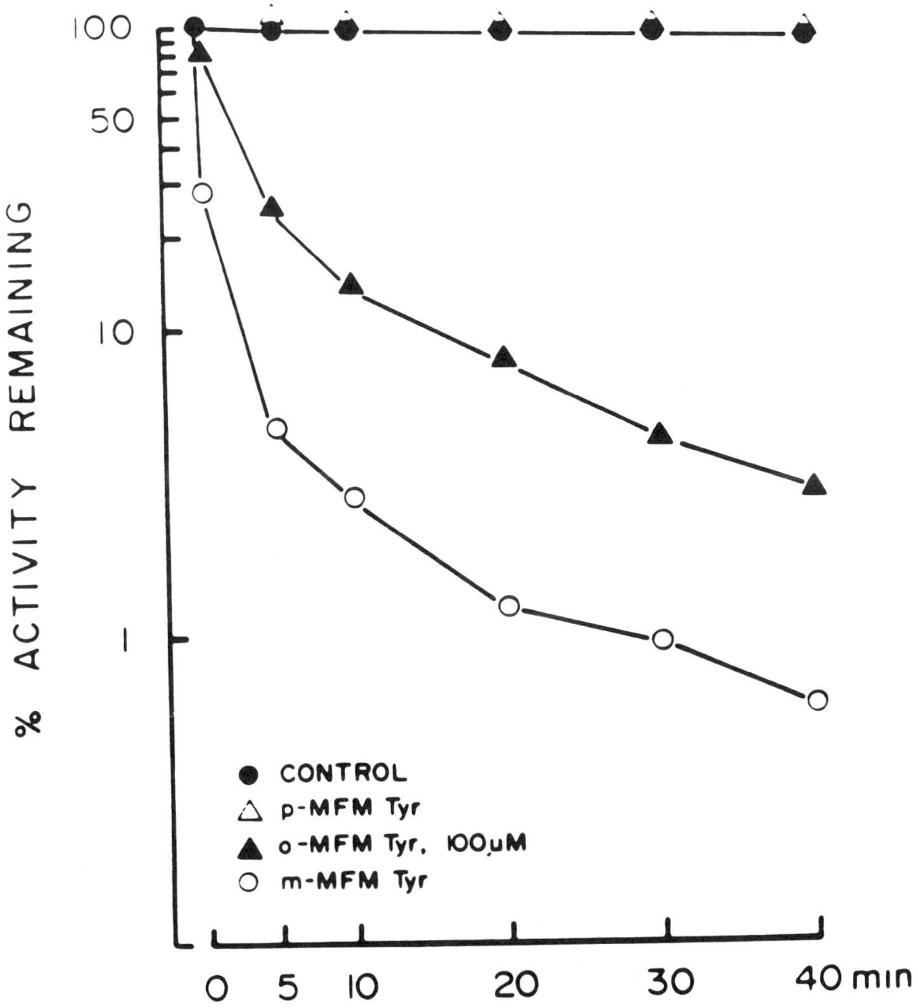

Fig. 1. *Inhibition of pig kidney AADC by o,m,p-analogues of α-monofluoromethyl tyrosine.*

Table I. Kinetic constants of α-substituted *p*-tyrosine analogues with TH isolated from beef adrenal.

SUBSTRATE	K_m (mM)	V_{max} (nmol/h/mg)	V_{max}/K_m
Tyr	0.060	920	15333
α-MT	0.107	100	934
MFMT	0.250	30	120
DFMT	∞	0.	-

hydroxylase (TH) in catecholaminergic neurons and tryptophan hydroxylase (TrpH) in serotoninergic neurons, are substrate and neuron specific. In addition, the hydroxylation represents the rate-limiting step in both cases and both enzymes are controlled by a range of short, medium and long term regulatory mechanisms which as we shall see later can be exploited positively (Lovenberg, this book, chapter 9). The concept requires a tyrosine analogue which itself does not inhibit AADC, is a reasonable substrate for TH and the resulting dopa analogue should be a potent inhibitor of AADC. As summarized above for MFMD, α-MFMT seemed to be a good candidate to meet these conditions.

AADC is known to efficiently decarboxylate *o*- and *m*-tyrosine but *p*-tyrosine is a very poor substrate. This helps our purpose as demonstrated in Figure 1 where we compare the time-dependent inhibitions of pig kidney AADC by the different α-fluoromethyl tyrosines; whilst the *o*- and *m*-analogues produce a marked rapid time-dependent inhibition, not reversed by dialysis, the *p*-analogue had no significant effect over the 40 min of incubation.

We then proceeded to evaluate the substrate activity of several α-substituted *p*-tyrosine analogues using partially purified TH from beef adrenals (Table I). We confirmed that α-methyl-*p*-tyrosine is a substrate of TH: both V_{max} and K_m are different and the rate of hydroxylation at low concentrations (V_{max}/K_m) is about 16% that of the natural substrate (note: all analogues are racemates, hence K_m should probably be divided by 2). Substitution of the methyl group by one fluorine atom more than doubles K_m and reduces V_{max} by a factor of 3, resulting in an ability to serve as substrate about 1/100 that of tyrosine. We could not detect any affinity of α-difluoromethyl-*p*-tyrosine for TH either as substrate or inhibitor. Such a profound effect of fluorine substitution on a carbon atom remote from the site of transformation is quite surprising. Nevertheless, as a result of these two experiments α-MFMT seems to have the required properties for our needs.

α-MFMT being a poor substrate of TH, we tested whether enough MFMD could be produced in brain homogenates to inhibit AADC. Homogenates of rat brain striata, a region rich in dopaminergic nerve terminals, were used. In the absence of the hydroxylase cofactor 6-methyltetrahydropterin, concentrations of α-MFMT up to 1 mM produced no time-dependent inhibition of AADC over 1 hour. Addition of the cofactor produced a time- and dose-dependent inhibition of the AADC present in the homogenate (50% inhibition at 0.25 mM after 1 hour). MFMD could be identified in the incubation mixture and addition of α-MT greatly reduced the rate of inhibition. These factors taken together suggest that inhibition of AADC is achieved through MFMD formed by hydroxylation of α-MFMT by the action of TH. In this experiment, α-MFMT-methyl-ester (α-MFMT-Me), to be used, later for *in vivo* experiments, had no effect presumably because of the absence of an appropriate hydrolase in the striatal extracts (Figure 2).

There is however a second hydroxylation system in the liver acting specifically on the methyl ester form (not on the amino acid). This hydroxylation is blocked by SKF 525A, but not by α-MT (or its Me ester) and is most probably due to an unspecific mixed function oxidase (not shown, see Hornsperger, 1984 for full details). All *in vivo*

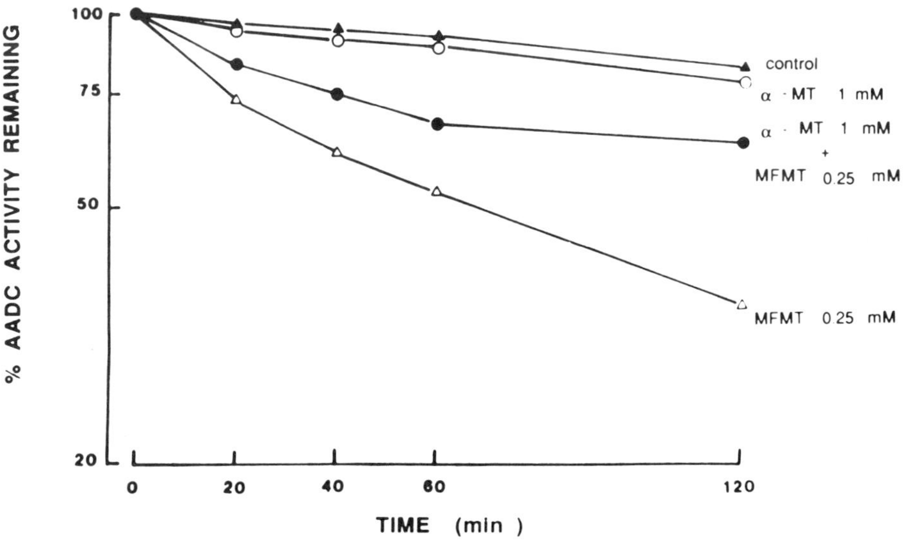

Fig. 2. *Factors influencing inhibition of AADC by α-MFMT derivatives in rat striata homogenates.*

work to be described now was performed on the methyl ester derivative of α-MFMT (α-MFMT-Me) as the parent amino acid (like most *p*-tyrosine analogues) was extremely insoluble in aqueous buffers and gave erratic biological results. Hydrolysis of the methyl ester represents a further step for the activation processes since we are now dealing with a pro-drug of a dual enzyme-activated inhibitor. α-MFMT-Me is equiactive by oral or intraperitoneal administration and gives reproducible results.

Figure 3 describes the dose-dependent inhibition of AADC in five organs of rats, 16 hours after administration of an oral dose of α-MFMT-Me. The most sensitive organ is the kidney, followed by liver, then adrenals and brain; finally, the heart enzyme is affected only at the highest dose of 200 mg/kg. It should be noted that there is significant enzyme activity left in brain and adrenals even after the highest dose.

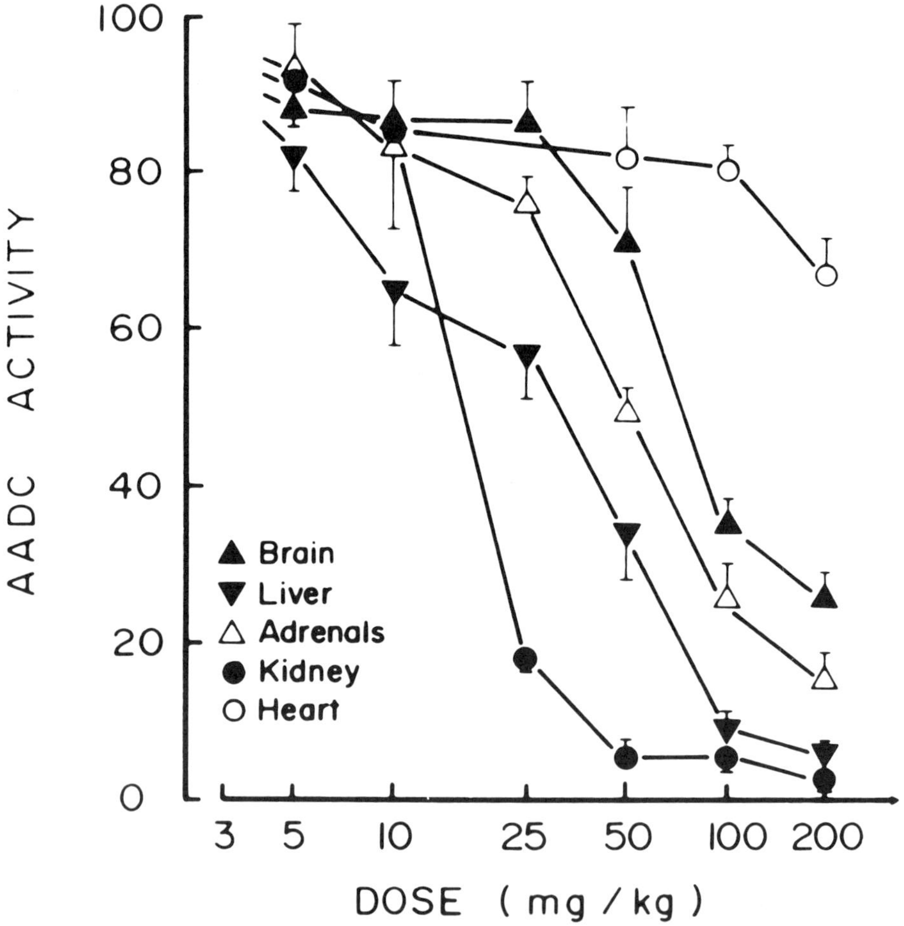

Fig. 3. *Dose-dependent inhibition of AADC in rat organs: 16 hr after oral dose of MFMT-Me. In this and all subsequent in vivo experiments, each value is the mean ±SEM of 5 animals.*

Inhibition of AADC in the brain is diminished if α-MT is coadministered with α-MFMT-Me (200 mg/kg; i.p. every 6 hours); this same treatment does not change the inhibition in liver, kidney and adrenals. In contrast, pretreatment of animals with SKF 525A (50 mg/kg) reduces significantly AADC inhibition in liver and kidney but not in the brain (Table II). Pretreatment of rats with apomorphine, a dopamine agonist known to down regulate TH activity decreases AADC inhibition in brain, while haloperidol (1 mg/kg, 30 min before α-MFMT-Me) increases the inhibition significantly (Table II).

As noticed before, inhibition of AADC in the brain responds to conditions regulating TH activity. However under no circumstance, was this enzyme activity totally suppressed at least at the whole brain level. Nevertheless, not only the new synthesis but also the concentrations of amines in the brain were profoundly affected by small doses of α-MFMT-Me. For instance, Figure 4 shows the dose dependent changes in amine metabolism in rat brain under the conditions of the experiment described in Figure 3.

As seen before in Figure 3, AADC activity in whole brain is decreased at doses of α-MFMT-Me higher than 25 mg/kg, but even at the highest dose is still of the order of 20% of control. While norepinephrine and the dopamine metabolite dopac are decreased in a smooth dose-dependent manner, dopamine levels remain unchanged until doses higher than 50 mg/kg are given. The accumulation of dopa, which indicates a blockade at the level of dopa decarboxylation, becomes significant at doses higher than 25 mg/kg. As outlined in a preliminary account (Jung *et al.*, 1984), the pattern of amines changes as related to AADC inhibition is totally different from that seen after

Table II. Influence of different chemical agents on AADC inhibition in rat organs (a: 100 mg/kg MFMT-Me, 16 hr; b: 200 mg/kg MFMT-Me, 4 hr).

	TISSUE / TREATMENT	AADC ACTIVITY (% CONTROL)		
		BRAIN	LIVER	KIDNEY
a)	MFMT-Me (100 mg/kg)	52 ± 5 %	4 ± 0.5 %	24 ± 6 %
	+ α-MT (3 x 200 mg / kg)	74 ± 8 %	5 ± 0.5 %	32 ± 9 %
b)	MFMT-Me (200 mg / kg)	50 ± 4 %	4 ± 0.5 %	4 ± 0.5 %
	+ SKF 525A (50 mg / kg)	52 ± 6 %	30 ± 4 %	25 ± 4 %
	+ APOMORPHINE (0.5 mg / kg)	64 ± 3 %	4 ± 0.5 %	4 ± 0.5 %
	+ HALOPERIDOL (1mg/kg)	40 ± 3 %	4 ± 0.5 %	5 ± 0.5 %

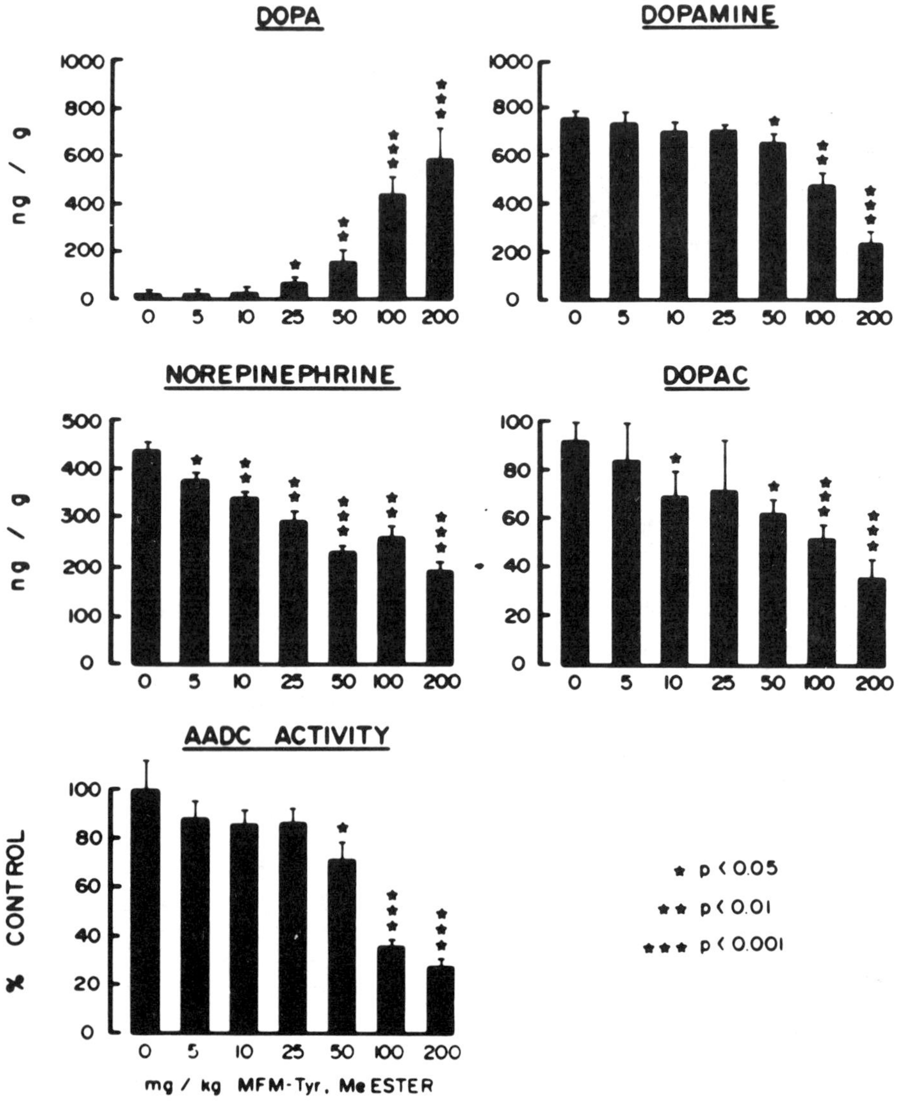

Fig. 4. *Dose-dependent effect of α-MFMT-Me on amine metabolism in rat brain. The drug was given orally, 16 hr before sacrifice of animals.*

MFMD administration. MFMD produces a parallel decrease of dopamine and norepinephrine and there is no significant change before near total inhibition of AADC.

Hornsperger, 1984 also describes the absolute selectivity of α-MFMT-Me towards the catecholaminergic system even after subchronic administration of doses up to 50 mg/kg/day. After the 6th dose, we could find no sign of a change in the serotoninergic system: no 5-HTP accumulation, no decrease of 5-HT or 5-HIAA. Under these same conditions, norepinephrine was depleted by 50% even at the daily 10 mg/kg dose; dopamine on the other hand was affected only at the 25 mg/kg/day dose (or higher). Dopa accumulation followed a smooth dose-response curve. It thus seems that α-MFMT-Me is able not only to distinguish between catecholaminergic and serotoninergic neurons, but also at small doses between noradrenergic and dopaminergic systems. The fact that norepinephrine concentrations are decreased by 50% at the small dose of 10 mg/kg/day and that increasing the daily dose by five times does not produce an additional effect is also worth mentioning (pools with different turnover rates?). If these selectivities could be confirmed by additional studies, is it conceivable that by the sole use of radiolabeled α-MFMT-Me at different doses, one could visualize all three aminergic systems separately?

We had seen in Table II, that AADC inhibition by α-MFMT-Me responds to conditions influencing TH activity. In this section we will demonstrate that the concentrations of amine, precursor and metabolite respond to these conditions even more exquisitely. As can be seen in Table III, apomorphine reduces the accumulation of dopa seen after α-MFMT-Me. On the contrary after haloperidol pretreatment, the accumulation of dopa seen after a single dose of α-MFMT-Me is increased dramatically, while the depletion of dopamine is increased and the accumulation of dopac normally produced by haloperidol is completely blocked.

Table III. Effect of chemicals changing dopamine turnover on the perturbation of dopamine metabolism by α-MFMT-Me (200 mg/kg; 4 hr; orally). Apomorphine 0.5 mg/kg (i.p.); haloperidol: 1 mg/kg (i.p.).

TREATMENT	DOPA (ng / g)	DA (ng / g)	DOPAC (ng / g)
CONTROL	0	670 ± 18	85 ± 7
MFMT - Me alone 200 mg / kg	181 ± 24 (1)	562 ± 22 (4)	78 ± 5
HALOPERIDOL alone 1 mg / kg	0	554 ± 27	224 ± 13 (6)
HALOPERIDOL + MFMT-Me	391 ± 34 (2)	327 ± 22 (5)	85 ± 10 (7)
APOMORPHINE alone 0.5 mg / kg	0	653 ± 16	85 ± 7
APOMORPHINE + MFMT-Me	121 ± 15 (3)	585 ± 24	75 ± 6

1-3, 4-5 : * $p < 0.05$ 1-2 : ** $p < 0.01$ 6-7 : *** $p < 0.001$

Numbers in parentheses refered to paired experiments to be compared to student's t-test.

Similarly the changes of norepinephrine metabolism due to α-MFMT-Me treatment can be modulated by factors affecting turnover of the noradrenergic system (or vice versa of course). We devised a reliable method to measure NE turnover based on concentration of MOPEG-SO_4 (4-hydroxy-3 methoxy phenyl ethylene glycol sulfated on the phenol function) (Hornsperger *et al.*, 1984). NE turnover is known to be increased by yohimbine (an α_2-antagonist) and physical stress such as cold restraint. The effect of these treatments combined with α-MFMT-Me are summarized in Table IV.

Both effects on the DA and the NE systems described in an acute experiment with a high dose of α-MFMT-Me can be reproduced in sub-chronic experiment with much lower doses (5-10 mg/kg/day). We have for instance administered α-MFMT-Me at 10 mg/kg/day for 2 weeks to rats. One group of rats received haloperidol 1 mg/kg daily, another group received haloperidol only with the last α-MFMT-Me dose. The animals were killed 4 hours after the 15th treatment. In animals receiving both α-MFMT-Me and haloperidol sub-chronically, the levels of dopac were the same as in untreated controls. In animals receiving haloperidol with the 15th dose of α-MFMT-Me the increase of dopac was less than in naive haloperidol controls but significantly higher than the previous group. Similar experiments were performed with chronic yohimbine and smaller doses of α-MFMT-Me with identical results. Our interpretation of this is that by chronic administration with small doses of α-MFMT-Me it should be possible

Table IV. Factors influencing NE turnover: α-MFMT-Me was given orally at a dose of 200 mg/kg; yohimbine (2.5 mg/kg) was given subcutaneously; stress was by immobilization at 4°C 4 hr. All animals were killed 6 hr after α-MFMT-Me.

TREATMENT	DOPA (ng / g)	NE (ng / g)	MOPEG-SO4 (ng / g)
CONTROL	0	375 ± 10	196 ± 10
MFMT - Me alone	160 ± 11	260 ± 10	118 ± 6
YOHIMBINEalone	0	334 ± 18	361 ± 18 (1)
MFMT - Me + YOHIMBINE	176 ± 26	192 ± 22	162 ± 6 (2)
STRESS alone	–	–	392 ± 9 (3)
MFMT- Me + STRESS	–	–	195 ± 3 (4)

1 - 2 , 3 - 4 : ** $p < 0.01$

to down regulate dopaminergic or noradrenergic systems whose activity has been abnormally increased. For instance, it has been suggested that the slow onset of neuroleptic action of haloperidol could be due to the stimulation of the dopaminergic system; clinical trials combining haloperidol and α-MT have been conducted with some success (Carlsson *et al.*, 1972). It is our understanding that they have been abandoned due to the difficulty of manipulating α-MT (high dose, short action, kidney toxicity). We believe that α-MFMT-Me could substitute for α-MT at much lower doses and with longer duration of action. It could also be envisaged that α-MFMT-Me could reduce withdrawal symptoms in humans addicted to various drugs affecting the DA and/or NE system. Indeed, we have conducted some experiments in rats implanted with morphine pellets (unpublished work). Twenty-four rats were implanted with two morphine pellets (75 mg free base) two days apart. Six of these rats (control group) were treated with vehicle (1% ascorbate, orally) started 1 day before the first implantation. A second group of 6 rats was treated with α-MFMT-Me 10 mg/kg/day, orally, for 6 days, starting 1 day before the first implantation and continuing for 2 days after the second implantation. A third group of 6 rats received clonidine 30 μg/kg s.c., 15 min before naloxone. The last group of 6 rats received only the morphine pellets. On day 2 after the second pellet, withdrawal syndrome was initiated with naloxone 3 mg/kg, s.c. The effects were observed over 15 min immediately after injection. The number of wet dog shakes (WDS), jumps, writhes were counted and various aspects of behaviour were scored. Each rat was weighed before naloxone and reweighed after 3 hours and sacrificed immediately after. Brain and heart were removed for analysis of norepinephrine and metabolites. MFMT-Me treated animals showed a tendency to reduce the number of naloxone precipitated jumps and WDS, and to reduce the behavioural scores although the effects were not statistically significant. Clonidine prevented the withdrawal syndrome to some extent. Naloxone precipitated morphine withdrawal induced hyperactivity of the noradrenergic system characterized by an increase in the concentration of MOPEG-SO4 in the brain. This hyperactivity could be entirely normalized after chronic treatment with MFMT-Me. Only partial normalization was obtained with clonidine.

Last but not least, it is well established that essential hypertension is accompanied by an increase in biogenic amine turnover. It is believed that α-methyldopa (see W. De Jong, this book, chapter 5) exerts its antihypertensive effect by being transformed in several steps into α-methylnorepinephrine a weak norepinephrine agonist. α-Methyl tyrosine has antihypertensive properties and would probably be used were it not for the problems mentioned before. We investigated the biochemical and pharmacological effects of α-MFMT-Me in Sprague Dawley (SD) and spontaneously hypertensive rats (SH) of the same age after sub-chronic administration (Table V).

It appears clearly that α-MFMT-Me has a significant effect on heart NE levels in SH rats but not is SD rats. This biological effect is accompanied by a dramatic reduction in blood pressure in SH rats: not only does α-MFMT-Me block further increase of blood pressure as seen in untreated control, but it even brings the level below the starting value.

Table V. Differential effects of α-MFMT-Me in SD and SH rats after sub-chronic administration. A dose of 10 mg/kg/day was given orally to SD rats for 15 days. The animals were killed 24 hr after the last dose. Results are the mean æ s.e.m. n=5 for SD rats, n=7 for SH rats. Blood pressure was measured in the femoral artery at the beginning and the end of the study in SH rats.

TISSUE	BRAIN		HEART		ARTERY	
	SD rats (n = 5)	SH rats (n = 7)	SD rats	SH rats	SD rats	SH rats
NE CONTROL	420 ± 19	520 ± 40	815 ± 49	450 ± 40	477 ± 16	1070 ± 100
NE TREATED	238 ± 19	160 ± 20	832 ± 79	100 ± 20	138 ± 21	350 ± 70
BLOOD PRESSURE (mm Hg) SH rats CONTROL	beginning 162 ± 10			end 196 ± 13		
BLOOD PRESSURE (mm Hg) SH rats TREATED	beginning 162 ± 14			end 118 ± 10**		

** $p < 0.001$

Our attempts to find a potential psychotropic effect of α-MFMT-Me have not been all too successful. There is no effect on spontaneous motor activity (10-50 mg/kg/day), nor on exploratory behaviour (4x10 mg/kg/day). α-MFMT-Me neither synergizes nor antagonizes chemically or physically induced convulsions and does not have analgesic properties. We could not detect any anxiolytic properties in the conditioned avoidance test. There was a hint of a positive effect in the Porsolt swimming test (antidepressant), but it was barely significant and would require a much larger experiment to confirm. In summary, we can see three areas of pharmacological activity: blood pressure regulation, possibly stress and selectively decreasing states of dopaminergic and noradrenergic hyperactivity (combination with neuroleptics, decreasing withdrawal symptoms).

II. α-MONOFLUOROMETHYLTRYPTOPHAN (α-MFMTrp)

Encouraged by the reasonable success encountered with α-MFMT, we attempted a parallel study to discover a means to selectively regulate serotoninergic activity. A preliminary account of this part of the work has appeared in (Schirlin *et al.*, 1988). In the first instance we shall look for a tryptophan analogue which does not inhibit AADC by itself but which is hydroxylated by tryptophan hydroxylase (TrpH) into a 5-hydroxy analogue which should be a potent inhibitor (preferentially time-dependent) of AADC.

α-Methyltryptophan is a commercial product, α-fluoromethyl and α-difluoromethyltryptophan were synthesized by a selective route (Schirlin *et al.*, 1988). Concentrations of all three compounds between 0.1 and 1 mM did not produce time-dependent decreases of hog kidney AADC activity when incubated in the presence of pyridoxal phosphate. The reaction in the absence of exogenous pyridoxal phosphate between AADC and α-methyl aromatic aminoacid is extremely complex and has been discussed elsewhere (Jung, 1986).

Tryptophan hydroxylase was prepared in crude or partially purified form from cerebral stem of rat brains and assayed with the different analogues, using 6-methyltetrahydropterin as cofactor (Schirlin *et al.*, 1988). The appearance of 5-hydroxy species was followed either by native fluorescence in acidic medium (Hamon *et al.*, 1977) or by HPLC with fluorescent detection, using authentic 5-hydroxytryptophan analogues as standards (Wagner). The kinetic constants of tryptophan and its analogues as measured under our conditions are summarized in Table VI. This work was done in collaboration with Dr. Michel Hamon, College de France.

Introduction of an α-methyl substituent on tryptophan decreases the affinity by a factor of four and halves the maximum rates of hydroxylation, resulting in a eight fold reduction of the rates of hydroxylation at low concentrations. Substitutions of the methyl group by one fluorine atom has a dramatic effect on K_m but not so much on V_{max}, the final result being a reduction of hydroxylation rate of 150 fold. Substitution of a second fluorine gives still measurable rates of hydroxylation due to the availability

Table VI. Kinetic constants of α-substituted tryptophane analogues with brain stem TrpH.

SUBSTRATE	K_m (mM)	V_{max} (nmol/20min/mg)	V_{max}/K_m
Trp	0.163	7.00	42.900
α- MeTrp	0.620	3.25	5.240
MFMTrp	7.500	2.55	0.340
DFMTrp	52.000	2.52	0.048

of very sensitive methods; again V_{max} is not affected and K_m is increased by a factor of 7. The overall effect of alkyl and fluoroalkyl substitution are not dissimilar from those described for tyrosine analogues. It is known that fluorine substitution in β position to an amine decreases dramatically the pKa of this amine (Lettre and Wolcke, 1967). As the α-amino group of these aromatic amino acids most certainly plays an important role in binding to the active site, this change in pKa would affect the affinity of the aminoacid for the hydroxylase.

Two conditions for the successful control of serotonin biosynthesis by α-fluoromethyl or difluoromethyltryptophan were therefore fulfilled: no direct action on AADC and some substrate activity on TrpH. It turned out however that the corresponding 5-hydroxy-tryptophan analogues are relatively poor inhibitors of AADC. For instance, at a concentration of 10μM, there was no detectable loss of activity over 40 min of incubation, MFMD in the same conditions irreversibly inhibits over 95% of enzyme activity within 5 min. At a 10 times higher concentration, α-difluorometh-yltryptophan and the monofluoromethyl analogues produce a 90% loss of enzyme activity within 10-20 min, i.e. both compounds have an intrinsic activity comparable to difluoromethyldopa rather than to MFMD. This is confirmed also *in vivo*, when administered to rats at doses of 100 mg/kg, the reduction of brain AADC activity is about

Table VII. Effect of α-substituted tryptophane analogues on brain 5-HT metabolism. Compounds were given i.p. at a dose of 100 mg/kg at 12, 24 and 36 hr. The animals were sacrificed at 42 hr and their brains analyzed for Trp, 5-HPT, 5-HT and 5-HIAA.

	5 - HTP (ng / g)	5 - HT (ng / g)	5 - HIAA (ng / g)
CONTROL	0 (1)	320 ± 19 (1)	183 ± 10 (4)
α - MeTrp	0	31 ± 3 (2)	28 ± 2 (5)
α MFMTrp	70 ± 17 (2)	223 ± 17 (3)	121 ± 9 (6)
α - DFMTrp	0	374 ± 18	190 ± 12

1-3, 4-6 :* $p < 0.05$ 1-2, 4-5 : *** $p < 0.001$

50%, this reduction being accompanied by a slight increase of brain 5-HTP concentration and a slight decrease of 5-HT (not shown).

To evaluate the potential of the non-hydroxylated analogues to undergo the dual enzyme-activation specifically, we administered 3 doses of 100 mg/kg at 12-hr intervals of each α-methyl, α-monofluoromethyl and α-difluoromethyltryptophan to rats. The animals were sacrificed six hours after the last dose. The changes in serotonin metabolism are reported in Table VII.

The effect of α-MeTrp on serotonin metabolism was known from the literature (Gal and Christiansen, 1975). There is a significant depletion of Trp, and the appearance of 5-0H MeTrp and α-Me-5-HT accompanied by decreases of 5-HT and 5-HIAA. Not shown, norepinephrine is also reduced by 45%. MFMTrp produces a significant accumulation of 5-HTP which indicates blockade of 5-HTP decarboxylation. There are also significant changes of 5-HT and 5-HIAA levels. Norepinephrine levels on the other hand are unchanged. None of this happens with the difluoromethyl analogue. In this experiment we have not searched for the formation of α-fluoromethylserotonin analogues (by analogy with α-methyl serotonin); there was no sign in earlier studies with the relevant dopa analogues that decarboxylation to the corresponding fluoro methylamine occurred to a significant extent (Maycock *et al.*, 1980).

In conclusion of this part, it seems that α-MFMTrp has the right properties *in vitro* and *in vivo* to selectively control 5-HT synthesis. The serotonin systems seems a bit more "sluggish" to respond; it would therefore be interesting to do a similar study under conditions of increased turnover.

III. BRAIN SELECTIVE MAO-A INHIBITOR AND CONCLUSION

To illustrate further the power of this concept of dual enzyme-activated inhibitors, we will briefly describe the very clever design of MDL 72,394 (β-fluoromethylene-*m*-tyrosine) as a brain selective MAO-A inhibitor by Palfreyman and his colleagues (Bey *et al.*, 1984a; McDonald *et al.*, 1986; McDonald *et al.*, 1984; Palfreyman *et al.*, 1985). [Full details of this concept can be found in Chapter 11 by Palfreyman *et al.*] It was known from previous work that 2-aryl-3-fluoroallylamines are very potent inhibitors of monoamine oxidase (MAO-A and B). Depending on the substitution of the aryl portion, impressive selectivity for A or B enzymes could be obtained (Bey *et al.*, 1984b). Thus, for instance, 2-[3-hydroxyphenyl]-3-fluoroallylamine, MDL 72,392, shows marked selectivity for the A form, i.e., the intraneuronal MAO. Inhibition of this enzyme in the brain is thought to be associated with the successful treatment of depression; unfortunately inhibition of this same enzyme is responsible for cardiovascular accidents (cheese effect). Palfreyman *et al.*, 1985, therefore, reasoned that if they could design a "bio-precursor," to protect this MAO inhibitor until it reaches the brain and there use an enzyme present in biogenic amine terminals, they could elegantly solve this apparently intractable problem. This was achieved with the aminoacid form of the MAO inhibitor. MDL 72,934 is an excellent substrate of AADC, is not by itself an inhibitor of MAO. When given to animals together with carbidopa to prevent decarboxylation in the periphery, profound inhibition of MAO-A in the brain can be obtained with little or no peripheral inhibition as measured by tyramine potentiation

(Di Francesco *et al.*, 1986). This and related compounds for MAO-B inhibition are now undergoing clinical trials (Palfreyman *et al.*, 1988); this class of compounds may well represent the beginning of a new era for monoamine oxidase inhibitors.

Not all attempts are as successful however. For instance, it had been suggested (McDonald *et al.*, 1985) that β-fluoromethylene glutamic acid could be decarboxylated by glutamic acid decarboxylase (GAD) into β-fluoromethylene-4-aminobutyric acid, itself a relatively weak inhibitor of 4-aminobutyric acid transaminase (GABA-T). Such a mechanism, would have been of great interest to restrict GABA-T inhibition to nerve terminals and increase selectively the neurotransmitter GABA pool while causing minimal perturbation in glial cells (Jung, 1981). Unfortunately, β-fluoromethylene glutamic acid is not decarboxylated by mammalian GAD. The concept remains appealing however, but we need to identify suitable GABA-T inhibitors. Our best GABA-T inhibitors are 4-fluoromethyl or 4-vinyl analogues and the corresponding glutamic acid analogues are actually inhibitors of GAD (Kuo and Rando, 1981; Bouclier and Jung, 1981).

In summary, in the present review, we have reported our largely unpublished work on α-MFMT as an attempt not only to highlight some of the neglected properties of this compound but also more generally to illustrate the advantages of using a helper enzyme to produce site- and organ-specific blockade of a metabolic pathway. If anything, the briefly described properties of MDL 72,394 emphasize this even more. The whole review should be considered as a call to creative chemists to exploit this possibility of dual, triple or even more stages of activation as an alternative to more physical means of drug targeting.

REFERENCES

Bey, P., Jung, M.J., Koch-Weser, J., Palfreyman, M.G., Sjoerdsma, A., Wagner, J., and Zreika, M. (1980). *Br. J. Pharmacol.* **70**, 571-576.

Bey, P., McDonald, I.A., and Palfreyman, M.G. (1984a). *J. Pharm. Pharmacol.* **36**, 38W.

Bey, P., Fozard, J., Lacoste, J.M., McDonald, I.A., Zreika, M., and Palfreyman, M.G. (1984b). *J. Med. Chem.* **27**, 9-10.

Borri-Voltattorni, C., Minelli, A., Vecchini, P., Fiori, A., and Turano, C. (1979). *Eur. J. Biochem.* **93**, 181-188.

Bouclier, M. and Jung, M.J. unpublished data.

Buri, P. and Gumma, A., eds. (1985). "Drug targeting." Amsterdam, Elsevier.

Carlsson, A., Persson, T., Roos, B.E., and Walinder, J. (1972). *J. Neural Transm.* **33**, 83-90.

DeJong, W. (1989). This book, pp. 31-41.

Di Francesco, G.F., Palfreyman, M.G., Spedding, M., and Zreika, M. (1986). *Br. J. Pharmacol.* **89**, 685P.

Fozard, J.R., Mohring, J., Palfreyman, M.G.; and Koch Weser, J. (1981). *J. Cardiovasc. Pharmacol.* **3**, 1038-1049.

Gal, E.M. and Christiansen, P.A. (1975). *J. Neurochem.* **24**, 89-95.

Gardner, C.R. and Alexander, J. (1985). In "Drug targeting" (P. Buri and A. Gumma, eds.), pp. 145-164. Amsterdam Elsevier.
Gottlieb, D. and Shaw, P., eds. (1967). "Antibiotics," Vol 1. New York, Springer-Verlag.
Hamon, M., Bourgoin, S., Artaud, F., and Hery, F. (1977). *J. Neurochem.* **28**, 811-818.
Hornsperger, J.M. (1984). Ph.D. Thesis, Universite Louis Pasteur, Strasbourg, France.
Hornsperger, J.M., Wagner, J., Hinkel, J.P., and Jung, M.J. (1984). *J. Chromatog.* **306**, 364-370.
Jung, M.J. (1981). In "Neurotransmitter interaction and compartmentation" (H. Bradford, ed.), pp. 329-344. New York, Plenum Press.
Jung, M.J. (1986). *Bioorganic Chem.* **14**, 429-433.
Jung, M.J., Hornsperger, J.M., Gerhart, F., and Wagner, J. (1984). *Biochem. Pharmacol.* **33**, 327-330.
Jung, M.J., Hornsperger, J.M., McDonald, I.A., Fozard, J., and Palfreyman, M.G. (1985). In "Drug targeting" (P. Buri and A. Gumma, eds.), pp. 165-178. Amsterdam, Elsevier.
Kollonitsch, J., Perkins, L.M., Patchett, A.A., Doldouras, G.A., Marburg, S., Duggan, D.E., Maycock, A.L., and Aster, S.D. (1978). *Nature* **274**, 906-908.
Kuo, D. and Rando, R.R. (1981). *Biochemistry* **20**, 506-511.
Lettre, H. and Wolcke, U. (1967). *Liebig's Ann. Chem.* **708**, 75-85.
Lovenberg, W. (1989). This book, pp. 101-119.
McDonald, I.A., Lacoste, J.M., Bey, P., Wagner, J., Zreika, M., and Palfreyman, M.G. (1984). *J. Am. Chem. Soc.* **106**, 3354-3356.
McDonald, I.A., Palfreyman, M.G., Jung, M.J., and Bey, P. (1985). *Tetrahedron Lett.* **26**, 4091.
McDonald, I.A., Lacoste, J.M., Bey, P., Wagner, J., Zreika, M., and Palfreyman, M.G. (1986). *Bioorganic Chem.* **14**, 103-118.
Maycock, A.L., Aster, S.D., and Patchett, A.A. (1980). *Biochemistry* **19**, 709-718.
Palfreyman, M.G., McDonald, I.A., Fozard, J.R., Mely, Y., Sleight, A.J., Zreika, M., Wagner, J., Bey, P., and Lewis, P.J. (1985). *J. Neurochem.* **45**, 1850-1860.
Palfreyman, M.G., McDonald, I.A., Bey, P., Schechter, P.J., and Sjoerdsma, A. (1988). *Prog. Neuropsychopharmacol. Biol. Psychiatry* **12**, 967-987.
Schirlin, D., Gerhart, F., Hornsperger, J.M., Hamon, M., Wagner, J., and Jung, M.J. (1988). *J. Med. Chem.* **31**, 30-36.
Wagner, J. Unpublished data.

11

TYROSINE AND TRYPTOPHAN ANALOGUES AS DUAL ENZYME-ACTIVATED INHIBITORS OF MONOAMINE OXIDASE

Michael G. Palfreyman

Merrell Dow Research Institute
Cincinnati, OH 45215

Monique Zreika

Merrell Dow Research Institute
Strasbourg Research Center
67084 Strasbourg Cedex, France

Ian A. McDonald

Merrell Dow Research Institute
Cincinnati, OH 45215

Monoamine oxidase (MAO, EC 1.4.3.4, monoamine: O_2 oxidoreductase [deaminating flavin-containing]), first described by Hare in 1928, is an important enzyme involved in the catabolism of both the neurotransmitter biogenic amines epinephrine, norepinephrine, dopamine and serotonin and the trace amines such as tyramine, tryptamine, phenethylamine, *N*-telemethylhistamine and octopamine (Blaschko, 1974). The enzyme is widely distributed throughout the body where it is principally associated with the outer mitrochondrial membrane (Dostert *et al.*, 1989). The existence of more than one form of MAO became apparent when the irreversible inhibitor clorgyline was observed to selectively inhibit only a portion of the rat brain enzyme pool (Johnston, 1968). This enzyme form, now designated MAO-A, preferentially deaminates norepinephrine and serotonin. The remaining brain enzyme, MAO-B, was found to be selectively inhibited by *L*-deprenyl and to preferentially deaminate phenethylamine and benzylamine (Knoll and Magyar, 1972). It has been recently established that the two enzyme forms are distinct proteins arising from separate genes and have different tissue distribution, inhibitor sensitivity and substrate specificity (Murphy *et al.*, 1984; Bach *et al.*, 1988). The proteins, however, are closely related, having 70% sequence homology (Bach *et al.*, 1988).

Enzymes as Targets for Drug Design

Astute clinical observations of the mood elevating effects of the anti-tubercular agent iproniazid (Selikoff *et al.*, 1952) in 1952 were followed in the same year by Zeller's observation that this compound had an inhibitory effect on MAO (Zeller *et al.*, 1952). Controlled clinical trials with iproniazid in depression were subsequently undertaken (Loomer *et al.*, 1957; Kline, 1958) and soon the introduction of several other MAO inhibitors into clinical practice led to their widespread use. However, from 1963, enthusiasm for MAO inhibitors waned as reports of life-threatening hypertensive crises associated with the ingestion of tyramine-containing foods such as cheese began to appear (Blackwell, 1963).

In recent years there has been a resurgence of interest in the clinical use of MAO inhibitors for a number of reasons. It is now well established that MAO inhibitors are effective in a variety of affective disorders (Youdim and Finberg, 1983; Robinson, chapter 18, this book, 1989). In addition, a better understanding of the enzyme sub-types, their tissue distribution and the availability of the newer, selective inhibitors has given impetus to look for other therapeutic indications. Finally, a reappraisal of the dangers associated with the ingestion of tyramine-containing foods has suggested that the risk is less severe than commonly believed and that dietary control is an effective strategy for significantly reducing the incidence of hypertensive accidents (Ilett *et al.*, 1980). There is, however, a need for improved MAO inhibitors. In particular, an inhibitor which could be used without dietary restrictions would be especially valuable.

Although the evidence is indirect, it is generally believed that MAO inhibitors produce their antidepressant effects by inhibition of MAO in the brain, thereby increasing the concentration of norepinephrine and/or serotonin (Murphy *et al.*, 1984) which, in turn, leads to down-regulation of adrenergic and/or serotonergic function. Inhibition of MAO-A in peripheral sympathetic nerves and tissues such as the gut, liver, lungs and blood vessels can increase the bioavailability of ingested tyramine present in foods such as cheese, resulting in concentrations that can cause release of norepinephrine. The ensuing vasoconstrictor and hypertensive response forms the basis of the so called "cheese-effect" (Youdim and Finberg, 1983; Ilett *et al.*, 1980). Obviously, an MAO inhibitor whose action was largely restricted to the brain would be expected to have a minimal risk of these cardiovascular problems.

In chapter 10 in this book, Jung *et al.*, have discussed their elegant studies directed towards the dual-enzyme activated inhibition of aromatic *L*-amino acid decarboxylase (AADC) and the inherent advantage that such inhibitors provide in terms of site selectivity. In this chapter we wish to demonstrate how a conceptually similar approach can lead to very selective inhibition of brain MAO.

I. SELECTIVE INHIBITION OF BRAIN MAO WITH (E)-β-FLUOROMETHYLENE-*m*-TYROSINE

Whereas AADC was the target enzyme in Jung's work, we relied upon this enzyme to liberate an MAO inhibitor from an appropriately-substituted amino acid precursor (Figure 1). Furthermore, in the presence of a peripherally-acting AADC inhibitor, we hoped to exclusively restrict decarboxylation to the brain. To test the validity of this

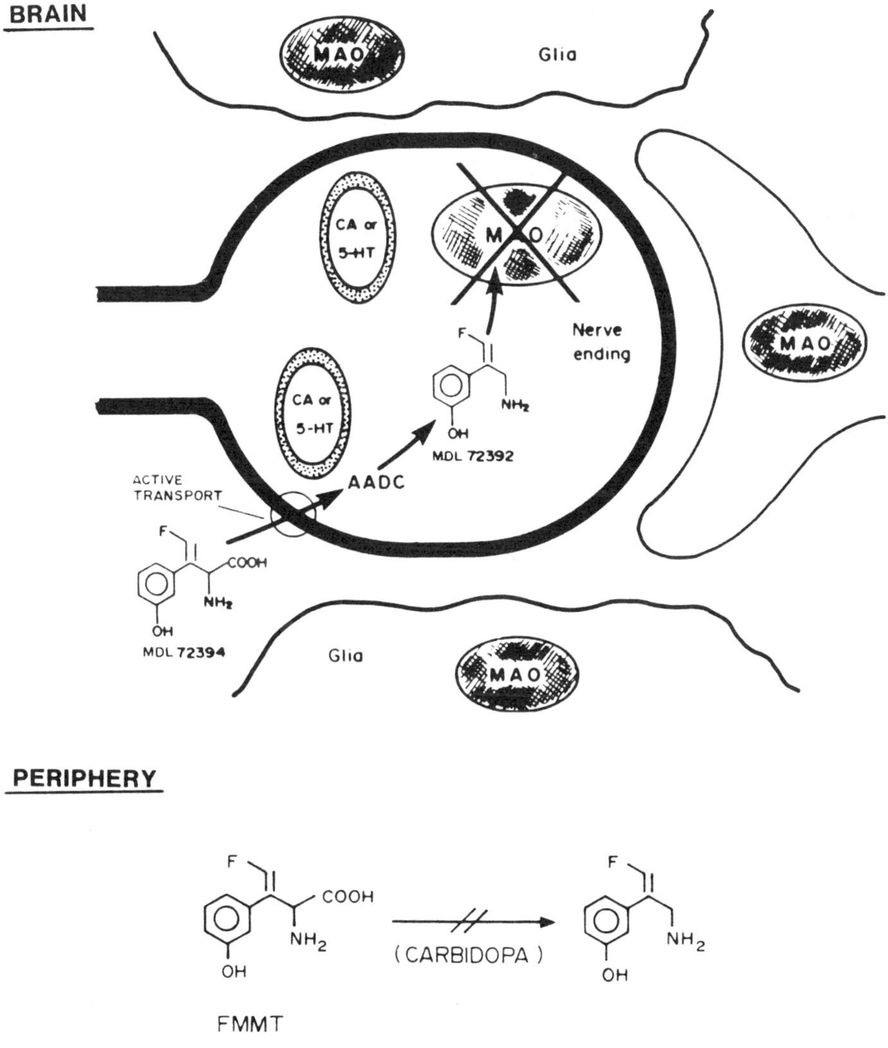

Fig. 1. *Application of the concept of dual enzyme activated inhibition to the design of brain selective inhibitors of MAO.*

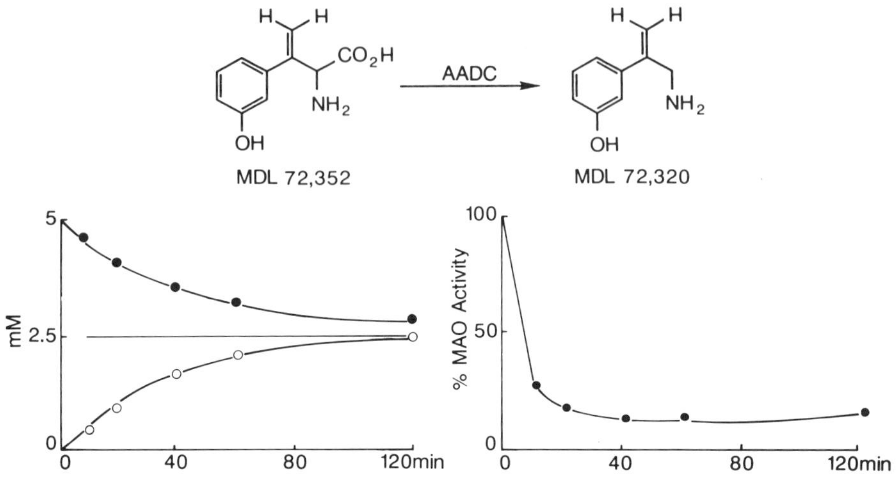

Fig. 2. *AADC mediated decarboxylation of MDL 72,352. The left hand side of the figure shows the disappearance of the amino acid (●) and the appearance of the amine, MDL 72,320 (○), as a function of time. Aliquots taken from the incubation medium were diluted to 500 μM (based on MDL 72,352) and tested as inhibitors of MAO using tyramine as substrate (right hand side).*

idea, we synthesized β-methylene-*m*-tyrosine (MMT, MDL 72,352), a known compound that Chari and Wemple (1979) had designed as a potential mechanism-based *inhibitor* of AADC. Rather than inhibit AADC, we were delighted to discover that MMT was decarboxylated by the partially purified hog kidney enzyme (McDonald *et al.*, 1986) as shown in Figure 2 and the product, β-methylene-*m*-tyramine (MDL 72,320), as expected, was a weak inhibitor of MAO. With the knowledge that fluoromethylene substituted phenethylamine analogues were much more potent inhibitors of MAO than the non-fluorine containing analogues, we synthesized and tested (*E*)-β-fluoromethylene-*m*-tyrosine (FMMT, MDL 72,394). This compound was efficiently decarboxylated be AADC (Figure 3A) to yield the very potent, time-dependent, irreversible, MAO-A selective inhibitor (*E*)-β-fluoromethylene-*m*-tyramine (MDL 72,392; Figure 3, B and C). Decarboxylation of FMMT was totally blocked by 50 μM of the irreversible AADC inhibitor α-mono-fluoromethyldopa (MFMD, MDL 71,963) and aliquots from the incubation medium failed to inhibit MAO. This confirmed that decarboxylation is essential for MAO inhibition and that FMMT is not an MAO inhibitor *per se* (Palfreyman *et al.*, 1983; McDonald *et al.*, 1984; Palfreyman *et al.*, 1985). Finally, HPLC separation of the enantiomers of FMMT afforded small amounts of pure *L* and *D* isomers (Wagner *et al.*, 1987) enabling us to demonstrate that only the *L* isomer is decarboxylated by AADC as would be expected (Figure 4).

When administered to rats, FMMT was found to be a very potent inhibitor of MAO with inhibition in the brain and duodenum being greater than in heart and liver (Figure 5A, Figure 6). Significant elevations of the concentrations of norepinephrine, dopa-

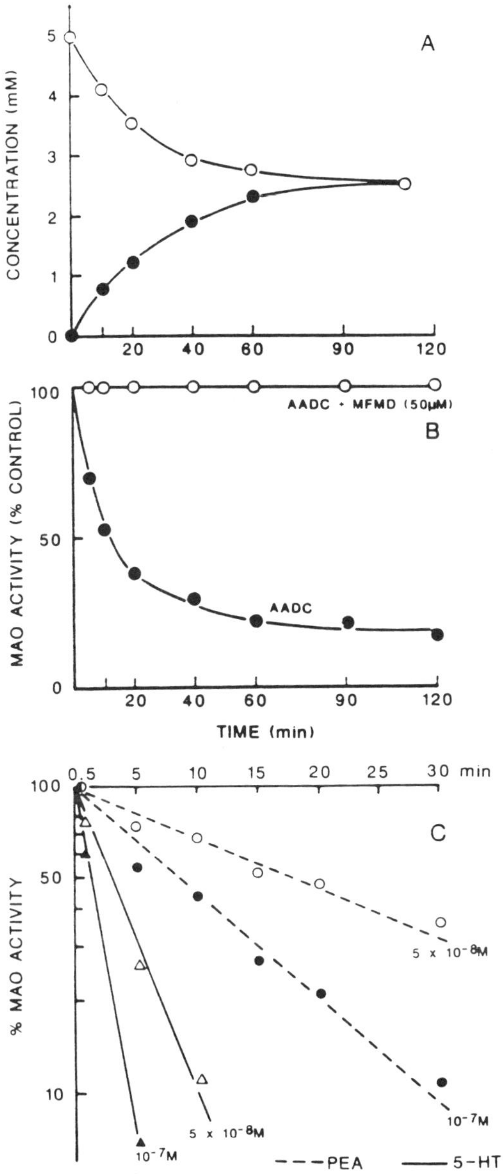

Fig. 3. *Incubation of racemic MDL 72,394 with AADC leads to the formation of a potent, MAO-A selective inhibitor. (Panel A) Disappearance of the amino acid (○) and appearance of the amine, MDL 72,392 (●) were monitored by HPLC. The decarboxylation was blocked with MFMD (MDL 71,963). (Panel B) Aliquots from the incubation medium either with MFMD (○) or without (●) were tested for inhibition of MAO. (Panel C) Time dependent inhibition of MAO-A (●,▲) and MAO-B (○,△) with MDL 72,392.*

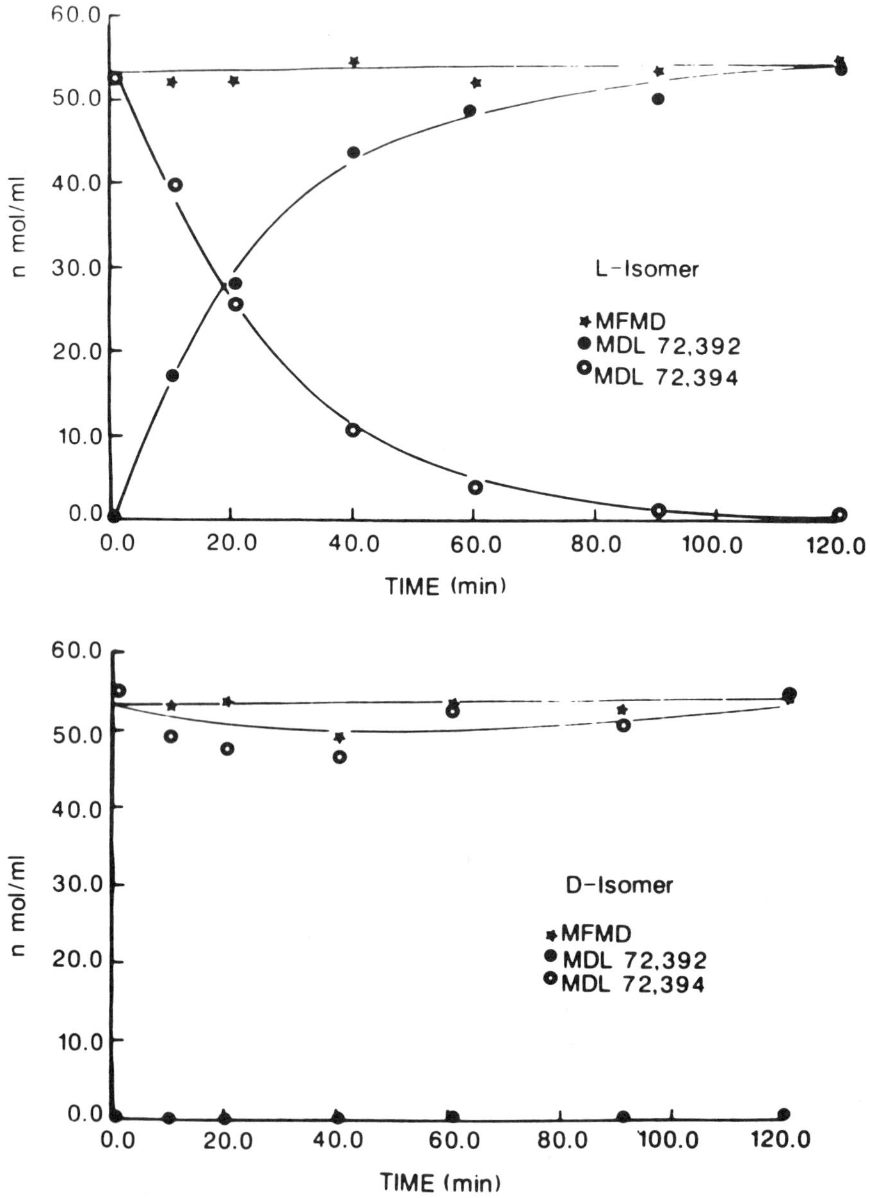

Fig. 4. *Incubation of the L and D isomers of (E)-β-fluoromethylene-m-tyrosine with AADC. In the top panel the concomitant disappearance of the L-amino acid MDL 72,394 and appearance of the amine MDL 72,392 is shown. This conversion is blocked by the AADC inhibitor MFMD. In the bottom panel, the D-isomer of MDL 72,394 is not decarboxylated by AADC.*

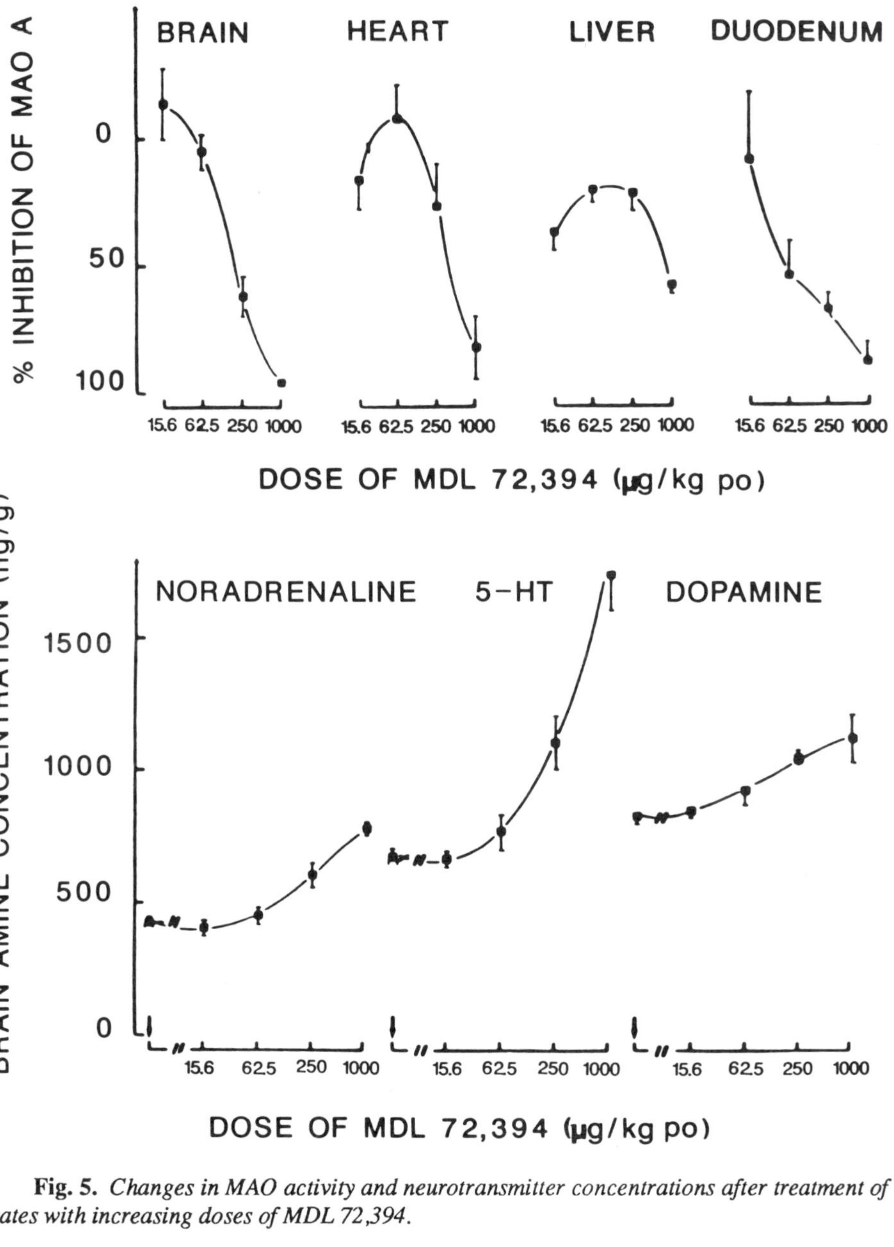

Fig. 5. *Changes in MAO activity and neurotransmitter concentrations after treatment of rates with increasing doses of MDL 72,394.*

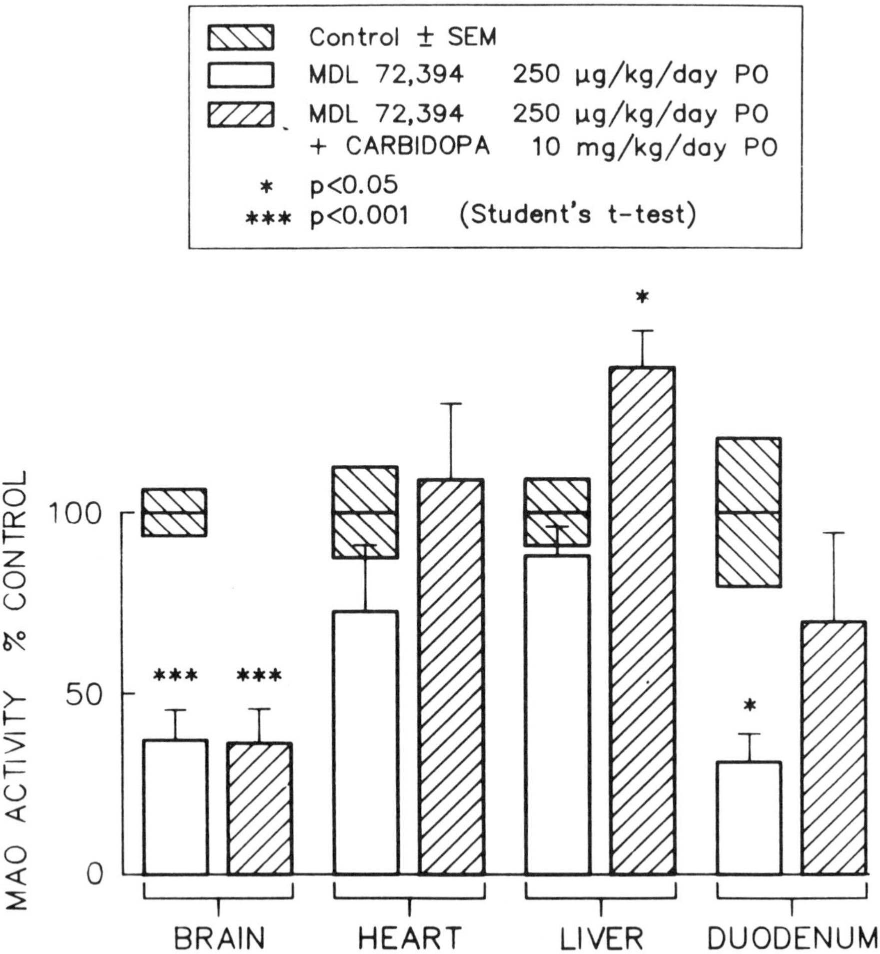

Fig. 6. *Changes in MAO activity in various organs after chronic treatment of rats either with MDL 72,394 alone or co-administered with carbidopa.*

mine and 5-hydroxytryptamine were measured in the brain (Figure 5B). Decarboxylation by AADC was also an essential step in the *in vivo* activation of FMMT because pretreatment with high doses of MFMD abolished the MAO inhibitory activity (Palfreyman *et al.*, 1985).

Since the major site of decarboxylation of orally administered *L*-Dopa is the gastrointestinal tract with up to 70% being metabolized in these tissues, one would extrapolate that co-administration of FMMT with a decarboxylase inhibitor would markedly enhance its bioavailability. Moreover, a secondary consequence of this combination would be that once FMMT had entered the systemic circulation, more unchanged drug would be available to reach the CNS. This is exactly what was seen when FMMT was co-administered with carbidopa to rats; brain MAO inhibition was enhanced, the dose to achieve this inhibition was reduced and inhibition of MAO was lower in peripheral tissues (Figure 6, Table I). Concomitantly, the extent to which the cardiovascular effects of tyramine were potentiated, relative to the effects of FMMT given alone, were markedly reduced (Table I).

Although much of the initial evaluation of FMMT was undertaken in rats, this species has very high decarboxylase activity in peripheral tissues and in this respect does not resemble man. Dogs, on the other hand, have low decarboxylase activity (Sasahara *et al.*, 1980) and were considered a more appropriate species for further testing of the MAO pro-drug.

A preliminary study was conducted in dogs (Davies *et al.*, 1984) receiving oral daily doses of 0.1 mg/kg FMMT, alone or combined with 10 mg/kg carbidopa, for four days. Three hours after the last dose, the animals were surgically prepared with isolated jejunal loops, tyramine was then introduced into the lumen of the loop and the degree of metabolism was determined according to Ilett *et al.* (1980). MAO activity and monoamine levels were determined at the end of the experiment. FMMT alone did not alter tyramine metabolism (Davies *et al.*, 1984) or inhibit MAO in the intestine but did substantially inhibit MAO activity in the liver and cerebral cortex and caused an elevation of cortical norepinephrine concentrations (Table II). On the other hand,

Table I. Effect of subacute (5 day) oral dosing with MDL 72,394, either alone or combined with carbidopa, on central MAO activity, central monoamine concentrations, and tyramine potentiation in the rat. Increases in norepinephrine and serotonin, and inhibition of MAO, were statistically significant ($p < 0.05$, Student's 't' test) for all treatments except serotonin after 0.0625 mg/kg/day MDL 72,394. Values represent the mean ± S.E.M. of 5-7 determinations. [a]Tyramine potentiation was calculated from the degree of shift of a curve of dose versus tachycardia to intraduodenally administered tyramine (0.312 - 50 mg/kg) when compared to animals pretreated with saline or carbidopa.

DOSE OF MDL 72,394 (mg/kg-day)	DOSE OF CARBIDOPA (mg/kg-day)	MAO ACTIVITY % CHANGE FROM CONTROL	NOREPINEPHRINE CONCENTRATION % CHANGE FROM CONTROL	SEROTONIN CONCENTRATION % CHANGE FROM CONTROL	DEGREE OF TYRAMINE POTENTIATION[a]
0.0625	0	-35 ± 7	+23 ± 3	+14 ± 6	4.0
0.0625	10	-76 ± 2	+118 ± 3	+95 ± 3	2.5
0.25	0	-62 ± 8	+77 ± 4	+114 ± 9	7.1

Table II. Tyramine metabolism, MAO inhibition and brain monoamine concentrations in dogs dosed for four days with MDL 72,394. [a]MAO activity was assessed using both ^{14}C-tyramine and ^{14}C-serotonin as substrates.

DOSE OF MDL 72,394 (mg/kg-day)	DOSE OF CARBIDOPA (mg/kg-day)	% OF RADIOACTIVITY IN VENOUS EFFLUENT		MAO ACTIVITY % OF CONTROL[a]			CORTEX MONOAMINE CONCENTRATIONS (% OF CONTROL)	
		TYRAMINE	P-HYDROXY PHENYLACETIC ACID	INTESTINE	LIVER	CORTEX	NE	5HT
0	0	13.2	74.2	100	100	100	100	100
0.1	0	13.2	72.4	108	66.8	61.6	160	110
0.1	10	24.2	60.0	92.1	110.2	77.6	135	80

carbidopa co-administration slightly enhanced the amount of unchanged tyramine reaching the portal circulation, prevented FMMT from inhibiting the liver enzyme and tended to reduce the MAO inhibitory effect in the brain (Table II). In retrospect, it is probable that this dose of carbidopa does penetrate the CNS and inhibit the decarboxylation of FMMT. Therefore, further experiments were undertaken using lower doses of carbidopa (2 mg/kg p.o.) and, in this instance, the cardiovascular responses to tyramine also were measured (Palfreyman *et al.*, 1988). Treatments with 0.1 mg/kg/day of FMMT for 4 days with 2 mg/kg/day carbidopa did not alter the tyramine sensitivity as measured by the mean blood pressure divided by heart rate ratio following oral tyramine challenge. FMMT alone at 0.5 mg/kg only marginally enhanced tyramine sensitivity compared to 2 mg/kg tranylcypromine which produced a dramatic pressor response at doses of tyramine estimated to be about 30-fold less than the amount needed to elicit the same response in the MDL 72,394/carbidopa treated animals (Figure 7).

Cardiovascular responses to intravenously administered tyramine were also determined in these MAO treated dogs (Palfreyman *et al.*, 1988). As we had seen with the oral tyramine challenge, combined administration of FMMT with carbidopa did not enhance the effects of tyramine (Figure 8). FMMT, given alone, produced only a marginal increase in the responses to tyramine. What was particularly striking, however, was that this potentiation was only seen at the higher doses of tyramine. Tranylcypromine treatments, as expected, increased the sensitivity to tyramine over the whole dose-response curve (Figure 8). All three treatment schedules markedly inhibited brain MAO activity (64 - 99% inhibition, Table III) and decreased the plasma concentration of the norepinephrine metabolite, dihydroxyphenylglycol (Table IV). These data would suggest that FMMT, even without the added protection of a peripherally acting decarboxylase inhibitor, will be less likely to produce the "cheese-effect" as compared to a classic, non-selective MAO inhibitor such as tranylcypromine. Because the relative activities of the decarboxylase and monoamine oxidases in peripheral and central tissues of the dog are closer in activity to those seen in man we feel confident that the favorable profile of FMMT in this species bodes well for therapeutic application.

FMMT *per se* has surprising selectivity for brain MAO. At this stage, we are uncertain as to the reasons for this selectivity although several explanations are plausible. Firstly, FMMT is a close structural analogue of tyrosine and we were able

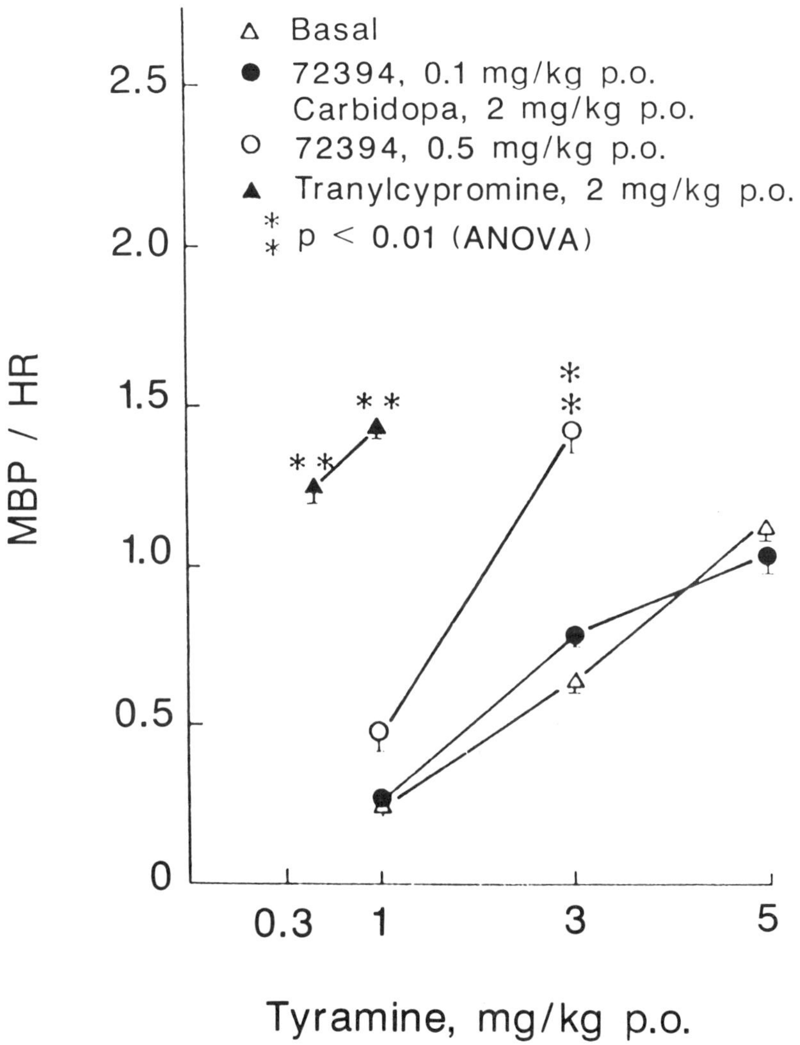

Fig. 7. *Mean blood pressure (MBP) divided by heart rate (HR) responses to increasing doses of oral tyramine in dogs treated with either MDL 72,394 alone, MDL 72,394 and carbidopa, or tranylcypromine.*

Table III. Effect of MAO inhibitors on MAO activity in the dog brain. The data is presented as % control activity ± S.E.M. [a]$p<0.01$ and [b]$p<0.05$ when compared to control by analysis of variance with Dunnett's 't' test.

	Control (n = 4)	MDL 72,394, 0.5 mg/kg-day p.o. (n = 3)	MDL 72,394, 0.1 mg/kg-day p.o. and carbidopa 2 mg/kg-day p.o. (n = 3)	Tranylcypromine 2 mg/kg-day p.o. (n = 3)
MAO-A	100 ± 15	25 ± 0.5[a]	48 ± 7[a]	3 ± 0.2[a]
MAO-B	100 ± 23	14 ± 0.3[a]	36 ± 3[b]	1 ± 0.2[a]

to demonstrate that it is transported into the brain on the amino acid carrier that also transports tyrosine (Palfreyman *et al.*, 1985). The second point is more subtle. AADC is predominently located in monoaminergic nerves in the brain and is found in peripheral sympathetic nerves, the gut, liver and kidney (Jaeger *et al.*, 1984). However, this enzyme is not necessarily in juxtaposition to the MAO enzyme in all these peripheral tissues so that the free amine (MDL 72,392) may not readily reach all MAO-containing tissues compartments. In contrast, in the brain the two enzymes may be more suitably located relative to each other. From lesioning studies with 6-hydroxydopamine and following microinjections of MFMD into dopamine-containing nerves of the brain, it was unambiguously established that FMMT depended on intraneuronal AADC for its activation to a potent MAO inhibitor (Palfreyman *et al.*, 1985; Sleight *et al.*, 1985). Thus, sufficient MAO enzyme could remain intact in the periphery ensuing rapid metabolism of any ingested tyramine at doses of FMMT which adequately inhibit the brain enzyme. A second level of selectivity may arise from the potential of the actual MAO inhibitor, (*E*)-β-fluoromethylene-*m*-tyramine (MDL 72,392), to be taken up into monoamine-containing nerves by the amine uptake mechanism. Recent experiments by Fagervall and Ross (1989), in which the doses of FMMT needed to inhibit MAO both *intra* and *extra* neuronally were determined, demonstrated that preferential MAO inhibition occurred within noradrenergic, and to a lesser extent, dopaminergic nerve terminals. This selectivity was seen following repeated as well as acute treatment with FMMT (Table V). They proposed that if MDL 72,392 were to be formed outside of monoamine nerves it would be a substrate for the norepinephrine and dopamine uptake systems, but probably not for the serotonin uptake system. It is possible that this neuronal selectivity may further explain why FMMT has little propensity to potentiate the cardiovascular effects of tyramine even in the absence of a peripheral decarboxylase inhibitor. The fact that FMMT is a substrate for the amino acid uptake system and that the decarboxylated metabolite is a

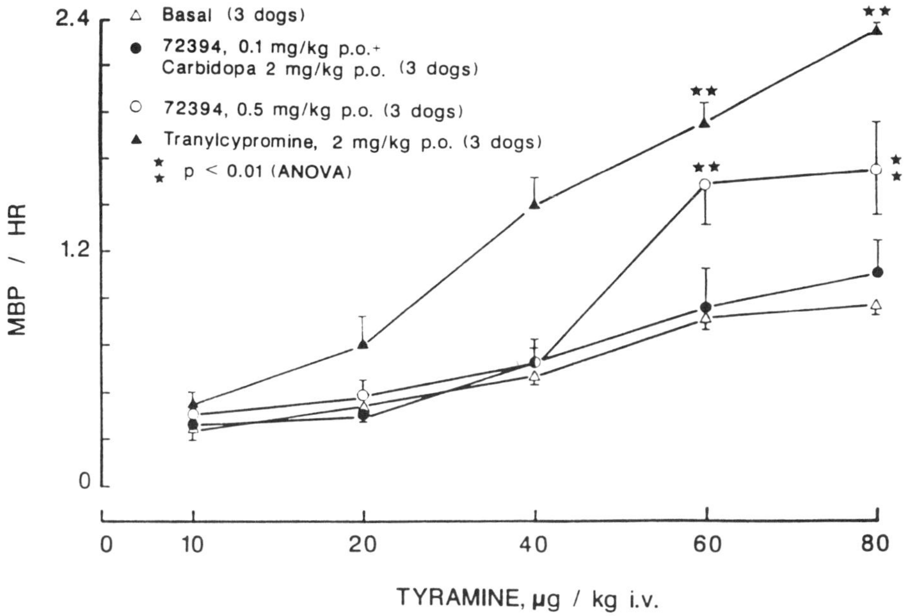

Fig. 8. *Mean blood pressure (MBP) divided by heart rate (HR) responses to increasing doses of intravenous tyramine in dogs treated with either MDL 72,394 alone, MDL 72,394 and carbidopa, or tranylcypromine.*

substrate for the catecholamine uptake system may also help to explain the remarkable potency of the compound as an MAO inhibitor.

Chronic administration to experimental animals of FMMT, in common with all effective antidepressants (Murphy *et al.*, 1984; Hollister, 1986), was shown to lead to an adaptive reduction in the number and/or function of certain monoamine receptors, suggesting antidepressant potential (Table VI). Moreover, as would be predicted for a potent MAO inhibitor, FMMT is also active in acute tests of antidepressant activity

Table IV. Concentration of dihydroxyphenylglyol (DHPG) in dog plasma. [a]$p<0.01$ vs. control.

TREATMENT	DHPG ng/ml ± s.e.m.
Control	1.16 ± 0.1
FMMT 0.1 mg/kg; Carbidopa 2 mg/kg	0.27 ± 0.1[a]
FMMT 0.5 mg/kg	0.43 ± 0.2[a]
Tranylcypromine 2 mg/kg	0.34 ± 0.1[a]

Table V. Calculated ED_{50} values for MDL 72,394 to inhibit MAO activity inside and outside various monoaminergic synaptosomes. [a]Data from Fagervall and Ross, 1989. All ED_{50} values were calculated from raw data, and are expressed as mg/kg of MDL 72,394 needed to produce a 50% reduction in MAO activity. Dosage ranges of MDL 72,394 evaluated were 0.1-1 mg/kg for acute drug studies and 0.025-0.1 mg/kg for repeated drug studies. In all experiments, 5 rats received each dose of MDL 72,394 that was evaluated. [b]For acute drug administration, MDL 72,394 was given subcutaneously, and animals were killed 24 hours later. For repeated drug administration, MDL 72,394 was given orally combined with carbidopa (20 mg/kg) for 7 days, and animals were killed 24 hours after their last dose.

	ED_{50} IN[a]	ED_{50} OUT[a]
A. ACUTE DRUG ADMINISTRATION[b]		
Noradrenergic Synaptosomes	0.24	0.40
Dopaminergic Synaptosomes	0.20	0.25
Serotonergic Synaptosomes	0.50	0.42
B. REPEATED DRUG ADMINISTRATION[b]		
Noradrenergic Synaptosomes	0.016	0.035
Dopaminergic Synaptosomes	0.010	0.020
Serotonergic Synaptosomes	0.048	0.027

such as reserpine hypothermia where pronounced antagonism was seen at 1 mg/kg (Palfreyman and Fozard, 1984).

Preliminary evaluation of FMMT in man has confirmed that inhibition of MAO occurs in the range of 1 - 16 mg and that these doses are not associated with changes in the sensitivity to the blood pressure elevating effects of intravenously infused tyramine (Palfreyman *et al.*, 1988; Davies *et al.*, 1986). Phase II clinical studies are now in progress to assess the antidepressant effect of this interesting inhibitor of MAO.

II. (E)-β-FLUOROMETHYLENE-5-HYDROXYTRYPTOPHAN AS A DUAL ENZYME-ACTIVATED INHIBITOR OF MAO

Because AADC will also decarboxylate 5-hydroxytryptophan to 5-hydroxytryptamine it was obviously of interest to synthesize the corresponding β-fluoromethylene-substituted amino acid. Enough material was prepared of the very sensitive (*E*)-β-fluoromethylene-5-hydroxytryptophan (FM-5-HTP, MDL 73,095) for testing as a substrate for AADC *in vitro* (Palfreyman *et al.*, 1989).

FM-5-HTP was decarboxylated at about 10% of the rate of 5-HTP and the amine formed, presumably (*E*)-β-fluoromethylene-5-hydroxytryptamine, inhibited MAO (Figure 9). Preliminary testing *in vivo*, in combination with carbidopa, indicated that as little as 0.1 mg/kg p.o. of FM-5-HTP significantly inhibited rat brain MAO. We believe that this MAO inhibitor has the potential to show selectivity for MAO within serotonergic nerves of the brain.

Table VI. Receptor modifications induced by chronic treatment with FMMT. Rats were treated with FMMT orally for 14 or 21 days and determinations made 72 hrs after the last dose. [a]NE and isoproterenol stimulated adenylate cyclase. [b]5-HT mediated decrease in adenylate cyclase. [c]Nitrendipine binding site.

RECEPTOR	RESPONSE	REFERENCE
α-Adrenergic		
α_1	No effect	Palfreyman, et al, 1986
α_2 post synaptic	↓	"
α_2 on NE terminals	No effect	"
α_2 on 5-HT terminals	No effect	"
β-Adrenergic		
β	↓	"
Adenylate cyclase[a]	↓	"
Serotonin		
5-HT_{1A}	↓	Palfreyman, et al, 1986 Sleight, et al, 1988a
5-HT_{1B}	No effect	Palfreyman, et al, 1986 Sleight, et al, 1989
Adenylate cyclase[b]	↓	Sleight, et al, 1988a Sleight, et al, 1988b
5-HT_2	↓	Palfreyman, et al, 1986
Calcium channel		
L-type[c]	No effect	Palfreyman, et al, 1986

III. CONCLUSION

In this chapter we have shown how it is possible to selectively inhibit MAO enzyme pools in the brain so that the peripheral enzyme is left intact in sufficient amounts to efficiently metabolize any ingested pressor amines, such as tyramine. The key to the success of this approach was to use dual enzyme-activated inhibitors whose structures were modelled as closely as possible to those of the natural substrates of the enzymes AADC and MAO. By taking this approach the inhibitors were able to benefit from both the amino acid and amine facilitated transport mechanisms that Nature has created to handle the monoamine neurotransmitters with greatest efficiency.

It is our conviction that this approach can be applied to enzyme systems beyond those involved in biogenic amine metabolism. This should lead not only to greater specificity for the target enzyme, but also may allow for the selective inhibition of those pools of enzyme associated with the particular therapeutic target.

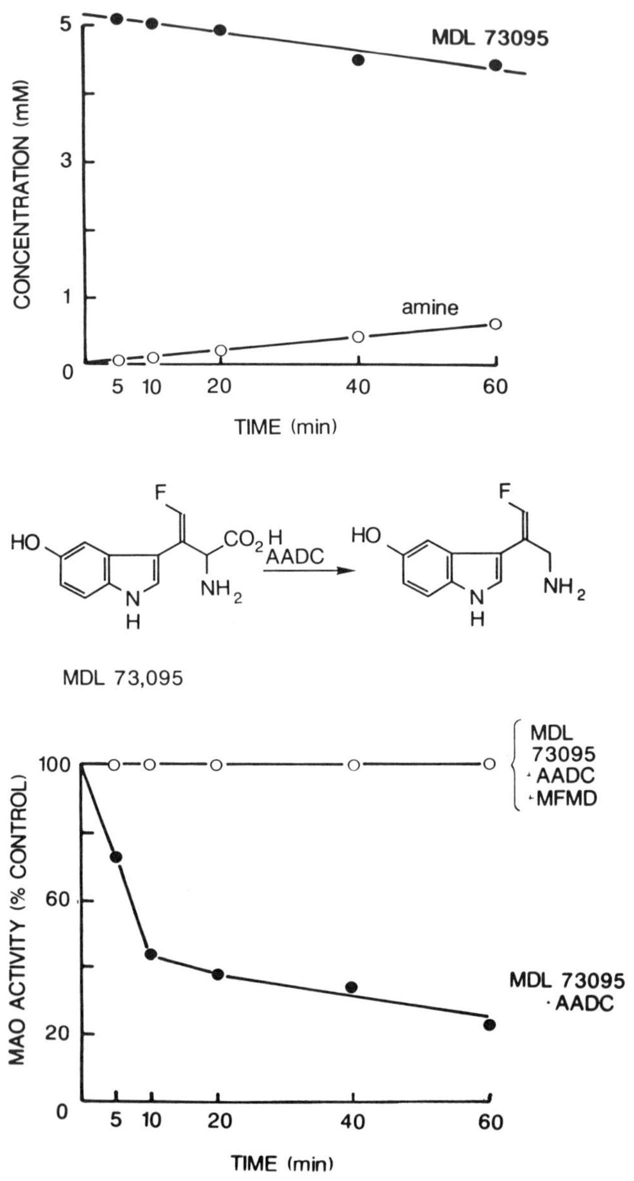

Fig. 9. *AADC mediated decarboxylation of racemic MDL 73,095 (top panel) leading to the formation of an MAO inhibitor (bottom panel).*

ACKNOWLEDGEMENTS

The authors express their gratitude to numerous co-workers who have contributed to studies described and cited in this chapter and to Mary Dooley for help in the preparation of the manuscript. In particular, we wish to thank Michel Jung and his co-workers for valuable discussion.

REFERENCES

Bach, A.W.J., Lan, N.C., Johnson, D.L., Abell, C.W., Bembenek, M.E., Kwan, S.W., Seeburg, P.H., and Shih, J.C. (1988). *Proc. Natl. Acad. Sci. U.S.A.* **85**, 4934-4938.

Blackwell, B. (1963). *Lancet* **2**, 849-851.

Blaschko, M. (1974). *Rev. Physiol. Biochem. Pharmacol.* **70**, 83-148.

Chari, R.V.J. and Wemple, J. (1979). *Tetrahedron Lett.* **21**, 111-114.

Davies, D.S., Yasuhara, H., Boobis, A.R., and George, C.F. (1984). In "Monoamine oxidase and disease" (K.F. Tipton, P. Dostert, and M. Strolin Benedetti, eds.), pp. 443-448. Orlando, FL, Academic Press.

Davies, D.S., Brown, M.J., Dollery, C.T., Pigache, R., and Orwin, J. (1986). *Acta Pharmacol. Toxicol.* **59** (Suppl. 5, Pt. 1), Abstr. 36, p. 42.

Dostert, P.L., Strolin Benedetti, M., and Tipton, K.F. (1989). *Med. Res. Rev.* **9**, 45-89.

Fagervall, I. and Ross, S.B. (1989). *J. Neurochem.* **52**, 467-471.

Hare, M.L.C. (1928). *Biochem. J.* **22**, 968-979.

Hollister, L.E. (1986). *Annu. Rev. Pharmacol. Toxicol.* **26**, 23-37.

Ilett, K.F., George, C.F., and Davies, D.S. (1980). *Biochem. Pharmacol.* **29**, 2551-2556.

Jaeger, C.B., Ruggiero, D.A., Albert, V.R., Park, D.H., Joh, T.H., and Reis, D.J. (1984). *Neuroscience* **11**, 691-713.

Johnston, J.P. (1968). *Biochem. Pharmacol.* **17**, 1285-1297.

Kline, N.S. (1958). *J. Clin. Exp. Psychopathol.* **19**, 72-78.

Knoll, J. and Magyar, K. (1972). *Adv. Biochem. Psychopharmacol.* **5**, 393-408.

Loomer, H.P., Saunders, J.C., and Kline, N.S. (1957). *Psychiatr. Res. Rep. Am.Psychiatr. Assoc.* **8**, 129-141.

McDonald, I.A., Lacoste, J.M., Bey, P., Wagner, J., Zreika, M., and Palfreyman, M.G. (1984). *J. Am. Chem. Soc.* **106**, 3354-3356.

McDonald, I.A., Lacoste, J.M., Bey, P., Palfreyman, M.G., and Zreika, M. (1985). *J. Med. Chem.* **28**, 186-193.

McDonald, I.A., Lacoste, J.M., Bey, P., Wagner, J., Zreika, M., and Palfreyman, M.G. (1986). *Bioorg. Chem.* **14**, 103-118.

Murphy, D.L., Garrick, N.A., Aulakh, C.S., and Cohen, R.M. (1984). *J. Clin. Psychiatry* **45**, 37-43.

Palfreyman, M.G. and Fozard, J.R. (1984). Unpublished data.

Palfreyman, M.G., McDonald, I.A., Zreika, M., and Fozard, J.R. (1983). *Prog. Neuro-Psychopharmacol. Biol. Psychiatry Suppl.*, Abstr. 369.

Palfreyman, M.G., McDonald, I.A., Fozard, J.R., Mely, Y., Sleight, A.J., Zreika, M., Wagner, J., Bey, P., and Lewis, P.J. (1985). *J. Neurochem.* **45**, 1850-1860.

Palfreyman, M.G., Mir, A.K., Kubina, M., Middlemiss, D.N., Richards, M., Tricklebank, M.D., and Fozard, J.R. (1986). *European J. Pharmacol.* **130**, 73-89.

Palfreyman, M.G., McDonald, I.A., Bey, P., Schechter, P.J., and Sjoerdsma, A. (1988). *Prog. Neuropsychopharmacol. Biol. Psychiatry* **12**, 967-987.

Palfreyman, M.G., McDonald, I.A., Zreika, M., Collard, J.N., Wagner, J., Bey, P., and Sjoerdsma, A. (1989). In "Amino Acids" (G. Lubec, G.A. Rosenthal, eds.) The Netherlands, ESCOM Science Publisher BV. In press.

Robinson, D.S. (1989). This book, pp. 233-257.

Sasahara, K., Nitanai, T., Habara, T., Morioka, T., and Nakajima, E. (1980). *J. Pharm. Sci.* **69**, 261-265.

Selikoff, I.J., Robitzek, E.H., and Ornstein, G.G. (1952). *Q. Bull. Seaview Hosp.* **13**, 17-26.

Sleight, A.J., Redfern, P.H., and Palfreyman, M.G. (1985). *Br. J. Pharmacol.* **86** (Suppl.), 480P.

Sleight, A.J., Marsden, C.A., and Palfreyman, M.G. (1988a). *Br. J. Pharmacol.* **93** (Suppl.), 95P.

Sleight, A.J., Marsden, C.A., Palfreyman, M.G., Mir, A.K., and Lovenberg, W. (1988b). *Eur. J. Pharmacol.* **154**, 255-261.

Sleight, A.J., Smith, R.J., Marsden, C.A. and Palfreyman, M.G. (1989). *Neuropharmacology* **28**, 477-480.

Wagner, J., Wolf, E., Meintzelmann, B., Zreika, M., and Palfreyman, M.G. (1987). International Symposium on Chiral Separations, Sept. 3-4, University of Surrey, U.K. (Abstr.)

Youdim, M.B.M. and Finberg, J.P.M. (1983). *Mod. Probl. Pharmacopsychiatry* **19**, 63-74.

Zeller, E.A., Barsky, J., Fouts, J.R., Kirchheimer, W.F., and Van Orden, L.S. (1952). *Experientia* **8**, 349-350.

12

INHIBITION OF POLYAMINE BIOSYNTHESIS AND FUNCTION AS AN APPROACH TO DRUG DESIGN

Anthony E. Pegg

Departments of Cellular and Molecular Physiology and of Pharmacology
Milton S. Hershey Medical Center
Pennsylvania State University
Hershey, PA 17033

Peter P. McCann
Albert Sjoerdsma

Merrell Dow Research Institute
Cincinnati, OH 45215

I. INTRODUCTION

This chapter is a brief account of recent research involving the synthesis of inhibitors of polyamine production and their potential as pharmacological agents. In order to place this work in perspective, summaries of the reactions involved in maintaining cellular polyamine levels and in the incorporation of polyamines into cellular components are also included. The design and use of inhibitors of polyamine biosynthesis is a very rapidly growing field and this chapter is not intended to provide a totally comprehensive coverage of this literature.

II. OUTLINE OF POLYAMINE BIOSYNTHESIS AND DISPOSITION

A. Biosynthesis and Interconversion

In mammalian cells the polyamine biosynthetic pathway, which is illustrated in Figure 1, is now thought to be well understood (reviewed by Tabor and Tabor, 1984; Pegg and McCann, 1982, 1988; Pegg, 1986). Spermidine and spermine are formed by the sequential addition of aminopropyl groups to putrescine. These reactions are catalyzed by two distinct enzymes, spermidine synthase and spermine synthase, that require decarboxylated S-adenoslymethionine as the aminopropyl donor. This nucleoside is formed by the action of S-adenosylmethionine decarboxylase [AdoMetDC][1], a pyruvoyl enzyme. Putrescine is formed by the action of ornithine decarboxylase [ODC], a pyridoxal phosphate-dependent enzyme.

These reactions are the only ones leading to the *de novo* formation of polyamines in the mammalian cells. However, it should be noted that there is a specific and regulatable membrane transport system for putrescine, spermidine and spermine and that these compounds can therefore be taken up from extracellular sources. Furthermore, spermine and spermidine can be converted back into putrescine by the combined actions of spermidine/spermine-N^1-acetyltransferase [SAT] (polyamine acetylase) and polyamine oxidase. The former enzyme transforms spermidine to its N^1-acetyl derivative which is then cleaved by polyamine oxidase to generate putrescine and 3-acetamidopropanol (Figure 1). Similar reactions starting with spermine convert it into spermidine. Putrescine, N^1-acetylspermidine and, under some circumstances, spermidine are readily excreted from mammalian cells (Wallace, 1987; Pegg *et al.*, 1988a, 1989a). The presence of serum oxidases may lead to the destruction of these extracellular amines but the possibility of a significant supply of polyamines arising from other cells or from dietary sources cannot be ruled out.

In general, the same reactions described above for the synthesis of polyamines are found in non-mammalian sources but there are some significant differences (Tabor and Tabor, 1985; Pegg and McCann, 1988). Plants and many bacteria also contain arginine decarboxylase [ADC]. This provides an alternate route to putrescine via the production of agmatine which is then converted to putrescine by agmatinase or agmatine deiminase. Many unicellular organisms do not contain spermine synthase and, under most conditions, produce very little, if any, spermine.

Unusual polyamines are found in a variety of fungi, algae, some plants and exotic microorganisms including extreme thermophiles, some cyanobacteria and certain archaebacteria. These compounds are probably produced via similar reactions but in only a few cases have these enzymes been characterized and this work is outside the scope of this chapter.

[1]*Abbreviations: AdoMetDC, S-adenosylmethionine decarboxylase; ODC, ornithine decarboxylase; ADC, arginine decarboxylase; SAT, spermidine/spermine-N^1-acetyltransferase; AdoMet, S-adenosylmethionine; DFMO, α-difluoromethylornithine, eflornithine, Ornidyl® (MDL 71,782); MAP, (2R, 5R)-δ-methylacetylenicputrescine (MDL 72,175); MGBG, methylglyoxal bis(guanylhydrazone); MAOEA, 5'-deoxy-5'[N-methyl-N-(aminooxyethyl)]-aminoadenosine; MHZPA, 5'-deoxy-5'-[N-methyl-N-(3-hydrazinopropyl)]-aminoadenosine; AdoDato, S-adenosyl-1,8-diamino-3-thiooctane; AdoDatad, S-adenosyl-1,12-diamino-3-thio-9-azadodecane; MTA, 5'-methylthioadenosine; $AdoS^+(CH_3)_2$, S-methyl-5'-methylthioadenosine [dimethyl (5'-adenosyl)sulfonium perchlorate]; DFMA, α-difluoromethylarginine (MDL 71,897).*

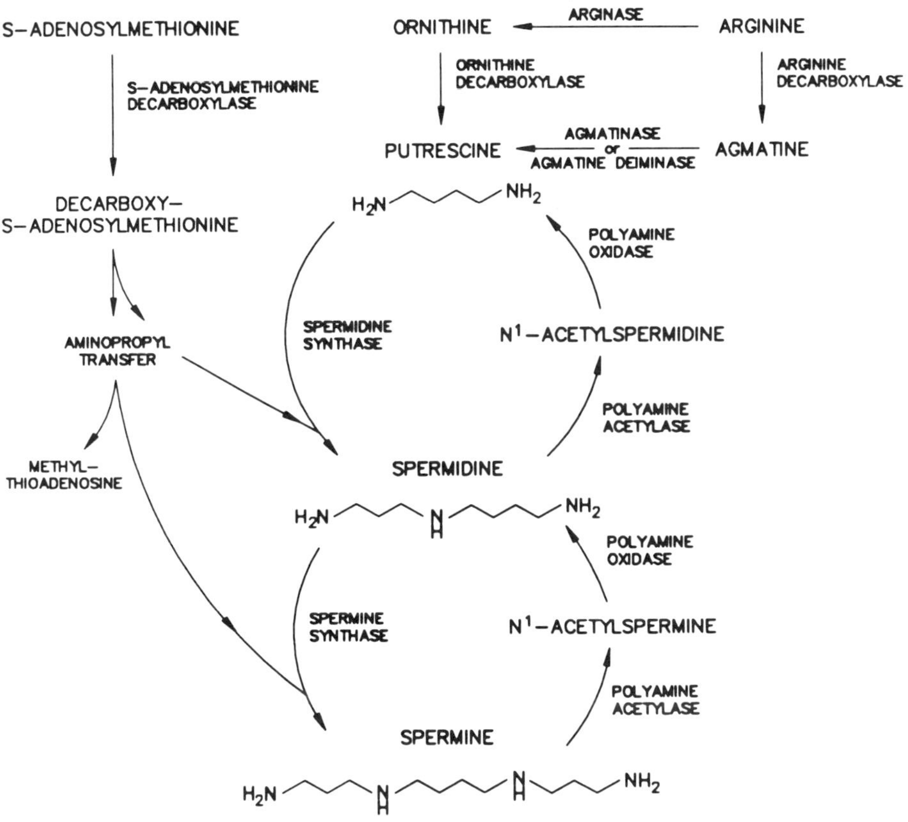

Fig. 1. *Schematic overview of polyamine metabolism.*

B. Incorporation of Polyamines into Other Cellular Molecules

One function of polyamines is to act as precursors for certain post-translational modifications of proteins. Although in only a few cases is there any indication of the physiological significance of these reactions, it has been found that transglutaminase-catalyzed incorporation of polyamines into both intracellular and extracellular proteins does take place in higher eukaryotes (Williams-Ashman and Canellakis, 1979; Williams-Ashman, 1984; Beninati *et al.*, 1988).

A second example of the role of polyamines as precursors of protein modification is the formation of hypusine in the protein synthesis initiation factor eIF-4D (Murphy and Gerner, 1987; Chen and Dou, 1988; Park, 1988; Park and Wolff, 1988). Hypusine, which is N^1-(4-amino-2-hydroxybutyl)lysine, is formed by the reaction of an internal lysine residue in this protein. The first step in this modification requires spermidine as a precursor to form deoxyhypusine. Putrescine must be converted into spermidine in order to take part in this reaction although the part of the molecule incorporated into the hypusine involves only the 1,4-diaminobutane fragment of the spermidine. 1,3-Diaminopropane is released in stoichiometric amounts. Hypusine is not present in bacteria but has been found in many mammalian cells and in yeast as a part of this protein. Its function is not known but the extreme conservation of a reaction pathway involving a number of specific enzymes leading to hypusine in eIF-4D suggests that it must play an important part in some physiological process.

In *E. coli*, spermidine can act as precursor of N^1-monoglutathionyl-spermidine [γ-*L*-glutamyl-*L*-hemicystinyl-glycyl-spermidine] (Tabor and Tabor, 1984, 1985). The signficance of this spermidine containing complex, which can account for more than 85% of the spermidine in the cells when the bacteria enter the stationary growth phase, is unknown. This complex is also present in trypanosomatids but these organisms also contain trypanothione [N^1, N^8-bis(glutathionyl)spermidine] (Fairlamb *et al.*, 1985; Pegg and McCann, 1988; Shim and Fairlamb, 1988). Trypanothione is a unique component of trypanosomatids where it plays an important role as an antioxidant defense mechanism by acting as a substrate for an NADPH-dependent flavoprotein disulfide reductase.

III. INTERFERENCE WITH POLYAMINE FORMATION AND FUNCTION IN MAMMALIAN CELLS

A. Inhibitors of Ornithine Decarboxylase [ODC]

The first clear demonstration that the activity of ODC was essential for mammalian cell growth was provided by the use of the competitive inhibitor α-methylornithine (Mamont *et al.*, 1976). Because this compound is not sufficiently potent for most purposes, the majority of the important studies on this topic over the past decade have been carried out using α-difluoromethylornithine (DFMO, eflornithine, Ordinyl®, MDL 71782), a compound synthesized at the now Merrell Dow Research Institute in Strasbourg, France (Bey *et al.*, 1987). Early biochemical and pharmacological studies in this area have been extensively reviewed (Sjoerdsma, 1981; Pegg and McCann, 1982; Sjoerdsma and Schechter, 1984; McCann *et al.*, 1987). Exhaustive work on the role of polyamines in the growth of normal and malignant mammalian cells has been carried out using DFMO, which acts as an enzyme-activated/mechanism based irreversible inhibitor of ODC (Sjoerdsma, 1981; Sjoerdsma and Schechter, 1984; Palfreyman *et al.*, 1987, Bey *et al.*, 1987), and some related inhibitors which have somewhat improved properties (Sjoerdsma and Schechter, 1984; Bey *et al.*, 1987). The use of DFMO, as a biochemical and pharmacological tool and as a treatment for various types of cancer, has been reviewed extensively (Pegg and McCann, 1982; Sjoerdsma and Schechter, 1984; Pegg, 1986, 1988; Porter and Sufrin, 1986; Sunkara, *et al.*, 1987; Danzin and Mamont, 1987; Porter and Jänne, 1987; Schechter *et al.*, 1987).

In general, these studies show that treatment of cells in culture with DFMO is very effective in depleting putrescine and spermidine but that there is only a small decline, if any, in spermine and a great increase in the content of decarboxylated AdoMet. The use of more active inhibitors such as (2R,5R)-δ-methyl-α-acetylenicputrescine [MAP, MDL 72,175] produced greater depletion of spermine in some experiments (reviewed by Pegg, 1988; Danzin and Mamont, 1987; Porter and Jänne, 1987), but even with this compound significant amounts of spermine remain in most treated cells. In order to deplete spermine, it appears that inhibitors of other enzymes in the polyamine biosynthetic pathway or the stimulation of spermine breakdown may be needed.

The synthesis of compounds which are claimed to inhibit more than one of the enzymic steps in polyamine biosynthesis has been reported (Khomutov *et al.*, 1985; Nakashima *et al.*, 1987; Hibasami *et al.*, 1988a). Although this is a potentially attractive idea, since a more complete block of the pathway could then be achieved with a single drug, the possibility that compounds can be designed with sufficient specificity and potency towards these enzymes and lack inhibitory properties towards other enzymes seems remote. It is likely that at the intracellular levels needed to maintain selectivity, such compounds will only inhibit the polyamine biosynthetic enzyme that is most sensitive. For example, although 1-aminooxy-3-aminopropane is an inhibitor of ODC, AdoMetDC and spermidine synthase (Khomutov *et al.*, 1985) *in vitro*, its effects on mammalian cells in culture are clearly due only to inhibition of ODC (Hyvönen *et al.*, 1988; Poulin *et al.*, 1989).

B. Inhibitors of S-adenosylmethionine Decarboxylase (AdoMetDC)

Early studies of the effects of blocking the AdoMetDC reaction were carried out using the compound methylglyoxal bis(guanylhydrazone) [MGBG]. MGBG is a potent competitive inhibitor of AdoMetDC and is capable of preventing the formation of decarboxylated AdoMet in exposed cells. Although these experiments did suggest that polyamines were important for the proliferation of mammalian cells, these results are not conclusive because, as discussed in several reviews (Pegg and McCann, 1982; Pegg, 1986, 1988; Porter and Jänne, 1987), MGBG is not specific and may have a variety of effects on cellular physiology unrelated to interference with polyamine synthesis. Furthermore, MGBG can be considered to be a polyamine analog and enters cells via the polyamine transport system (see below). The reversal of the effects of MGBG by addition of spermidine may therefore be due to the prevention of its uptake or binding to key intracellular sites rather than to replacement of cellular spermidine.

Although a variety of other inhibitors of AdoMetDC have been reported (summarized in Kolb *et al.*, 1982 and Pegg, 1983a) only recently have compounds with sufficient potency and specificity become available. Secrist and Khomutov and their respective colleagues have described the synthesis of analogues of AdoMet which contain a reactive aminooxy-or hydrazino- group. Among the most interesting compounds were 5'-deoxy-5'[N-methyl-N-(aminooxyethyl)]-aminoadenosine [MAOEA] and 5'-deoxy-5'[N-methyl-N-(3-hydrazinopropyl)]-aminoadenosine [MHZPA] (Secrist, 1987; Pegg *et al.*, 1988b) and S-(5'-deoxy-5'-adenosyl)methylthioethylhydroxylamine (Artamonova *et al.*, 1986). These compounds act as irreversible inhibitors of AdoMetDC. Although the exact nature of the adduct responsible for this inhibition is not yet known, they presumably act by binding to the substrate site and then forming a covalent bond with the pyruvate prosthetic group of the enzyme. Since this pyruvate is bound to the enzyme in an amide linkage and is formed from a serine residue within the polypeptide chain during the processing of proenzyme, it cannot be replaced and the enzyme is permanently inactivated.

Exposure of a variety of mammalian cells to these inhibitors has been found to produce the expected changes in cellular polyamine content and to inhibit cell growth. For example, in a detailed study, either MAOEA or MHZPA lead to a reduction of spermidine and spermine, decarboxylated AdoMet and 5'-methylthioadenosine [MTA] and a massive accumulation of putrescine in L1210 cells (Pegg *et al.*, 1988b). Cell growth was reduced once spermidine and spermine became depleted, indicating that putrescine cannot fully substitute for the higher polyamines in L1210 cells. Addition of either spermidine or spermine restored the normal growth rate. The combination of DFMO and MAOEA completely prevented any increase in polyamines in cultures of L1210 or HT29 cells (Pegg *et al.*, 1988b, 1989a). These results establish that inhibition of AdoMetDC is an appropriate target for the design of drugs which interfere with cell growth by preventing the synthesis of polyamines. Unfortunately these derivatives of AdoMetDC may not have sufficient biological stability for convenient use *in vivo*. However, more active and more stable inhibitors may soon be forthcoming.

Another approach has been to synthesize a variety of cogeners of MGBG including some which are also likely to form covalent adducts at the enzyme active site (Pegg, 1983a; Hibasami *et al.*, 1988a,b). Although several compounds have been reported to be more active than MGBG (Elo *et al.*, 1986, 1988; Nakashima *et al.*, 1987; Hibasami *et al.*, 1988a), it is not clear that these compounds have any better specificity than MGBG itself and these compounds also may act more as toxic polyamine analogues than as inhibitors of AdoMetDC. Despite this, it is possible that some of these compounds may be useful anti-tumor agents (Elo *et al.*, 1988; Hibasami *et al.*, 1988b).

Recent studies involving cloning of the mammalian AdoMetDC cDNA and the *in vitro* translation of the mRNA for this enzyme have provided more details of its structure and biosynthesis (Pajunen *et al.*, 1988). The mammalian enzyme is a dimer of two pairs of subunits with approximate molecular weights of 31,000 and 7,000. These subunits are formed by the cleavage of a proenzyme chain (M.W., 38,000) in a reaction which forms the pyruvate prosthetic group at the amino terminal end of the larger subunit. The cleavage occurs at the sequence -Leu-Ser-Glu-Ser-Ser-Met-Phe- forming -Leu-Ser-Glu at the carboxyl end of the small subunit and Pyruvate-Ser-Met-Phe- at the amino terminus of the large subunit (Stanley *et al.*, 1989). The processing/cleavage step may be an autocatalytic reaction of the proenzyme chain and is accelerated in the presence of putrescine. Since this step is essential for the production of the active sites of the enzyme, it provides an attractive target for the future design of therapeutically useful inhibitors of AdoMetDC. The AdoMetDC from non-mammalian species also contain a covalently bound pyruvate prosthetic group which is formed from a proenzyme in a similar way (Tabor and Tabor, 1987; Anton and Kutney, 1987; Kashiwagi *et al.*, 1988) but the amino acid sequences involved are sufficiently different that the design of species specific inhibitors might be possible.

C. Aminopropyltransferase Inhibitors

Spermidine synthase from mammalian cells is strongly inhibited by the multisubstrate analog, S-adenosyl-1,8-diamino-3-thiooctane [AdoDato] which was designed by Coward and colleagues for this purpose (Tang *et al.*, 1980; Coward and Pegg, 1987). It is also inhibited by low concentrations of cyclohexylamine which acts as a powerful competitive inhibitor with respect to putrescine [K_i of about 0.2 μM] (Hibasami *et al.*, 1980; Batchelor *et al.*, 1986). This surprising finding has recently been clarified and extended by the finding that n-butylamine is also a potent inhibitor of spermidine synthase with a K_i of 0.5 μM (Baillon *et al.*, 1989). Similarly, methylthiopropylamine (Hibasami *et al.*, 1987) and 2-mercaptoethylamine (Hibasami *et al.*, 1988c) were reported to be competitive inhibitors with K_i values of 5 μM and 0.3 μM respectively. It appears that these compounds act as analogues of the putrescine substrate. Certain other compounds related to cyclohexylamine are also effective in blocking the spermidine synthase reaction and recently *trans*-4-methylcyclohexylamine was found to have a K_i of 0.04 μM (Shirahata *et al.*, 1988).

AdoDato, which is the most potent inhibitor of isolated spermidine synthase *in vitro*, does enter cells and reduce the synthesis of spermidine in cultured mammalian cells (Pegg *et al.*, 1982; Sherman *et al.*, 1986). However, the effectiveness of the compound is limited by a compensatory rise in the activity of AdoMetDC which increases the supply of decarboxylated AdoMet. The main effect of exposing mammalian cells to AdoDato is therefore to favor the use of the available decarboxylated AdoMet for the synthesis of spermine rather than spermidine. The cellular polyamine pools are perturbed by a decrease in spermidine and an increase in spermine. Only after exposure to high levels of AdoDato was there any effect on cell growth but this inhibition of proliferation was prevented by the provision of exogenous spermidine. Treatment of rats with AdoDato did eliminate the rise in spermidine which occurs after partial hepatectomy (A.E. Pegg and J.K. Coward, unpublished observations) but there has been little additional experimental work with this compound in animals because of the cost of synthesis of the amount of material needed. Cyclohexylamine [incorrectly identified as dicylohexylamine (Batchelor *et al.*, 1986)] did reduce polyamine levels in mouse tumors (Ito *et al.*, 1982). Both methylthiopropylamine and 2-mercaptoethylamine (Hibasami *et al.*, 1987; 1988c) and a compound described as N-chlorosulfonyl- dicyclohexylamine (Nakashima *et al.*, 1986) were found to deplete spermidine levels and to act as antiproliferative agents in cultured human lymphoid leukemia Molt 4B cells. The effects of these compounds and of the more potent *trans*-4-methylcyclohexylamine in tumor bearing animals will be of interest.

The synthesis of specific spermine synthase inhibitors has been of particular importance since (a) spermine levels in many cells are not greatly affected by the use of inhibitors of ODC and (b) the physiological importance of spermine synthase, which is a ubiquitous component of mammalian cells but is absent from many microorganisms, is not yet known.

Although spermine synthase is strongly inhibited by its reaction product, MTA, and can be shown to be inhibited *in vitro* by a variety of analogues of this compound (Pegg, 1983b), these substances were not useful inhibitors in the cell. The reasons for this are firstly that some of the compounds are rapidly degraded by MTA phosphorylase, which normally acts to reduce MTA levels to values at which spermine synthesis is not affected, and secondly that MTA analogues have a wide variety of deleterious effects on cells that are not related to interference with polyamine metabolism.

The first compound shown to be a useful inhibitor of spermine synthase was S-methyl-5'-methylthioadenosine [$AdoS^+(CH_3)_2$] which was an inhibitor of the enzyme *in vitro* and led to a selective depletion of spermine in cultured SV 3T3 cells (Tang *et al.*, 1980; Pegg and Coward, 1985). When added to the culture medium at levels of 50 μM or of less, $AdoS^+(CH_3)_2$ brought about a decline in spermine and an increase in spermidine that is consistent with the blockage of the polyamine biosynthetic pathway at the spermine synthase step. At higher concentrations of the nucleoside there was also a decrease in spermidine, which is in accord with the observation that it was a competitive inhibitor of AdoMetDC (Kolb *et al.*, 1982; Pegg *et al.*, 1986). A variety of other nucleosides were tested for inhibition of spermine synthase but no compounds more active than $AdoS^+(CH_3)_2$ were found until the synthesis of S-adenosyl-1,12-diamino-3-thio-9-azadodecane [AdoDatad] (Coward and Pegg, 1987; Woster *et al.*, 1989). This compound, which is structurally related to AdoDato by the addition of an aminopropyl group, was designed as a multisubstrate analog inhibitor of spermine synthase and was found to be an extremely potent and specific inhibitor of this enzyme. The specificity of AdoDato for spermidine synthase and of AdoDatad for spermine synthase is quite remarkable.

Exposure of cell cultures (SV 3T3, L1210 or HT 29) to AdoDatad at concentrations of 50 μM or less lead to a complete inhibition of spermine synthesis and an increase in spermidine (Pegg *et al.*, 1989b). The AdoDatad was converted to a toxic product by the action of serum oxidases but, when this was prevented by the addition of aminoguanidine, there was no effect on cell growth with concentrations of AdoDatad that exerted a maximal effect on spermine levels. Higher concentrations of AdoDatad did affect growth but this effect was not reversed by polyamines. The lack of effect of AdoDatad-induced spermine depletion on cell proliferation, is a similar result to that found when $AdoS^+(CH_3)_2$ was used to prevent spermine synthesis (Pegg and Coward, 1985). These results suggest that spermine synthesis is not needed for the proliferation of these cells but it should be noted that the compensatory increase in spermidine may allow for cell growth.

The increase in spermidine is due both to the lack of conversion of spermidine into spermine and to an increase in the activity of AdoMetDC and, consequently, in the formation of decarboxylated AdoMet which is used as a substrate by spermidine synthase. The rise in the cellular content of decarboxylated AdoMet which occurs in cells treated with ODC or aminopropyltransferase inhibitors accounts for the paradoxical finding that when cells are treated with a combination of AdoDato and either Adodatad or $AdoS^+(CH_3)_2$ the cellular content of spermidine and spermine is normal (Pegg *et al.*, 1986, 1989b).

Very recently, it has been found that several N-alkylated-1,3-diaminopropane derivatives are inhibitors of spermine synthase, being competitive inhibitors with the spermidine substrate (Baillon *et al.*, 1989). The most active of these was *N*-butyl-1,3-

diaminopropane which has a K_i value of about 0.01 μM. This compound prevents the synthesis of spermine in HTC cells (Baillon *et al.*, 1989), L1210 cells, HT 29 cells and SV 3T3 cells (Pegg *et al.*, 1989c) when added at 50 μM concentrations to the culture medium. There was no effect on the growth of HTC cells under these conditions confirming the results with the nucleoside aminopropyltransferase inhibitors discussed above. *N*-Butyl-1,3-diaminopropane may be a particularly useful compound for the prevention of spermine synthesis since it is simple to prepare, and its action is competitive with respect to the spermidine substrate and is therefore not affected by the compensatory increase in decarboxylated AdoMet. However, its specificity has not yet been fully investigated and preliminary experiments indicate that it also acts as an inducer of spermidine/spermine-N^1-acetyltransferase (Pegg 1989c). Such induction could actually help to bring about the depletion of spermine by increasing the conversion of spermine and spermidine into putrescine via the acetylase/oxidase pathway.

D. Polyamine Analogues and the Role of Polyamine Transport

Despite the fact that the polyamine biosynthetic pathway is a ubiquitous cellular component and that all cells therefore have the capacity to synthesize polyamines, it is now firmly established that cells also contain a membrane transport system permitting the uptake of extracellular polyamines (Pegg and McCann, 1982; Pegg, 1986; Byers *et al.*, 1987; Kakinuma *et al.*, 1988; Heston and Charles, 1988; Gawel-Thompson and Greene, 1988; Byers and Pegg, 1989). This uptake system is regulated to respond to the intracellular polyamine content in some way which is not yet well understood at the biochemical level. When polyamines are reduced by exposure to inhibitors, the activity of this system is increased and, conversely, in the presence of high levels of polyamines the activity is reduced.

This transport system has considerable importance in reducing the effectiveness of polyamine biosynthesis inhibitors used as therapeutic agents since exogenous polyamines provided in the diet, by intestinal microorganisms, or by cell turnover can be very effectively scavenged by the induced transport system present in cells with depleted polyamines. It has been demonstrated that DFMO is a more active antitumor agent against L1210 cells in mice when the L1210 cell line used has a mutation preventing polyamine transport (Persson *et al.*, 1988). Also, the efficacy of DFMO against rodent tumors is enhanced when salvage of polyamines is prevented (Claverie *et al.*, 1987) and reduced when intestinal diamine oxidase which degrades the polyamines is blocked (Kallio *et al.*, 1984).

Calmodulin antagonists have recently been reported to influence polyamine transport (Heston and Charles, 1988) but, at present, no specific inhibitors of the transport process have been identified. Such compounds may be useful adjuncts to therapy with polyamine biosynthesis inhibitors. A considerable number of polyamine analogues are also substrates for the transport system and these compounds act as competitive inhibitors with regard to the natural polyamines. There is some evidence that there may actually be a number of different carrier systems with overlapping specificities (Byers *et al.*, 1987). At least one of these systems is Na^+ dependent and can be inhibited by substances affecting Na^+ flux (Kakinuma *et al.*, 1988; Heston and Charles, 1988; Gawel-Thompson and Greene, 1988). More detailed biochemical characterization of the transport system and the factors responsible for its regulation are clearly needed in order to facilitate the development of inhibitors. The progress of this work has been hampered by the low concentration of the carrier protein present in mammalian cell membranes, but mutant cells lacking the transport system have been obtained by exposure to a mutagen followed by selection for resistance to MGBG which enters cells on the polyamine transport system (Mandel and Flintoff, 1978; Rodrigues *et al.*, 1987; Persson *et al.*, 1988; Heaton and Flintoff, 1988). Transfection of human DNA into such CHO cell mutants has been used successfully to obtain hamster cells expressing the human polyamine transport protein and these cells would be a suitable tool for cloning the relevant genes (Byers *et al.*, 1989).

The toxicity of MGBG is clearly related to the accumulation of this compound at very high levels within the cell. The high levels of MGBG lead to a number of effects including a loss of active mitochondria and interference with macromolecular synthesis. This accumulation is brought about by the polyamine transport system for which MGBG is a good substrate. Moreover, MGBG does not appear to mimic the effects of natural polyamines and a number of other polyamine analogues in leading to the down regulation of the polyamine transport system (Porter and Jänne, 1987; Byers and Pegg, 1989). The system therefore continues to operate and transport MGBG into the cell despite the very high intracellular concentration of the drug. This lack of repression of the polyamine transport system may be an important factor in the toxicity of MGBG which has not been widely appreciated. The uptake of MGBG is increased by pretreatment of cells with inhibitors of polyamine biosynthesis such as DFMO because such pretreatment increases the activity of the transport system (Porter and Jänne, 1987). Such combinations have therefore been suggested as a way of increasing the uptake of MGBG into tumor cells. However, this increased accumulation may not be sufficiently selective to improve the therapeutic value of MGBG without increasing its already distressing toxicity to unacceptable levels.

An interesting series of polyamine analogues containing various N-alkyl substituents have been synthesized by Bergeron and colleagues (Porter and Bergeron, 1988; Bergeron *et al.*, 1988, 1989). Many of these compounds are recognized by the polyamine transport system and are taken up by it. Of particular interest are the bis(ethyl) compounds in which the ethyl groups are present on the terminal nitrogen atoms. It was found that the exposure of L1210 cells to N^1,N^8-bis(ethyl)spermidine reduced cellular ODC levels and decreased the intracellular content of putrescine, spermidine and spermine in L1210 cells (Porter *et al.*, 1987a; Porter and Bergeron, 1988). The corresponding spermine analog, N^1,N^{12}-bis(ethyl)spermine was even more active in repressing ODC and also reduced AdoMetDC content and brought about the almost complete loss of all of the cellular polyamines including putrescine (Porter *et al.*, 1987b). These effects are presumably due to the analogues being recognized as polyamines by the physiological systems which regulate ODC and AdoMetDC levels (Pegg *et al.*, 1988c). However, at least in L1210 cells, the bis(ethyl) derivatives were not able to support cell growth (Porter *et al.*, 1987b). These compounds therefore acted as inhibitors of cell growth in L1210 cells and in a number of other cell lines. They were particularly active as inhibitors of the growth of human undifferentiated large cell lung carcinoma cells and N^1,N^{12}-bis(ethyl)spermine appears to be particularly promising for the chemotherapy of such tumors (Casero *et al.*, 1987, 1989a).

Recently, examination of the properties of a series of compounds differing from N^1,N^{12}-bis(ethyl)spermine in the number of methylene groups has indicated that N^1,N^{14}-bis(ethyl)homospermine is even more active against the growth of L1210 cells (Bergen *et al.*, 1988) and that N^1,N^{11}-bis(ethyl)norspermine is more active against human melanoma lines (C.W. Porter, personal communication). These results raise the possibility that some selectivity towards the growth of certain cell types may be achieved by minor changes in structure. It should be pointed out that the mechanism by which the bis(ethyl) polyamines interfere with cell growth is not at all well understood. Although these compounds do lead to the loss of the normal cellular polyamines and may well affect growth because of this loss, they could also affect the cell in other ways. A dramatic fall in mitochondria and mitochondrial DNA is produced by exposure of L1210 cells to N^1,N^{12}-bis(ethyl)spermine (Bergen *et al.*, 1988).

It is also noteworthy that the mechanism by which the bis(ethyl) spermine derivatives deplete cellular polyamines involves not only the repression of ODC and AdoMetDC which shuts off polyamines synthesis but also involves the excretion of cellular polyamines from the cell. When HT 29 human colon carcinoma cells were treated with N^1,N^{12}-bis(ethyl)spermine there was a rapid and virtually complete efflux of putrescine, N^1-acetylspermidine and spermidine from the cell (Pegg *et al.*, 1989a). This efflux contributes significantly to the fall in intracellular polyamines and accounts for the fact that exposure to N^1,N^{12}-bis(ethyl)spermine is actually more effective in reducing the polyamine content of these cells than the combination of DFMO and MAOEA in amounts sufficient to inhibit ODC and AdoMetDC completely. The mechanism by which the efflux of the natural polyamines occurs is not yet known but it may be due to a physiological process which keeps cellular polyamine levels constant. When the total polyamine concentration in the cell including the analog is expressed on a nitrogen equivalences basis, the total cationic charge of the intracellular polyamines is conserved in cells treated with these bis(ethyl) spermine derivatives (Bergeron *et al.*, 1989).

Another interesting facet of the biochemistry of responses to the bis(ethyl) polyamines is that these compounds are extremely powerful inducers of SAT (Erwin and Pegg, 1986; Libby *et al.*, 1989; Byers and Pegg, 1989; Casero *et al.*, 1989b). Increases of several hundred fold in SAT content occur in a variety of mammalian cells within a few hours of exposure to these compounds. The exact magnitude of the stimulation varies with the cell type but in general the order of potency of inducers of SAT is N^1,N^{11}-bis(ethyl)-norspermine > N^1,N^{12}-bis(ethyl)spermine > N^1,N^{14}-bis (ethyl)homospermine. Particularly large increases in SAT were brought about by N^1,N^{12}-bis(ethyl)spermine in the undifferentiated human large cell lung carcinoma [NCI H82] cells and this increase may be relevant to the sensitivity of these cells to the cytotoxic effects of this agent (Casero *et al.*, 1989b). The induction of SAT may play a role in the rapid loss of polyamines from cells treated with these agents since the acetylation facilitates the conversion of spermine (which is not excreted) to spermidine and putrescine which are lost from the cell (Pegg *et al.*, 1989a). Irrespective of the mechanism of SAT induction, the extremely high levels of SAT which amount to about 1 part in 300 in the NCI H82 cells and about 1 part in 800 in CHO cells treated with these agents (R. Pakala, R.A. Casero and A.E. Pegg, unpublished observations) will facilitate the cloning of cDNAs for this enzyme.

Bitonti *et al.*, (1989a) have demonstrated that a bis(benzyl)polyamine analogue [N,N'-bis{3-[(phenylmethyl)amino]propyl}1,-7-diaminoheptane](MDL 27,695) depletes polyamines in rat hepatoma cells by repression of ODC and AdoMetDC activities and completely abolishes cell proliferation at μM concentrations. It appears that this repression is brought about by the free-amine counterpart since the addition of polyamine oxidase inhibitor which prevents the metabolism of the benzyl derivative the free-amine completely abolished the activity. The lipid soluble bis(benzyl) polyamine analogues which are rapidly taken up by the cells and then efficiently converted into the free-amine therefore represent useful pro-drugs which lead to the loss of polyamine biosynthetic capacity after intracellular metabolism. This metabolism may require the activity of polyamine oxidase and the compounds may have differential effects on cells according to their polyamine oxidase activity (Bitonti *et al.*, 1989a,b). It is also possible that the bis(benzyl) polyamine analogues exert effects related to their binding to intracellular sites normally occupied by polyamines.

A series of fluorinated analogues of putrescine and spermidine has been synthesized at the Merrell Dow Research Institute in Strasbourg, France and tested for their effects on isolated enzymes and cellular polyamine content and growth (Sarhan *et al.*, 1987; Baillon *et al.*, 1988; Dezeure *et al.*, 1988; Seiler *et al.*, 1988; Mamont *et al.*, 1989). As might be expected from the powerful electron withdrawing effect of fluorine atoms the position of the substitution has a striking effect on the biochemistry associated with these compounds. Thus, 2-fluoroputrescine is a good substrate for spermidine synthase with a very high affinity [K_m of about 12 μM] and 2,2-difluoroputrescine has an even greater affinity but acts as an inhibitor at higher concentrations (Dezeure *et al.*, 1988). Similarly, replacement of the hydrogens by fluorines at the methylene of the 7-position increases the affinity for spermine synthase but causes a pH-dependent substrate inhibition of the enzyme. However, *gem*-difluorine substituents reduce the affinity if present at the 6-position and abolish it if present attached to the 2-methylene position (Baillon *et al.*, 1988). The addition of 7,7-difluorospermidine to HTC cells blocks spermine synthesis either directly or via the formation of 6,6-difluorespermine. Combined treatment with 7,7-difluorospermidine and the ODC inhibitors, DFMO or MAP causes a loss of viability of HTC cells and tumor regression in rodents (Mamont *et al.*, 1989).

Another interesting possible use for the fluorine-containing polyamines is in tumor imaging studies using ^{19}F-NMR spectroscopy (Hull *et al.*, 1988). The 2-fluoro- and 2,2-difluoroputrescine derivatives may be of particular value for this purpose since they are converted into the higher polyamines in tumor tissues (Dezeure *et al.*, 1988).

A series of polyamine derivatives consisting of 5 achiral, *gem*-dimethylspermidines and two analogous tetramethylated spermines were examined by Nagarajan *et al.* (1988). None of the compounds was an effective substrate for spermine synthase and only one, 8,8-dimethylspermidine, was a substrate for spermidine/spermine-N^1-acetyltransferase. However, all of the spermidine derivatives and 1,1,12,12-tetramethylspermine were able to support the growth of DFMO-treated SV 3T3 cells. In contrast to the result with 1,1,12,12-tetramethylspermine, 3,3,10,10-tetramethylspermine did not restore growth of these polyamine-depleted cells.

E. Inhibition of Hypusine Formation

The synthesis of hypusine requires the reaction in which the 4-aminobutyl group of spermidine is transferred to the ε-amino group of the lysine acceptor present in eIF-4D followed by a second reaction in which the deoxyhypusine formed is hydroxylated. It is apparent that hypusine formation cannot take place in cells that are prevented from synthesizing spermidine. Several groups have now described cell free systems for studying hypusine formation which can be used to determine the content of the precursor, non-hypusinylated, form of eIF-4D and it is clear that this material accumulates in cells treated with DFMO (Murphey and Gerner, 1987; Park, 1988; Park and Wolff, 1988; Chen and Dou, 1988).

It is possible that the absence of the completed hypusinylated eIF-4D contributes to the reduction of growth in these cells but it should be noted that 5,5-dimethyl spermidine, which cannot take part in the deoxyhypusine formation reaction, was able to support the growth of DFMO-treated SV 3T3 cells (Nagarajan *et al.*, 1988). The accumulation of non-hypusinylated eIF-4D was confirmed in these cells and it was found that none of the other *gem*-dimethylspermidine derivatives described above were substrates for the deoxyhypusine synthase (A.E. Pegg and B. Ganem, unpublished observations). Thus, it appears that hypusine formation is not needed for the growth of these cells but it remains possible that a very small fraction of the eIF-4D is converted to the hypusine-containing derivative and that this is sufficient for the normal growth.

A potent inhibitor of hypusine formation would be a useful tool to answer this question. Although a variety of compounds have been described which are capable of inhibiting this reaction including guazatine (Murphey and Gerner, 1987), a number of spermidine analogues, including *sym*-norspermidine (Park and Wolff, 1988), the *gem*-dimethylspermidines described above, and the alkynyl derivative of spermidine [N-(3-aminopropyl-1,4-diamino-2-butyne] (A.E. Pegg and B. Ganem, unpublished observations) were inhibitory, none of these compounds were sufficiently powerful to be likely inhibitors of hypusine formation. It is known that 1,3-diaminopropane, a product of the reaction is a reasonably good inhibitor (Park and Wolff, 1988) and it is possible that some derivatives of this compound can be developed for this purpose. It may be recalled that for some time before the availability of satisfactory ODC inhibitors such as DFMO, 1,3-diaminopropane was used to study the function of polyamines because of its ability to reduce cellular ODC content (Jänne *et al.*, 1983). It is possible that some of the effects of 1,3-diaminopropane, which is known to lead to inhibition of protein synthesis in mammalian cells, may be due to interference with the production of hypusine.

IV. INTERFERENCE WITH POLYAMINE FORMATION AND FUNCTION IN PLANTS

A. Inhibition of Plant Ornithine and Arginine Decarboxylases

In plants, as well as a number of procaryotic organisms such as bacteria, there is an alternate pathway for the synthesis of putrescine via the decarboxylation of arginine to form agamatine which then yields putrescine. The presence of this second pathway for putrescine makes experimental procedures for inhibiting the biosynthesis of putrescine more complex. A further complicating experimental factor is that changes in the measured activities of these two enzymes in plants may not actually reflect similar changes in the polyamine levels (Birecka *et al.*, 1985). However, as recently discussed by Slocum and Galston (1987), ODC and ADC are differentially induced in plants and serve distinctly different roles in cell division versus growth and stress responses. In spite of these complexities, a number of mostly descriptive experiments have been done in a variety of plant systems with various specific and less-specific inhibitors of polyamine biosynthesis.

Inhibition of ODC in plants and specific tissues has resulted in effects primarily on actively dividing cells. Initial experiments with DFMO in cultured tobacco cells showed inhibition of cell division and subsequent unusual cell enlargement (Berlin & Forche, 1981). The specificity of similar effects of DFMO in blocking early stages of tomato fruit development was shown by reversal with putrescine (Cohen *et al.*, 1982). Other experiments, however, showed that some plants had ODC activities which were insensitive to DFMO, particularly oats and potato (reviewed in Smith, 1985), similar to the insensitivity of the *E. coli* ODC to DFMO noted earlier (Kallio and McCann, 1981). Later experiments, more biochemically oriented, found that DFMO also inhibited the division of tobacco ovary tissue *in vivo* and established a relationship between use of the inhibitor, polyamine biosynthesis inhibition and subsequent polyamine depletion (Slocum and Galston, 1985). Recently DFMO-resistant tobacco cell lines have been isolated with the plants grown from these cells having developmental abnormalities and the expected high putrescine levels (Hiatt and Malmberg, 1988).

ADC, in general, seems to be induced more for vegetative growth and in response to stress conditions. The addition of difluoromethylarginine [DFMA, MDL 71,897 (see Bey *et al.*, 1987)] first described as an irreversible inhibitor of bacterial ADC (Kallio *et al.*, 1981), to oat leaf cells and protoplasts *in vitro* completely inhibited a tremendous rise in ADC and putrescine caused by osmotic stress (Flores and Galston, 1982). Subsequent experiments also showed that inhibition of ADC in such oat protoplasts greatly improved their viability and the improvement was directly related to decreased putrescine and polyamine levels (Tiburcio *et al.*, 1986).

Perhaps one of the most dramatic effects of ADC inhibition was demonstrated by Feirer *et al.* (1984) wherein DFMA completely blocked embryogenesis in wild carrot cells, demonstrating an absolute requirement for *de novo* polyamine biosynthesis in a strict differentiation event. This work was almost simultaneously confirmed by Fienberg *et al.* (1984).

Inhibition of ADC and putrescine synthesis was also shown to modulate specific growth effects on gibberellin-induced growth in pea seedlings (Kaur-Sawhney *et al.*, 1986). In tobacco cells, blocking ADC decreased the levels of putrescine and putrescine-derived alkaloids (nornicotine and nicotine), previously thought to arise from putrescine formed by ODC (Tiburcio and Galston, 1986; Burtin *et al.*, 1989). The lowering of these nicotine alkaloids, in turn, caused promotion of root organogenesis (Tiburcio *et al.*, 1987; Burtin *et al.*, 1989).

A linkage of plant senescence and lowered polyamine levels was indicated by inhibition of ADC with DFMA in carnations, confirming a relationship between ethylene formation and availability of amino propyl groups derived from S-adenosylmethionine (Roberts *et al.*, 1984).

A complexity in the use of DFMA in plants was recently described by Slocum *et al.* (1988) wherein DFMA was found to be metabolized to DFMO by arginase in tobacco and mammalian cells as well. However, the recent synthesis and development of potent agmatine-based ADC inhibitors by the Merrell Dow Research Institute (Bitonti *et al.*, 1987) permits selective inhibition without any subsequent conversion to an ODC inhibitor, since agmatine is not a substrate for arginase.

B. Inhibition of Plant S-adenosylmethionine Decarboxylase and Spermidine Synthase

Less experimental work has been done with S-adenosylmethionine decarboxylase in plants compared to mammalian cells, although the enzyme has been purified from tobacco by use of an affinity column with the AdoMetDC inhibitor MGBG, which was also used to induce a stabilized enzyme in intact cells (Hiatt *et al.*, 1986). MGBG-resistant mutants had previously been isolated from cultured tobacco cells which were subsequently shown to overproduce the AdoMetDC peptide (Malmberg and McIndoo, 1983). AdoMetDC is actually a particularly important enzyme in plants because its substrate AdoMet is a precursor for both ethylene, a senescene-inducing plant hormone, and spermidine or spermine formation. Several recent studies have shown that inhibition of either pathway [ethylene or polyamine] impacts on the other (Slocum and Galston, 1987).

Spermidine synthase has been purified from Chinese cabbage (Sindhu and Cohen, 1984) and was shown to be inhibited by cyclohexylamine *in vivo* in Chinese cabbage protoplasts with a subsequent rapid depletion of spermidine accompanied by a 10-fold increase in decarboxylated AdoMet and increased spermine levels (Greenberg and Cohen, 1985). Cyclohexylamine was also found to deplete spermidine and consequently reduce growth and embryogenesis in cultured aspen cells and wild carrot cells, similar to the effects seen earlier with DFMA (Feirer *et al.*, 1985).

V. INHIBITION OF POLYAMINE FORMATION AND FUNCTION IN MICROORGANISMS EXCLUDING PROTOZOA

A. Phytopathogenic Fungi

Unlike bacteria (discussed below) most yeast and filamentous fungi synthesize putrescine only via ODC. Thus, unlike plants, the phytopathogenic fungi (or any fungi) contain no ADC (Davis and Paulus, 1983; Tabor and Tabor, 1985; Birecka *et al.*, 1986). It was therefore logical to test DFMO against several of these species. Rajam and Galston (1985) demonstrated that at relatively low concentrations, DFMO depletes polyamines and strongly inhibits the growth of four phytopathogenic fungi; *Botrytis* sp., *Botrytis cinera*, *Rhizoctonia solani*, and *Monilinia fructiola*. Subsequently Rajam *et al.* (1985) demonstrated that a spray containing 0.5% DFMO completely protected pinto bean plants infected with bean rust fungus, *Uromyces phaseoli*, a remarkable finding. DFMO also completely inhibited ODC activity in the corn blight fungus *Helminthosporium maydis* and completely arrested sporulation (Birecka *et al.*, 1986). Other studies have shown that DFMO retards mycelial growth of *Verticicillium dahliae* (tomato wilt) at low concentrations, and such effects are clearly reversed by putrescine (Mussell *et al.*, 1987). Rust fungus in broad beans (Walters, 1986), wood decay fungus (Illman, 1987) and wheat rust (Balass and Cohen, 1988) were completely controlled by low concentrations of DFMO, generally sprayed on infected plants.

B. Zoophilic Yeast and Fungi

Relatively little has been done in terms of polyamine enzyme inhibition in these yeast and fungi. Yeast ODC is sensitive to DFMO (Tyagi *et al.*, 1982) and ^{14}C-DFMO has been used to titrate the amout of ODC present in *Saccharomyces cervisiae* and *S. uvarum* (Poso and Pegg, 1983). DFMO will deplete polyamines in three yeast species that are human pathogens, *Candida albicans*, *C. tropicalis* and *C. parapsilosis*, although *C. tropicalis* was significantly more susceptible to the growth inhibitory effects of DFMO than the other two species (Pfaller *et al.*, 1987). Two dermatophyte species of fungi, *Microsporum* and *Trichophyton*, are also quite sensitive to DFMO (Boyle *et al.*, 1988) and are 25-fold more sensitive to α-monofluoromethyldehydroornithine-methylester, a pro-drug inhibitor of ODC, which presumably is taken into the cells at a faster rate than DFMO, analogous to what has been reported in African trypanosomes (Bacchi *et al.*, 1987).

C. Bacteria

A detailed review of the effects of polyamine biosynthesis inhibitors in bacteria was published recently (Bitonti and McCann, 1987). In essence, bacteria have both ODC and ADC, and inhibition of either enzyme alone has been shown to induce the other to maintain a fixed intracellular level of putrescine (Kallio *et al.*, 1982; Bitonti *et al.*, 1982). Even though most bacteria contain only putrescine and spermidine and not spermine, it still has been quite difficult to manipulate polyamine levels in these cells and demonstrate any effects on growth and cell division.

Among the early problems encountered was the fact that many bacteria, e.g. *E. coli, Klebsiella pneumoniae*, have ODC activities that are totally insensitive to DFMO (Kallio and McCann, 1981). Other inhibitors, also synthesized by the Merrell Dow Research Institute, were found to irreversibly inactivate these "unusual" bacterial ODC activities and included α-monofluoromethylputrescine, (MDL 71,864; Kallio *et al.*, 1982) and α-monofluoromethylornithine, (MDL 71,919; Bitonti *et al.*, 1982).

α-Difluoromethylarginine [DFMA, MDL 71,897] was first described as an irreversible inhibitor of bacterial ADC by Kallio *et al.* (1981), and even more potent arginine and agmatine analogs were later synthesized as irreversible inhibitors of both bacterial and plant ADC's (Bitonti *et al.*, 1987).

S-adenosylmethionine decarboxylase has not really been successfully inhibited intracellularly in bacteria although *in vitro* MGBG will inhibit *E. coli* AdoMetDC. Other analogs of AdoMet as well as pentamidine and Berenil® are also *in vitro*, i.e. cell-free, inhibitors of bacterial AdoMetDC (reviewed in Bitonti and McCann, 1987).

Inhibition of spermidine synthase has been as much more successful approach in bacteria. Cyclohexylamine was shown to be a potent inhibitor of *E. coli* and *Pseudomonas aeruginosa* as was AdoDato, a transition-state analog of the spermidine synthase reaction (Pegg *et al.*, 1983). The latter compound had no intracellular effect on bacterial spermidine levels, presumably due to lack of uptake into the cell. However, cyclohexylamine inhibited the growth of both *E. coli* and *Pseudomonas aeruginosa* due specifically to depletion of intracellular spermidine (Bitonti *et al.*, 1982).

Ethambutol, an antitubercular drug, was reported to inhibit spermidine synthase from *Mycobacteria* spp. but not from other bacteria which correspondingly are not killed by the drug (Paulin *et al.*, 1985). It also appears that the ODC, AdoMetDC and ADC enzymes of *Mycobacterium bovis* are essentially insensitive to previously described inhibitors such as DFMO, α-monofluoromethylornithine, MGBG, Berenil®, and DFMA (Paulin *et al.*, 1987).

Overall the best inhibition of growth by polyamine biosynthesis inhibitors in bacteria were seen with combinations of α-monofluoromethylornithine, DFMA, and cyclohexylamine in *E. coli*, *Pseudomonas aeruginosa*, and *Serratia marcescens* i.e., simultaneously inhibiting ODC, ADC, and spermidine synthase (Bitonti *et al.*, 1982; Bitonti and McCann, 1987). Although the growth inhibitory effects were highly significant, they were by no means as pronounced as in a number of fungi and protozoa and thus, it seems unlikely at this point, that the currently available polyamine biosynthesis inhibitors will be useful as potential therapy for bacterial diseases.

VI. INHIBITION OF POLYAMINE FORMATION AND FUNCTION IN PARASITIC PROTOZOA

Of all the microorganisms reviewed, the most extensive use of polyamine inhibitors, primarily DFMO, has been against a number of parasitic protozoa. This subject has been extensively reviewed by Sjoerdsma *et al.* (1984). Sjoerdsma and Schechter (1984), Bacchi and McCann (1987) with the most recent clinical aspects specifically discussed by Sjoerdsma (1987) and Schechter *et al.* (1987) and by Schechter and Sjoerdsma, chapter 14, this book.

A. African Trypanosomes

The initial studies were done in African trypanosomes wherein Bacchi *et al.* (1980) found that DFMO would cure acute infections of the veterinary parasite, *Trypanosoma brucei brucei*, in mice. This remarkable finding rapidly led to clinical studies by the Merrell Dow Research Institute for the dramatic, successful therapeutic use of DFMO in what would have been fatal cases of drug-resistant, late-stage, human sleeping sickness in Africa (Sjoerdsma and Schechter, 1984; Sjoerdsma *et al.*, 1984; Schechter *et al.*, 1987; Schechter and Sjoerdsma, chapter 14, this book).

This initial finding of the effect of DFMO in trypanosomes led to numerous studies on the role of polyamines in trypanosomal growth and metabolism. Although the African trypanosome is very sensitive to the growth inhibitory effects of DFMO, most probably due to its rapid doubling time and lack of spermine (Bacchi and McCann, 1987), the trypanosomal enzyme has similar kinetic sensitivity to DFMO as mammalian enzyme (Bitonti *et al.*, 1985). Furthermore, as in mammalian cells, DFMO enters the trypanosome by passive diffusion and not by facilitated transport (Bitonti *et al.*, 1986a). An intact immune system appears necessary for DFMO cures (Bitonti *et al.*, 1986b), suggesting that inhibition of polyamine metabolism also interferes with the changing of antigenic determinants on the parasite membrane (Bitonti *et al.*, 1988). In addition, polyamine depletion caused by DFMO causes major biochemical and morphological alterations of entire trypanosome populations, shifting them from parasitic blood stream forms to nonreplicating "short stumpy" forms, a specific differentiation event which is independent of a chronologically earlier block of bloodstream form replication (Giffin *et al.*, 1986).

Another quite unusual aspect of polyamine metabolism in trypanosomes is a novel spermidine-containing cofactor that is obligatory for trypanosomatid glutathione reductase activity. This cofactor, trypanothione, is uniquely present in trypanosomatids (Fairlamb *et al.*, 1985). DFMO-induced spermidine depletion greatly reduced the amount of trypanothione in *T. b. brucei*, indicating another likely target for the effects of DFMO on trypanosome infections (Fairlamb *et al.*, 1987). In a related approach, several substituted naphthoquinones have been shown to act as substrates for the trypanosomal glutathione reductase, in place of trypanothione, and subsequently inhibit the enzyme with a consequent reduction of parasite growth (Henderson *et al.*, 1988).

Much recent work has focused on molecular aspects of the trypanosomal enzyme isolated from *T. b. brucei*. Initial attempts to isolate the enzyme from gene amplified mutants were unsuccessful, as the DFMO resistant organisms isolated did not have an amplified ODC gene but contained elevated levels of ornithine that competed with DFMO for ODC binding (Bellofatto *et al.*, 1987). Use of a mouse cDNA for ODC ultimately allowed cloning and sequencing of ODC from *T. b. brucei* (Phillips *et al.*, 1987) and subsequent purification, characterization, and expression in *E. coli* (Phillips *et al.*, 1988). It is apparent that the trypanosomal ODC has a significantly longer half life than the mammalian enzyme (Phillips *et al.*, 1987) and was recently measured to be about 6 hours (Bacchi *et al.*, 1989). Although very homologous to the mammalian enzyme, the trypanosomal ODC lacks a specific 36 amino acid carboxy terminal peptide containing a "PEST", i.e., proline-glutamate-serine-threonine, region which subjects proteins to rapid degradation (Phillips *et al.*, 1987).

Other ODC inhibitors synthesized at or in the Merrell Dow Research Institute (Sjoerdsma and Schechter, 1984; Bey *et al.*, 1987) have been tested against trypanosomes, particularly α-monofluoromethyldehydroornithine methylester (MDL 72,403), the ethyl ester of the same compound (MDL 72,430), and MAP (Bitonti *et al.*, 1985). The α-monofluoromethyldehydroornithine methylester is particularly potent and shows greater synergism than DFMO when combined with suramin (Bacchi *et al.*, 1987a) or 9-deazainosine (Bacchi *et al.*, 1987b).

The other trypanosomal enzymes of the polyamine pathway have been much less studied than ODC. However, Berenil®, pentamidine, and MGBG all inhibit the AdoMetDC of *T. b. brucei* even though such enzymatic inhibition probably has no *in vivo* effect in terms of the antiprotozoal activity of these drugs (Bitonti *et al.*, 1986). The spermidine synthase from *T. b. brucei* is markedly sensitive to cyclohexylamine and AdoDato, although neither has any effect on the trypanosomal infections in mice as might be expected since intracellular putrescine and spermidine levels do not change in the parasites (Bitonti *et al.*, 1984).

B. Other Trypanosomatids

With regard to other trypanosomatids, DFMO will very effectively inhibit the growth of *Leishmania donovani* promastigotes (Kaur *et al.*, 1986), and, unlike the African trypanosome, *Leishmania donovani* cells resistant to DFMO seem to be ODC gene amplified (Coons *et al.*, 1989). However, DFMO has no effect *in vitro* or *in vivo* on the intracellular-infective cousin of *Leishmania* spp., *Trypanosoma cruzi*, the causative agent of American trypanosomiasis (Chagas' disease) (Kierszenbaum *et al.*, 1987a), although DFMO did affect normal mammalian macrophage ability to bind or ingest the parasite (Kierzsenbaum *et al.*, 1987b).

However, DFMA, the previously described specific inhibitor of arginine decarboxylase (an enzyme not found in mammalian cells), markedly inhibited the infection of macrophages by *T. cruzi* and replication of the parasites as well (Kierszenbaum *et al.*, 1987a). This would imply a specific arginine decarboxylase in *T. cruzi* which could be a unique therapeutic target for the now incurable Chagas' disease.

C. Sporozoea

Many of the Sporozoea class of protozoa are sensitive to DFMO including veterinary *Eimeria* spp. parasites (Hanson *et al.*, 1982). However, some of the most interesting findings have been with the *Plasmodia* spp., agents of malaria, the most prevalent parasitic disease in the world today. DFMO will inhibit the replication of the human parasite *P. falciparum* in culture, and will cure exoerythrocytic infections of *P. berghei* in mice and completely block the sporogenous cycle of *P. berghei* in mosquitos (reviewed in Bacchi and McCann, 1987). DFMO, while apparently not curative for erythrocytic infection of *P. berghei*, does appreciably reduce parasitemia, and in combination with a bis(benzyl)polyamine (MDL 27,695) (thought to interfere with some functioning of natural polyamines) will cure *P. berghei* infected mice (Bitonti *et al.*, 1989b).

Pneumocystis carinii, commonly thought to be related to the protozoal class Sporozoea, subclass Coccidia, has recently been described as being related to the fungi because of studies indicating somewhat homologous ribosomal RNA (Edman *et al.*, 1988). Whatever its biological origin, the pneumonia caused by *P. carinii* is a common cause of death in AIDS patients. Although DFMO was initially used successfully in clinical trials on patients (Sjoerdsma *et al.*, 1984; Sjoerdsma, 1987), subsequent experiments in animals (Hughes and Smith, 1984) showed no activity against *P. carinii* until the dose of DFMO was increased by a factor of two (Clarkson *et al.*, 1988). These studies (Clarkson *et al.*, 1988) and *in vitro* cell culture experiments (Cushion *et al.*, 1985), clearly demonstrated that *P. carinii* growth is inhibited by DFMO and its effects can be reversed by putrescine. Although a recent study found no detectable activity of ODC in *P. carinii* (Pesanti *et al.*, 1988), when concentrated pellets of *P. carinii* were utilized to measure polyamine metabolism, DFMO inhibited 90% of the putrescine formation, i.e. ODC activity (Lipschik *et al.*, 1989).

VII. INHIBITION OF POLYAMINE BIOSYNTHESIS AND VIRUSES

Numerous studies have been done with polyamine biosynthesis inhibitors as antiviral agents, particularly against such DNA viruses as herpes simplex, human cytomegalovirus, and vaccina. DFMO, α-monofluoromethyldehydroornithine methylester, and MGBG, among others, have been shown to be active against these targets (reviewed in Tyms and Williamson, 1987 and Tyms *et al.*, 1988). As polyamine biosynthesis is clearly required for viral replication, future studies need to be done to establish the potential use of inhibitors as antiviral therapy.

VIII. COMMENT

It is apparent that inhibitors of ODC and ADC, such as DFMO and DFMA, widely utilized as tools to study the function of polyamines in animal, plants, and microbial cells have or will have definite use as drugs or therapeutic treatments for a variety of clinical, veterinary and phytopathological purposes. It seems very likely, as well, that new approaches to designing compounds that affect polyamine biosynthesis and function, will prove to be similarly useful.

ACKNOWLEDGEMENTS

Research on polyamines in A.E. Pegg's laboratory is supported by grants CA-18138, CA-37606 and GM-26290 from the National Institutes of Health. All compounds mentioned with MDL (Merrell Dow Lepetit) numbers were synthesized by chemists at or in the Merrell Dow Research Institute.

REFERENCES

Anton, D.L. and Kutney, R. (1987). *J. Biol. Chem.* **262,** 2817-2822.

Artamonova, E.Y., Zavalova, L.L., Khomutov, R.M., and Kohomutov, A.R. (1986). *Biorg. Khim.* **12,** 206-212.

Bacchi, C.J. and McCann, P.P. (1987). In "Inhibition of polyamine metabolism: biological significance and basis for new therapies" (P.P. McCann, A.E. Pegg, and A. Sjoerdsma, eds.), pp. 317-344. San Diego, Academic Press.

Bacchi, C.J., Nathan, H.C., Hutner, S.H., McCann, P.P., and Sjoerdsma, A. (1980). *Science* **210,** 332-334.

Bacchi, C.J., Nathan, H.C., Clarkson, A.B., Jr., Bienen, E.J., Bitonti, A.J., McCann, P.P., and Sjoerdsma, A. (1987a). *Am. J. Trop. Med. Hyg.* **36,** 46-52.

Bacchi, C.J., Berens, R.L., Nathan, H.C., Klein, R.S., Elegbe, I.A., Rao, K.V.B., McCann, P.P., and Marr, J.J. (1987b). *Antimicrob. Agents Chemother.* **31,** 1406-1413.

Bacchi, C.J., Garofalo, J., Santana, A., Hannan, J.C., Bitonti, A.J. and McCann, P.P. (1989). *Exp. Parasitol.* **68.** In press.

Baillon, J.G., Mamont, P.S., Wagner, J., Gerhart, F., and Lux, P. (1988). *Eur. J. Biochem.* **176,** 237-242.

Baillon, J.G., Kolb, M., and Mamont, P.S. (1989). *Eur. J. Biochem.* **179,** 17-21.

Balass, O. and Cohen, Y. (1988). *Phytoparasitica.* **16,** 208.

Batchelor, K.W., Smith, R.A., and Watson, N.S. (1986). Biochem. J. **233,** 307-308.

Bellofatto, V., Fairlamb, A.H., Henderson, G.B., and Cross, G.A.M. (1987). *Mol. Biochem. Parasitol.* **25,** 227-238.

Beninati, S., Piacentine, M., Cocuzzi, E.T., Autuori, F., and Folk, J.E. (1988). *Biochim. Biophys. Acta* **952,** 325-333.

Bergeron, R.J., Neims, A.H., McManis, J.S., Hawthorne, T.R., Vinson, J.R.T., Bortell, R., and Ingeno, M.J. (1988). *J. Med. Chem.* **31,** 1183-1190.

Bergeron, R.J., Hawthorne, T.R., Vinson, J.R.T., and Beck, D.E., Jr., Ingeno, M.J. (1989). *Cancer Res.* **49,** 2959-2964.

Berlin, J. and Forche, E. (1981). *Z. Pflanzenphysiol.* **101,** 277-282.

Bey, P., Danzin, C., and Jung, M. (1987). In "Inhibition of polyamine metabolism: biological significance and basis for new therapies" (P.P. McCann, A.E. Pegg, and A. Sjoerdsma, eds.), pp. 1-32. San Diego, Academic Press.

Birecka, H., Bitonti, A.J., and McCann, P.P. (1985). *Plant Physiol.* **79,** 515519.

Birecka, H., Garraway, M.O., Baumann, R.J., and McCann, P.P. (1986). *Plant Physiol.* **80,** 798-800.

Bitonti, A.J. and McCann, P.P. (1987). In "Inhibition of polyamine metabolism: biological significance and basis for new therapies" (P.P. McCann, A.E. Pegg, and A. Sjoerdsma, eds.), pp. 259-275. San Diego, Academic Press.

Bitonti, A.J., McCann, P.P., and Sjoerdsma, A. (1982). *Biochem. J.* **208,** 435-441.

Bitonti, A.J, Kelly, S.E., and McCann, P.P. (1984). *Mol. Biochem. Parasitol.* **13,** 21-28

Bitonti, A.J., Bacchi, C.J., McCann, P.P., and Sjoerdsma, A. (1985). *Biochem. Pharmacol.* **34,** 1773-1777.

Bitonti, A.J., McCann, P.P., and Sjoerdsma, A. (1986a). *Biochem. Pharmacol.* **35**, 331-334.

Bitonti, A.J., Bacchi, C.J., McCann, P.P., and Sjoerdsma, A. (1986b). *Biochem. Pharmacol.* **35**, 351-354.

Bitonti, A.J., Dumont, J.A., and McCann, P.P. (1986c). *Biochem. J.* **237**, 685-689.

Bitonti, A.J., McCann, P.P., and Sjoerdsma, A. (1987a). *Exp. Parasitol.* **64**, 237-243.

Bitonti, A.J., Casara, P.J., McCann, P.P., and Bey, P. (1987b). *Biochem. J.* **242**, 69-74.

Bitonti, A.J., Cross-Doersen, D.E., and McCann, P.P. (1988). *Biochem. J.* **250**, 295-298.

Bitonti, A.J., Bush, T.L., and McCann, P.P. (1989a). *Biochem. J.* **257**, 769-774.

Bitonti, A.J., Dumont, J.A., Bush, T.L., Edwards, M.L., Stemerick, McCann, P.P., and Sjoerdsma, A. (1989b). *Proc. Natl. Acad. Sci. U.S.A.* **86**, 651-655.

Boyle, S.M., Sriranganathan, N., and Cordes, D. (1988). *J. Med. Vet. Mycol.* **26**, 227-235.

Burtin, D., Martin-Tanguy, J., Paynot, M., and Rossin, N. (1989). *Plant Physiol.* **89**, 104-110.

Byers, T.L. and Pegg, A.E. (1989). *Am. J. Physiol.* In press.

Byers, T.L, Kameji, R., Rannels, D.E., and Pegg, A.E. (1987). *Am. J. Physiol.* **252**, C663-C669.

Byers, T.L., Wechter, R., Nuttall, M., and Pegg, A.E. (1989). *Biochem. J.* In press.

Casero, R.A., Jr., Go, B., Theiss, H.W., Smith, J., Baylin, S.B., and Luk, G.D. (1987). *Cancer Res.* **47**, 3964-3967.

Casero, R.A., Jr., Ervin, S.J., Celano, P., Baylin, S.B., and Bergeron, R.J. (1989a). *Cancer Res.* **49**, 639-643.

Casero, R.A., Jr., Celano, P., Ervin, S.J., Porter, C.W, Bergeron, R.J., and Libby, P. (1989b). *Cancer Res.* **49**. In press.

Chen, K.Y. and Dou, Q.P. (1988). *FEBS Lett.* **229**, 325-328.

Clarkson, A.B., Jr., Williams, D.E., and Rosenberg, C. (1988). *Antimicrob. Agents Chemother.* **32**, 1158-1163.

Claverie, N., Wagner, W., Knödgen, B., and Seiler, N. (1987). *Anticancer Res.* **7**, 765-772.

Cohen, E., Arad, S., Heimer, Y.M., and Mizrahi, Y. (1982). *Plant Physiol.* **70**, 540-543.

Coons, T., Hanson, S., Bitonti, A.J., McCann, P.P., and Ullman, B. (1989). *Mol. Biochem. Parasitol.* In press.

Coward, J.K. and Pegg, A.E. (1987). *Adv. Enzyme Regul.* **26**, 107-113.

Cushion, M.T., Stanforth, D., Linke, M.J., and Walzer, P.D. (1985). *Antimicrob. Agents Chemother.* **28**, 796-801.

Danzin, C. and Mamont, P.S. (1987). In "Inhibition of polyamine metabolism: biological significance and basis for new therapies" (P.P. McCann, A.E. Pegg, and A. Sjoerdsma, eds.), pp. 141-164. San Diego, Academic Press.

Davis, R.H. and Paulus, T.J. (1983). *Methods Enzymol.* **94**, 112-116.

Dezeure, F., Sarhan., S., and Seiler, N. (1988). *Int. J. Biochem.* **20**, 1299-1312.

Edman, J.C., Kovacs, J.A., Masur, H., Santi, D.V., Elwood, H.J., and Sogin, M.L. (1988). *Nature* **334**, 519-522.

Elo, H. Mutikainen, I., Alhonen-Hongisto, L., Laine, R., Jänne, J., and Lumme, P. (1986) *Z. Naturforsch.* **41c**, 851-855.
Elo, H. Mutikainen, I., Alhonen-Hongisto, L., Laine, R., and Jänne, J. (1988). *Cancer Lett.* **41**, 21-30.
Erwin, B.G. and Pegg, A.E. (1986). *Biochem. J.* **238**, 581-587.
Fairlamb, A.H, Blackburn, P., Ulrich, P., Chait, B.T., and Cerami, A. (1985). *Science* **227**, 1485-1487.
Fairlamb, A.H., Henderson, G.B., Bacchi, C.J. and Cerami, A. (1987). *Mol. Biochem. Parasitol.* **24**, 185-191.
Feirer, R.P., Mignon, G., and Litvay, J.D. (1984). *Science* **223**, 1433-1435.
Feirer, R.P., Wann, S.R., and Einspahr, D.W. (1985). *Plant Growth Regulation* **3**, 319-327.
Fienberg, A.A., Choi, J.H., Lubich, W.P., and Sung, Z.R. (1984). *Planta.* **162**, 532-539.
Flores, H.E. and Galston, A.W. (1982). *Science* **217**, 1259-1261.
Gawel-Thompson, K. and Greene, R.M. (1988). *J. Cell Physiol.* **136**, 237-246.
Giffin, B.F., McCann, P.P., Bitonti, A.J., and Bacchi, C.J. (1986). *J. Protozool.* **33**, 238-243.
Greenberg, M.L., and Cohen, S.S. (1985). *Plant Physiol.* **78**, 568-575.
Hanson, W.L., Bradford, M.M., Chapman, W.L., Jr., Waits, V.B., McCann, P.P., and Sjoerdsma, A. (1982). *Am. J. Vet. Res.* **43**, 1651-1653.
Heaton, M.A., and Flintoff, W.F. (1988). *J. Cell Physiol.* **136**, 133-139.
Henderson, G.B., Ulrich, P., Fairlamb, A.H., Rosenberg, I., Pereira, M., Sela, M., and Cerami, A. (1988). *Proc. Natl. Acad. Sci. U.S.A* **85**, 5374-5378.
Heston, W.D. and Charles, M. (1988). *Biochem. Pharmacol.* **37**, 2511-2514.
Hiatt, A. and Malmberg, R.L. (1988). *Plant Physiol.* **86**, 441-446.
Hiatt, A.C., McIndoo, J., and Malmberg, R.L. (1986). *J. Biol. Chem.* **261**, 1293-1298.
Hibasami, H., Tanaka, M., Nagai, J., and Ikeda, T. (1980). *FEBS Lett.* **116**, 99-101.
Hibasami, H., Sakurai, M., Maekawa, S., and Nakashima, K. (1987). *Anticancer Res.* **7**, 1213-1216.
Hibasami, H., Sakurai, M., Maekawa, S., Murata, T., and Nakashima, K. (1988a). *Cancer Lett.* **39**, 121-127.
Hibasami, H., Tsukada, T., Maekawa, S., Sakurai, M. Shirakawa, S., and Nakashima, K. (1988b). *Cancer Chemother. Pharmacol.* **22**, 187-190.
Hibasami, H., Kawase, M., Tsukada, T., Maekawa, S., Sakurai, M., and Nakashima, K. (1988c). *FEBS Lett.* **229**, 243-246.
Hughes, W.T. and Smith, B.L. (1984). *Antimicrob. Agents Chemother.* **26**, 436-440.
Hull, W.E., Kunz, W., Port, R.E., and Seiler, N. (1988). *NMR Biomed.* **1**, 11-19.
Hyvönen, T., Alakuijala, L., Andersson, L., Khomutov, A.R., Khomutov, R.M., and Eloranta, T.O. (1988). *J. Biol. Chem.* **263**, 11138-11144.
Illman, B.L. (1987). *Phytopathol.* **77**, 1754.
Ito, H., Hibasami, H., Shimura, K., Nagai, J., and Hidaka, H. (1982). *Cancer Lett.* **15**, 229-235.
Jänne, J., Holtta, E., Kallio, A., and Kapyaho, K. (1983). *Spec. Top. Endocrinol. Metab.* **5**, 227-293.

Kakinuma, Y., Hoshino, K., and Igarashi, I. (1988). *Eur. J. Biochem.* **176,** 409-414.
Kallio, A. and McCann, P.P. (1981). *Biochem. J.* **200**, 69-75.
Kallio, A., McCann, P.P., and Bey, P. (1981). *Biochemistry* **20**, 3163-3166.
Kallio, A., McCann, P.P., and Bey, P. (1982). *Biochem. J.* **204**, 771-775.
Kallio, A., Nikula, P., and Jänne, J. (1984). *Biochem. J.* **218,** 641-644.
Kashiwagi, K., Taneja, S.K, Tabor, C.W., and Tabor, H. (1988). *J. Cell Biol.* **107,** 412a.
Kaur, K., Emmett, K., McCann, P.P., Sjoerdsma, A., and Ullman, B. (1986). *J. Protozool.* **33,** 518-521.
Kaur-Sawhney, R., Dai, Y.-R., and Galston, A.W. (1986). *Plant Cell Physiol.* **27,** 253-260.
Khomutov, R.M., Hyvonen, T., Karvonen, E., Kauppinen, L., Paalanen, T., Paulin, L., Eloranta, T., Pajula, R.-L., Andersson, L.C., and Poso, H. (1985). *Biochem. Biophys. Res. Commun.* **130,** 596-602.
Kierszenbaum, F., Wirth, J.J., McCann, P.P., and Sjoerdsma, A. (1987a). *Proc. Natl. Acad. Sci. U.S.A.* **84,** 4278-4282.
Kierszenbaum, F., Wirth, J.J., McCann P.P., and Sjoerdsma, A. (1987b). *Infect. Immun.* **55,** 2461-2464.
Kolb, M., Danzin, C., Barth, J., and Claverie, N. (1982). *J. Med. Chem.* **25,** 550-556.
Libby, P.R., Bergeron, R.J., and Porter, C.W. (1989). *Biochem. Pharmacol.* **38,** 1435-1442.
Lipschik, G.Y., Kovacs, J.A., Parillo, J.E., and Masur, H. (1989). *FASEB J.* **3,** A1192.
Malmberg, R.L. and McIndoo, J. (1983). *Nature* **305,** 623-625.
Mamont, P.S., Bohlen, P., McCann, P.P., Bey, P., Schuber, F., and Tardif, C. (1976). *Proc Natl. Acad. Sci. U.S.A.* **73,** 1626-1630.
Mamont, P.S., Claverie, N., and Gerhart, F. (1989). *Adv. Exp. Med. Biol.* **250,** 691-706.
Mandel, J.L. and Flintoff, W.F. (1978). *J. Cell Physiol.* **97,** 335-343.
McCann, P.P., Pegg, A.E. and Sjoerdsma, A., (eds.) (1987). "Inhibition of polyamine metabolism: biological significance and basis for new therapies." San Diego, Academic Press.
Murphey, R.J. and Gerner, E.W. (1987). *J. Biol. Chem.* **262,** 15033-15036.
Mussell, H., Osmeloski, J., Wettlaufer, S.H., and Weinstein, L. (1987). *Plant Dis.* **71,** 313-316.
Nagarajan, S., Ganem, B., and Pegg, A.E. (1988). *Biochem. J.* **254,** 373-378.
Nakashima, K. Tsukada, T., Hibasami, H., and Maekawa, S. (1986). *Biochem. Biophys. Res. Comm.* **141,** 718-722.
Nakashima, K., Hibasami, H., Tsukada, T., and Maekawa, S. (1987). *Eur. J. Med. Chem.* **22,** 553-558.
Pajunen, A., Crozat, A., Jänne, O.A., Ihalainen, R., Laitinen, P.H., Stanley, B., Madhubala, R., and Pegg, A.E. (1988). *J. Biol. Chem.* **263,** 17040-17049.
Palfreyman, M.G., Bey, P., and Sjoerdsma A. (1987). *Essays Biochem.* **23,** 28-81.
Park, M.H. (1988). *J. Biol. Chem.* **263,** 7447-7449.
Park, M.H. and Wolff, E.C. (1988). *J. Biol. Chem.* **263,** 15264-15269.
Paulin, L.G., Brander, E.E., and Poso, H.J. (1985). *Antimicrob. Agents Chemother.* **28,** 157-159.

Paulin, L.G., Brander, E.E., and Poso, H.J. (1987). *Experientia* **43**, 174-176.
Pegg, A.E. (1983a). *Methods Enzymol.* **94**, 239-247.
Pegg, A.E. (1983b). *Methods Enzymol.* **94**, 294-297.
Pegg, A.E. (1986). *Biochem. J.* **234,** 249-262.
Pegg, A.E. (1988). *Cancer Res.* **48**, 759-774.
Pegg, A.E. and Coward, J.K. (1985). *Biochem. Biophys. Res. Comm.* **133,** 82-89.
Pegg, A.E. and McCann, P.P. (1982). *Am. J. Physiol.* **243**, C212-C221.
Pegg, A.E. and McCann, P.P. (1988). *ISI Atlas Sci. Biochem.* **1**, 11-18.
Pegg, A.E., Tang, K-.C., and Coward, J.K. (1982). *Biochemistry* **21**, 5082-5089.
Pegg, A.E., Bitonti, A.J., McCann, P.P., and Coward, J.K. (1983). *FEBS Lett.* **155**, 192-196.
Pegg, A.E., Coward, J.K, Talekar, R.R., and Secrist, J.A., III. (1986). *Biochemistry* **25**, 4091-4097.
Pegg, A.E., Secrist, J.A., III, and Madhubala, R. (1988a). *Cancer Res.* **48**, 2678-2682.
Pegg, A.E., Jones, D.B., and Secrist, J.A., III. (1988b). *Biochemistry* **27**, 1408-1415.
Pegg, A.E., Madhubala, R., Kameji, T., and Bergeron, R.J. (1988c). *J. Biol. Chem.* **263,** 11008-11014.
Pegg, A.E., Wechter, R., Pakala, R., and Bergeron, R.J. (1989a). *J. Biol. Chem.* Submitted for publication.
Pegg, A.E., Wechter, R., Poulin, R., Woster, P.M., and Coward, J.K. (1989b). *Biochemistry.* In preparation.
Pegg, A.E., Wechter, R., Coward, J.K., Pakala, A., and Poulin, R. (1989c). *Biochem. J.* In preparation.
Pesanti, E.L., Bartlett, M.S., and Smith, J.W. (1988). *J. Infect. Dis.* **158**, 1137-1138.
Persson, L., Holm, I., Ask, A., and Heby, O. (1988). *Cancer Res.* **48**, 4807-4811.
Pfaller, M.A., Gerarden, T., and Riley, J. (1987). *Mycopathologia* **98**, 3-8.
Phillips, M.A., Coffino, P., and Wang, C.C. (1987). *J. Biol. Chem.* **262**, 8721-8727.
Phillips, M.A., Coffino, P., and Wang, C.C. (1988). *J. Biol. Chem.* **263**, 17933-17941.
Porter, C.W. and Bergeron, R.J. (1988). *Advan. Enzyme Regul.* **27**, 57-82.
Porter, C.W. and Jänne, J. (1987). In "Inhibition of polyamine metabolism: biological significance and basis for new therapies" (P.P. McCann, A.E. Pegg, and A. Sjoerdsma, eds.), pp. 203-248. San Diego, Academic Press.
Porter, C.W. and Sufrin, J.R. (1986). *Anticancer Res.* **6**, 525-542.
Porter, C.W., Berger, F.G., Pegg, A.E., Ganis, B., and Bergeron, R.J. (1987a). *Biochem. J.* **242,** 433-440.
Porter, C.W., McManis, J., Casero, R.A., and Bergeron, R.J., (1987b). *Cancer Res.* **47**, 2821-2825.
Poulin, R., Secrist, J.A. III, and Pegg, A.E. (1989). *Biochem. J.* Submitted for publication.
Poso, H. and Pegg, A.E. (1983). *Biochim. Biophys. Acta* **747**, 209-214.
Rajam, M.V. and Galston, A.W. (1985). *Plant Cell Physiol.* **26**, 683-692.
Rajam, M.V., Weinstein, L.H., and Galston, A.W. (1985). *Proc. Natl. Acad. Sci. U.S.A.* **82**, 6874-6878.

Roberts, D.R., Walker, M.A., Thompson, J.E., and Dumbroff, E.B. (1984). *Plant Cell Physiol.* **25**, 315-322.
Rodrigues, M.T., Palkonyay, L., Sircar, S., Fleurent, J., Dessureault, J., and Weber, J.M. (1987). *Biochem. Biophys. Res. Commun.* **147,** 675-681.
Sarhan, S., Knodgen, B., Gerhart, F., and Seiler, N. (1987). *Int. J. Biochem.* **19,** 843-852.
Schechter, P.J., Barlow, J.L., and Sjoerdsma, A. (1987). In "Inhibition of polyamine metabolism: biological significance and basis for new therapies" (P.P. McCann, A.E. Pegg, and A. Sjoerdsma, eds.), pp. 345-364. San Diego, Academic Press.
Secrist, J.A., III. (1987). *Nucleosides Nucleotides.* **6**, 73-84.
Seiler, N., Sarhan, S., Knodgen, B., and Gerhart, F. (1988). *J. Cancer Res. Clin. Oncol.* **114,** 71-80.
Sherman, M.L., Shafman, T.D., Coward, J.K, and Kufe, D.W. (1986). *Biochem. Pharmacol.* **35,** 2633-2636.
Shim, H. and Fairlamb, A.H. (1988). *J. Gen. Microbiol.* **134**, 807-817.
Shirahata, A., Morohoshi, T., and Samejima, K. (1988). *Chem. Pharm. Bull.* **36**, 3220-3222.
Sindhu, R.K. and Cohen, S.S. (1984). *Plant Physiol.* **74**, 645-649.
Sjoerdsma, A. (1981). *Clin. Pharmacol. Ther.* **30**, 3-22.
Sjoerdsma, A. (1987). In "Proceedings of the 10th International Congress of Pharmacology (Intl. Cong. Ser.)" (M.J. Rand, and C. Raper, eds.), pp. 643-652, Amsterdam, Elsevier.
Sjoerdsma, A. and Schechter, P.J. (1984). *Clin. Pharmacol. Ther.* **35**, 287-300.
Sjoerdsma, A., Golden, J.A., Schechter, P.J., Barlow, J.L., and Santi, D.V. (1984). *Trans. Assoc. Am. Physicians* **97**, 70-79.
Slocum, R.D. and Galston, A.W. (1985). *Plant Cell Physiol.* **26**, 1519-1526.
Slocum, R.D. and Galston, A.W. (1987). In "Inhibition of polyamine metabolism: biological significance and basis for new therapies" (P.P. McCann, A.E. Pegg, and A. Sjoerdsma, eds.), pp. 305-316. San Diego, Academic Press.
Slocum, R.D., Bitonti, A.J., McCann, P.P., and Feirer, R.P. (1988). *Biochem. J.* **255**, 197-202.
Smith, T.A. (1985). *Annu. Rev. Plant Physiol.* **36**, 117-143.
Stanley, B., Holm, I., and Pegg, A.E. (1989). *J. Biol. Chem.* In preparation.
Sunkara, P.S., Baylin, S.B., and Luk, G.D. (1987). In "Inhibition of polyamine metabolism: biological significance and basis for new therapies" (P.P. McCann, A.E. Pegg, and A. Sjoerdsma, eds.), pp. 121-140. San Diego, Academic Press.
Tabor, C.W. and Tabor, H. (1984). *Annu. Rev. Biochem.* **53**, 749-790.
Tabor, C.W. and Tabor, H. (1985). *Microbiol. Rev.* **49**, 81-99.
Tabor, C.W. and Tabor, H. (1987). *J. Biol. Chem.* **262**, 16037-16040.
Tang, K-.C., Pegg, A.E., and Coward, J.K, (1980). *Biochem. Biophys. Res. Commun.* **96**, 1371-1377.
Tiburcio, A.F. and Galston, A.W. (1986). *Phytochemistry* **25**, 107-110.
Tiburcio, A.F., Kaur-Sawhney, R., and Galston, A.W. (1986). *Plant Physiol.* **82**, 375-378.

Tiburcio, A.F., Kaur-Sawhney, R., and Galston, A.W. (1987). *Plant Cell Tissue Organ Culture* **9**, 111-120.

Tyagi, A.K., Tabor, H., and Tabor, C.W., (1982). *Biochem. Biophys. Res. Commun.* **109**, 533-540.

Tyms, A.S. and Williamson, J.D. (1987). In "Inhibition of polyamine metabolism: biological significance and basis for new therapies" (P.P. McCann, A.E. Pegg, and A. Sjoerdsma, eds.), pp. 277-304. San Diego, Academic Press.

Tyms, A.S., Williamson, J.D., and Bacchi, C.J. (1988). *J. Antimicrob. Chemother.* **22**, 403-427.

Ullman, B.L. (1987). *Phytopathology* **77**, 1754.

Wallace, H.M. (1987). *Med. Sci. Res. Biochem.* **15**, 1437-1440.

Walters, D.R. (1986). *New Phytol.* **104**, 613-619.

Williams-Ashman, H.G. (1984). *Mol. Cell. Biochem.* **58**, 51-61.

Williams-Ashman, H.G. and Canellakis, Z.N. (1979). *Perspect. Biol. Med.* **22**, 421-453.

Woster, P.M., Black, A.Y., Duff, K.J., Coward, J.K, and Pegg, A.E. (1989). *J. Med. Chem.* **32**, 1300-1307.

13

CONCEPTS FOR DERIVING SPECIFIC INHIBITORS OF POLYAMINE BIOSYNTHESIS-- HUMAN LUNG CANCER CELLS AS A MODEL SYSTEM

Robert A. Casero, Jr.
Stephen B. Baylin

The Johns Hopkins Oncology Center
Research Laboratories
The Johns Hopkins University School of Medicine
Baltimore, MD 21231

I. INTRODUCTION

The polyamines spermidine (Spd) and spermine (Spm), and their precursor diamine putrescine (Put), are naturally occurring polycationic compounds whose synthesis and metabolism are carefully controlled in all eukaryotic cells. The precise roles which these compounds play in growth and differentiation in normal and neoplastic cells are only now becoming apparent. In this review we will discuss the cellular effects of interference with the polyamine synthesis pathway by a number of agents. We will also discuss how the different major types of human lung cancers, in cell culture, provide a unique setting to study the differential sensitivities of cells to inhibitors of polyamine biosynthesis. We focus upon how these tumor cell systems may be utilized to elucidate more clearly how cells synthesize, regulate and require the polyamines for growth and survival. Finally, we suggest that work with lung cancer cells can help develop the rationale for using inhibitors of polyamine biosynthesis as antineoplastic agents.

II. POLYAMINE METABOLISM

It is not the intent of the current review to examine polyamine metabolism in detail. For such discussions the reader is directed to several excellent recent reviews (Pegg, 1986; Pegg, 1988; Tabor and Tabor, 1984). However, before any discussion of the effects of interference with polyamine metabolism can be entertained, some salient points must be stressed.

The polyamine metabolism pathway (Figure 1) begins with the production of ornithine by the breakdown of arginine. Putrescine is then formed from ornithine by

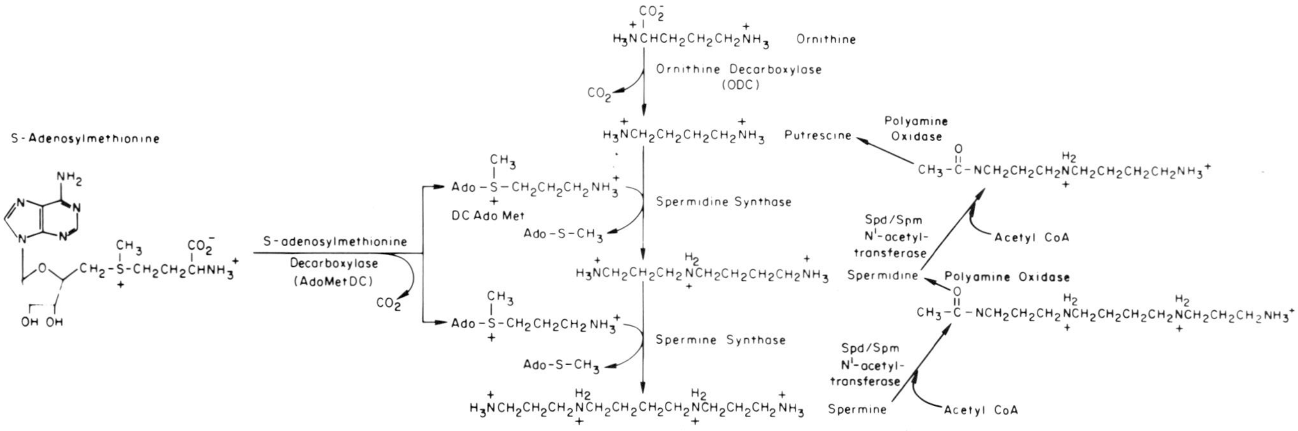

Fig. 1. *The polyamine biosynthetic pathway.*

the decarboxylation reaction catalyzed by the enzyme ornithine decarboxylase (ODC). The polyamines Spd and Spm are formed by the action of successive aminopropyltransferases, Spd synthase and Spm synthase. The aminopropyl donor for these latter reactions is decarboxylated S-adenosylmethionine which is formed from S-adenosylmethionine by the activity of the enzyme S-adenosylmethionine decarboxylase (AdoMetDC). Methylthioadenosine is the second product of each aminopropyltransferase reaction and the adenine and thiol moieties from this metabolite are salvaged for subsequent reutilization by a pathway beginning with methylthioadenosine phosphorylase. Although each of the above biosynthetic reactions are essentially irreversible, the back conversion of the polyamines to putrescine can occur by the actions of a two enzyme pathway. Spd/Spm N^1-acetyltransferase (SSAT), N^1-acetylates both Spd and Spm using acetyl coenzyme A as the acetyl donor. The acetylated polyamines then are substrates for the action of polyamine oxidase (FAD-requiring) which converts Spm to Spd, or Spd to putrescine, and 3-acetamidopropanal.

A. Polyamine Biosynthetic Enzymes

A hallmark of the polyamine pathway is its highly regulated nature. This property is underscored by the fact that 3 enzymes of the metabolic pathway are highly inducible. ODC, a pyridoxal-requiring enzyme, has been shown by several investigators (McCann, 1980; Russell and Snyder, 1969; Jänne and Raina, 1968) to be among the most rapidly and highly inducible mammalian enzymes. ODC responds to several stimuli including hormones, growth factors, and changes in cell density and serum factors (Tabor and Tabor, 1976). The recent cloning of the mouse and human ODC genes have allowed in depth studies of their regulation (Kahana and Nathans, 1984; Berger *et al.*, 1984; McConlogue *et al.*, 1984; Winqvist *et al.*, 1986). The induction of ODC occurs at the level of increased transcription which generally occurs within 1-2 hrs after stimulation (Kahana and Nathans, 1984; Berger *et al.*, 1984). The ODC protein also has a remarkably short half-life of less than 1 hr (Kontula *et al.*, 1984). The short half-life of the protein appears, in part, to result from signals inherent to amino acid sequences in the carboxy terminal region which are not required for enzyme activity, but which modulate stability. Ghoda *et al.* (1989) have recently provided evidence to substantiate such a role for these so called "PEST" sequences (Rogers *et al.*, 1986) by showing that mutations or deletions within one such region result in production of a very stable ODC protein. The above combination of a rapid induction and short half-life for ODC allow for fine tuned modulation of this enzyme activity and a resultant modulation of polyamine levels within cells.

Similar to ODC, AdoMetDC is also highly inducible and has a short half-life. Additionally, AdoMetDC is activated by Put thus coupling its activity to the action of ODC. AdoMetDC contains a covalently bound pyruvate co-factor that apparently is a product of self catalysis (Pegg and McCann, 1982; Shirahata and Pegg, 1986; Pajunen *et al.*, 1988). Recent cloning and sequence analysis of AdoMetDC by Pajunen *et al.* (1988) indicate that rat, bovine and human AdoMetDC undergo similar post-translational modification from a 38 Kd precursor to 32 Kd peptide associated with the formation of a required pyruvate co-factor which remains covalently linked to the active protein.

It is important to note that the polyamine synthetic pathway is essentially controlled by the above modulations in the activities of ODC and AdoMetDC. The subsequent synthetic steps are mediated by the aminopropyltransferases which, by contrast, are constitutive in nature. However, there is evidence that both Spd and Spm synthase are product inhibited and this property adds an additional level of control to the pathway (Coward and Pegg, 1987).

B. Spermidine/Spermine N^1-acetyltransferase and Polyamine Oxidase

The synthetic portion of the polyamine metabolic pathway has been emphasized for the past several years, both with regards to the study of its regulation and, as will be discussed below, with respect to efforts to synthesize compounds which deplete cells of polyamines. However, considerable interest has arisen recently in the study of the polyamine back-conversion pathway (Della Ragione and Pegg, 1982; Della Ragione and Pegg, 1983; Erwin and Pegg, 1986; Hibasami *et al.*, 1986; Libby *et al.*, 1989; Matsui and Pegg, 1982; Pegg *et al.*, 1985; Libby, 1980). As stated above, the back conversion is a two-step process, in which the final step is an oxidation by a action of a constitutive FAD-dependent polyamine oxidase. The rate-limiting step in this pathway appears to be the acetylation step of the polyamines, mediated by the enzyme SSAT. This enzyme is highly inducible in several cell types in response to many of the same stimuli as ODC (Erwin and Pegg, 1986; Libby *et al.*, 1989). In addition to these stimuli, SSAT is highly induced in response to exposure to the polyamines and several structural analogues (Erwin and Pegg, 1986; Hibasami *et al.*, 1986; Libby *et al.*, 1989; Casero *et al.*, 1989). It is speculated that modulation of SSAT activity, in addition to regulation of the biosynthetic enzymes, allows further cellular control of the intracellular concentration of polyamines. It is also possible that the acetylation step which changes polyamine charge, allows these molecules to be moved from one intracellular compartment to another. Finally, the magnitude of the induction of SSAT appears to be cell type specific and may play a role, as will be discussed in detail below, in determining specific cellular sensitivity to inhibitors of polyamine metabolism (Casero *et al.*, 1989).

III. INHIBITORS OF POLYAMINE METABOLISM

As is readily apparent from the above discussion, the polyamine metabolic pathway has several levels of regulation which allow most eukaryotic cell-types to rapidly alter intracellular polyamines during response to multiple stimuli. The full spectrum of these regulations for alterations in cell phenotype is not yet known. However, it is extremely clear that growth of cells, in general, is dependent upon intact polyamine biosynthesis. Therefore, the exquisitively sensitive control points in the polyamine pathway provide intriguing targets for therapeutic intervention in diseases characterized by disorders of proliferation and, especially, cancer.

A. Inhibitors of Ornithine Decarboxylase

Most efforts to synthesize compounds which interfere with polyamine metabolism have been directed to the biosynthetic enzymes, and most specifically, ODC (Danzin *et al,* 1983; Metcalf *et al.*, 1978). The reagents resulting from this work, the patterns of ODC inhibition, and the phenotypic consequences in various cell types have been discussed in several comprehensive reviews (Pegg, 1986; Pegg and McCann, 1982; Sjoerdsma and Schechter, 1984). Of the many inhibitors of ODC examined, 2-difluoromethylornithine (DFMO) has been one of the most effective and extensively studied. Synthesized by Metcalf and colleagues (1978), DFMO is an enzyme activated irreversible inhibitor of ODC which is decarboxylated by the enzyme, forming a highly reactive species which alkylates ODC, presumably near the active site (Metcalf *et al.*, 1978). The use of DFMO has been critical for confirming the essential requirement of polyamines in growth and differentiation of many eukaryotic cell types (Pegg and McCann, 1982). In addition to being a biochemical tool for determining polyamine function, DFMO, as outlined in other chapters in this book, has been the prototypical compound for design of therapeutic agents targeted against polyamine synthesis (Sjoerdsma and Schechter, 1984).

The large number of studies which have indicated that polyamines are absolutely required for growth in virtually all eukaryotic cell types, normal and neoplastic, stimulated intense interest in the potential clinical use of DFMO and similar compounds in the treatment of cancer. This interest and the ensuing experimentation led to clinical trials of DFMO, as will be discussed below (Sjoerdsma and Schechter, 1984; Abeloff *et al.*, 1986; Schechter *et al.*, 1987; Luk *et al.*, 1981). Other inhibitors possessing mechanisms similar to DFMO, have been synthesized. Some of the newer compounds have specific advantages over DFMO with respect to K_i, transport into cells and biological half-life (Danzin *et al.*, 1983; Danzin *et al.*, 1982). One such compound (2R, 5R)-δ-methyacetylenicputrescine (R'RMAP) has been demonstrated to be somewhat more active in animal tumor models than DFMO (Danzin *et al.*, 1983).

B. Inhibitors of S-adenosylmethionine Decarboxylase

In contrast to the many specific inhibitors of ODC, very few inhibitors of AdoMetDC have been discovered, and certainly none with the utility of the enzyme activated irreversible inhibitors of ODC. Methylglyoxal bis(guanylhydrazone) (MGBG) was found to be a potent inhibitor of AdoMetDC (Williams-Ashman and Schenone, 1972). However, many other actions of MGBG including mitochondrial toxicity, stabilization of AdoMetDC, and induction/inhibition of Spd/Spm N^1-acetyltransferase make it difficult to determine what, if any, role the inhibition of AdoMetDC plays in the observed antitumor effects of MGBG (Pegg *et al.*, 1985; Porter and Sufrin, 1986). Secrist and colleagues, and Artamonova *et al.* have reported the synthesis of irreversible inhibitors of AdoMetDC which show promise for elucidating the specific role of this enzyme, but an apparent lack of stability of these compounds limits their usefulness *in vivo* (Artamonova *et al.*, 1986; Secrist, 1987; Pegg *et al.*, 1988).

C. Other Inhibitors of Polyamine Biosynthesis

While inhibitors of ODC have demonstrated the requirements of polyamines for cell proliferation, there are certain therapeutic disadvantages to directing inhibitors of polyamine biosynthesis towards ODC. First, ODC is rapidly induced and has a very short turnover time as discussed above. Consequently, new synthesis of ODC may rapidly overcome drug inhibition by effectively titrating out the inhibitor unless sufficient concentrations of the drug are continually presented. An analogous situation might be expected if similar inhibitors were available for AdoMetDC.

To avoid such problems, some investigators have designed inhibitors of other synthetic steps in the polyamine pathway, with most efforts directed at the aminopropyltransferases, the Spd and Spm synthases. Coward and co-workers have synthesized transition state analogues which act as highly specific inhibitors of the synthases (Coward and Pegg, 1987; Tang *et al.*, 1980). S-adenosyl-1,8-diamino-3-thiooctane (AdoDATO) and S-adenosyl-1,12-diamino-3-thio-9-azadodecane (Ado-DATAD) are specific inhibitors of Spd synthase and Spm synthase respectively, and should be extremely useful in providing tools to determine the precise cellular roles of the individual polyamines, Spd and Spm. The above two inhibitors have already helped verify that the level of each of these polyamines can strongly modulate the specific synthetic steps by which they are derived. For example, studies using AdoDATO have demonstrated that Spd is a major intracellular regulator of AdoMetDC (Coward and Pegg, 1987; Tang *et al.*, 1980).

D. Polyamine Analogues

The facts that the polyamines regulate their own synthesis, and are essential for cell growth, prompted studies to determine the precise structural requirements of these compounds which facilitate these events. Morris and colleagues, in a prokaryotic system (Jorstad *et al.*, 1980), and Porter, Bergeron and colleagues (Porter and Sufrin, 1986; Porter and Bergeron, 1983; Casero *et al.*, 1984), in eukaryotes, defined structural criteria for allowing cellular uptake of polyamine analogues and subsequent substitution of such compounds for polyamines in growth (Porter and Bergeron, 1983; Casero *et al.*, 1984). These studies suggested means by which compounds with structural similarities to the natural polyamines might be used to inhibit polyamine biosynthesis and thereby selectively influence cell processes dependent upon polyamines. Two basic requirements were inherent to this concept. First, the synthesized compounds must be transported into the cell via the polyamine transport system and be able to substitute for the polyamines in modulating the synthetic pathway. Second, to be effective inhibitors, the analogues must not substitute for the polyamines in supporting those functions required for cell growth (Porter and Bergeron, 1983).

A class of compounds which appears to fit the above two criteria are the bis(ethyl)polyamines (Figure 2) synthesized by Bergeron and co-workers (Bergeron *et al.*, 1984; Bergeron *et al.*, 1988; Bergeron *et al.*, 1982). These polyamine analogues have been shown to be rapidly transported into the cell, apparently via the polyamine transport system (Porter and Sufrin, 1986; Porter *et al.*, 1984). Once inside, the analogues down regulate ODC and cause a rapid decrease in all cellular polyamines

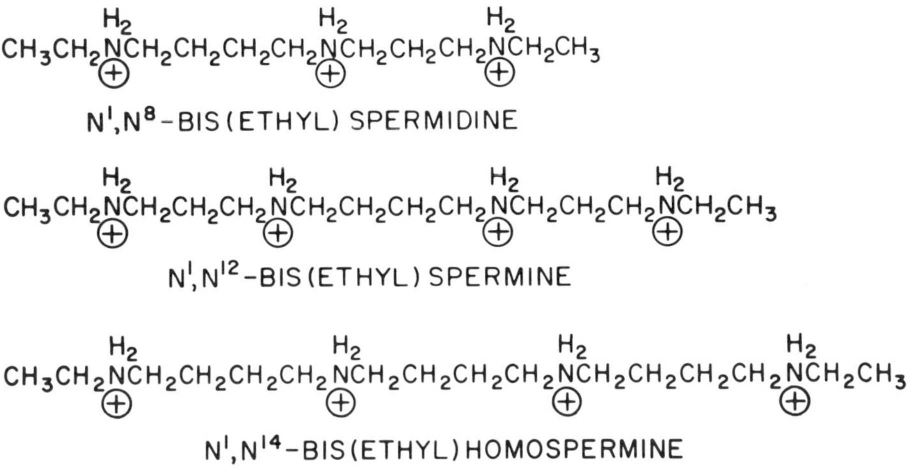

Fig. 2. *The bis(ethyl)polyamine analogues.*

including Spm (Porter *et al.*, 1985; Porter *et al.*, 1986; Porter *et al.*, 1987; Casero *et al.*, 1987; Casero *et al.*, 1989). Interestingly, these bis(ethyl) compounds do not appear to have significant direct effects on ODC when incubated with the enzyme *in vitro* (Porter *et al.,* 1987). Bis(ethyl)spermine (BESpm), in addition to completely down regulating ODC activity in most cell types, also significantly down regulates AdoMetDC activity (Porter *et al.*, 1987; Casero *et al.*, 1989). The observed decreases in enzyme activity are similar to those observed when cells are exposed to exogenous Spd or Spm, indicating that the analogues may regulate the enzymes in a manner similar to the natural polyamines. The mechanisms for this regulation may include classical product inhibition. However, Kameji and Pegg have shown that both the natural polyamines and the analogues may regulate the translation of ODC and AdoMetDC mRNA (Kameji and Pegg, 1987).

The bis(ethyl) compounds are readily transported and reach significant intracellular concentrations in most cell types tested. More importantly, they are apparently unable to functionally substitute for the polyamines with respect to supporting cell growth. These compounds then cause both a decrease in the biosynthesis of polyamines and an inhibition of cellular processes dependent upon polyamines.

IV. HUMAN LUNG TUMOR CELL SYSTEMS

In strategies for developing therapeutically useful inhibitors of polyamine biosynthesis, studies of individual cell systems will be particularly important. It is essential for these efforts, to understand how different cell types depend upon regulatory events in polyamine metabolism for critical functions and to assess cell specific phenotypic consequences of inhibiting these events. For such purposes, our group has found the spectrum of human lung cancer cell types, in culture, to be particularly useful model systems (Baylin and Gazdar, 1987). Before discussing our studies with these cells, it is necessary to review, briefly, the relevant biology of human lung cancer.

Lung cancer is now the leading cause in this country of death from neoplastic disease (see p. 7 of Silverberg and Lubera, 1987). Four major histologic categories of this cancer are recognized in patients. Small cell lung cancer (SCLC) comprises approximately 25% of diagnosed lung cancers. It is distinguished from the other types by its very early metastasizing potential, initially high rate of response to chemotherapy and/or radiotherapy (Mathews and Gazdar, 1981; Goodwin and Baylin, 1982; Carthew *et al,* 1985) and the presence of neuroendocrine phenotypic characteristics (Baylin, 1985). The other major types, large cell, adeno, and squamous lung carcinomas tend to metastasize slower, are amenable to cure by early surgical intervention, but are generally quite resistant to chemotherapy and radiotherapy.

A. Lung Tumor Phenotypes

The clinically observed differences between SCLC and non-SCLC lung cancers are generally preserved in cell culture, including differential responses to irradiation and drugs, and general segregation of neuroendocrine properties with SCLC. However, each of the major types of lung cancer may be related to one another through the complex cell differentiation events which underlie normal epithelial cell renewal in the bronchial epithelium. We and others have used the cell culture systems to explore the biology of these relationships.

Transitions between the SCLC and non-SCLC lung cancer phenotypes are particularly important to understanding the cell biology of lung cancer, to elucidating some of the tumor progression events which produce inevitable therapy resistance in patients with SCLC, and to understanding all of our work with inhibitors of polyamine biosynthesis in lung cancer cells. In patients, mixtures between SCLC and non-SCLC lung cancer phenotypes have been observed, and especially between SCLC and large cell undifferentiated carcinoma (Goodwin *et al.*, 1983). In cell culture, a spectrum of SCLC subtypes has been observed which represents variable movements from SCLC towards the large cell phenotype (Baylin and Gazdar, 1987; Carthew *et al.,* 1985; Gazdar *et al.*, 1985; Carney *et al.*, 1985). Cells drifting from the typical SCLC cells ("classic type") have been designated as "variant" SCLC cells in culture. The "variant" phenotype is closely associated with amplification and high expression of the c- or n-myc oncogenes, loss of neuroendocrine features, and conversion to an irradiation resistant status. In recent gene insertion studies, we have determined that complementation events between high myc gene expression and expression of a mutated *ras* oncogene can convert SCLC to the large cell phenotype (Mabry *et al.,* 1988).

B. Phenotypic Determinants of Sensitivity to Inhibitors of Polyamine Biosynthesis

1. Lung cancer cells and DFMO

We have utilized our emerging understanding of the above lung cancer culture systems to study the biochemical and phenotypic effects of inhibiting polyamine biosynthesis. Each of the best inhibitors of polyamine biosynthesis, whether acting directly or indirectly to inhibit enzyme activity, have exerted a general response of cell cytostasis following polyamine depletion. Such effects of DFMO have been well described in multiple systems, and once the DFMO is removed or polyamines replaced, most cell types readily resume growth at a normal rate (Pegg and McCann, 1982; Danzin *et al.*, 1983). The first report that polyamines were actually required by particular cell types for survival was by Luk, *et al.* (1981) and involved a culture line of human small cell lung carcinoma (SCLC). In these cells, polyamine depletion by DFMO led, first to a decline in growth rate followed by a profound loss of cell viability. Further study indicated that these results were not generalized to the four major types of lung tumor cells discussed above, but were related specifically to the SCLC phenotype. Multiple lines of SCLC, representing the discussed various sub-classifications of this tumor in culture, respond cytotoxically to DFMO treatment. In striking contrast, non-SCLC lines representing the other phenotypes of human lung cancers all respond to DFMO treatment with the usually observed cytostasis or by actually growing slowly through the DFMO-induced blockade of polyamine biosynthesis (Luk *et al.*, 1982; Luk and Casero, 1987).

The above results *in vitro* with SCLC prompted studies in a nude mouse xenograph model which provided further evidence that DFMO treatment could be used to arrest and/or kill SCLC *in vivo* (Luk *et al.*, 1983; Luk *et al.*, 1986). The above cell culture and animal studies led to phase I and phase II clinical trials. DFMO in these patient studies, proved to be relatively well tolerated as compared to other chemotherapeutic agents. However, results with DFMO have not been encouraging in the treatment of human neoplastic disease in general, and SCLC in specific (Sjoerdsma and Schechter, 1984; Abeloff *et al.*, 1986). The reasons for the clinical ineffectiveness of this compound against human SCLC is not well understood, but may be associated with the relatively high concentrations which need to be maintained intracellularly in order to see effects *in vivo* and possibly because of rescue of affected cells by circulating polyamines provided via the diet or other sources. However, as reviewed further below, the discussed transitions between SCLC and the non-SCLC lung cancer types may also play a pivotal role in diminishing SCLC responsiveness to DFMO.

Despite the lack of clinical success of DFMO as an antineoplastic, our initial studies with human lung cancer cells in culture underscored the fact that different phenotypes of human cells can behave quite differently to polyamine depletion. The mechanisms underlying this differential sensitivity are important to understand, not only for continuing to explore the polyamine pathway as a target for cancer chemotherapy--but also, for elucidating the precise functional role(s) of the polyamines. To explore these areas further, we have continued to use the major types of human lung cancer cells in culture to study other inhibitors of polyamine biosynthesis. The results have demonstrated, further, a remarkable interaction between differences in cell phenotype and response to different classes of polyamine synthesis inhibitors. Also, new leads are emerging for use of these inhibitors in cancer treatment.

2. Lung cancer and the bis(ethyl)polyamine analogues

We have focused our recent studies on the bis(ethyl)polyamine analogues synthesized by Bergeron and colleagues (Bergeron *et al.*, 1984; Bergeron *et al.*, 1988; Bergeron *et al.*, 1982). These compounds, as previously discussed, are generally more effective than DFMO in depleting total polyamine pools, including Spm. However, in all cell types initially studied, the general phenotypic response to treatment with bis(ethyl)polyamines was similar to that observed with DFMO (Porter *et al.*, 1985; Porter *et al.*, 1986; Porter *et al.*, 1987), i.e., cell growth was arrested but cytotoxic effects were not observed. However, our results with human lung cancer cells have been strikingly different and also, exactly opposite to those we had obtained previously with DFMO. We initially investigated line NCI-H82 of SCLC, which dies rapidly upon exposure to DFMO (Luk *et al.*, 1982; Casero *et al.*, 1986) and NCI-H157, a large cell undifferentiated lung tumor which is completely resistant to the cytotoxic effects of DFMO and actually continues to grow, at a slow rate, in the presence of 5 mM DFMO (Luk *et al.*, 1982; Casero *et al.*, 1986). Surprisingly, the large cell line was found to be extremely sensitive to each of the polyamine analogues, BESpd, BESpm and bis(ethyl)homo-spermine (BEHSpm) (Casero *et al.*, 1987; Casero *et al.*, 1989). Significant cell loss was observed after 6 days of exposure to any of these agents at a concentration of 10 μM. BESpm was found to be the most active compound and, in addition to reducing intracellular polyamines and ODC activity, also significantly reduced the activity of AdoMetDC.

In contrast to the above results, the DFMO sensitive line of SCLC was found to be particularly resistant to any of the analogues tested. Significant growth inhibition was observed only after extended treatment and no cytotoxicity was observed even at concentrations of 100 μM BESpm (Casero *et al.*, 1989). Similarly, the polyamine pools and biosynthetic enzymes were only modestly inhibited as compared to the sensitive large cell line and other cell lines previously examined (Porter *et al.*, 1985; Porter *et al.*, 1986; Porter *et al.*, 1987; Casero *et al.*, 1987; Casero *et al.*, 1989). It is interesting to note that the large cell line was also more sensitive than the SCLC line, in terms of ODC and AdoMetDC inhibition, following exposure to the natural polyamines, Spd and Spm (Casero *et al.*, 1989).

The contrasting results with the effects of DFMO and the bis(ethyl)polyamines in the large cell and SCLC lines of human lung cancer cells underscore, further, that significant differences exist between cells of different phenotypes with respect to regulation of, and requirements for, the polyamines. They also emphasize that the potential use of polyamine biosynthesis inhibitors and analogues as chemotherapeutic agents may be closely related to the effects of these compounds against specific types of cells. Large cell undifferentiated lung cancer is a virulent and difficult tumor to treat and the bis(ethyl)polyamine compounds are being carefully considered for their treatment potential in this disease.

3. Induction of SSAT in response to the bis(ethyl)polyamines

Work with the polyamine analogues and lung cancer cells is suggesting molecular mechanisms underlying differential cell responsiveness to these compounds. As discussed earlier, most strategies to inhibit polyamine metabolism, have focussed on the forward biosynthetic pathway. In our studies comparing sensitivity of SCLC and non-SCLC lung cancer cells to the bis(ethyl)polyamines we observed that intracellular polyamines were depleted more rapidly in non-SCLC cells than would be expected from decreases of ODC or AdoMetDC activity alone. Pegg and co-workers (Della Ragione and Pegg, 1983; Della Ragione and Pegg, 1982; Pegg *et al.*, 1985), and Libby *et al.* (1988), and others (Hibasami *et al.*, 1986; Takenoshita *et al.*, 1984; Shinki *et al.*, 1985) have investigated the regulation of the back conversion pathway (Figure 1) and effects on polyamine pool levels. It has been observed that SSAT is highly induced in L1210 and L6 myoblast cells treated with the bis(ethyl) compounds and this enzyme is also known to be inducible by the natural polyamines. We, therefore, studied the relationship between the response of human lung cancer cells to the bis(ethyl) compounds and levels of the enzyme activity, SSAT (Casero *et al.*, 1989).

We initially examined the basal and induced levels of SSAT in both bis(ethyl)polyamine insensitive and sensitive lung cancer cell lines. In the insensitive SCLC line, NCI-H82, exposure to either the natural polyamines or the bis(ethyl)polyamines did not result in substantial induction of SSAT at concentrations up to 100 μM. However, when the sensitive large cell line, NCI-H157, was exposed to any of the analogues, a remarkable induction of SSAT was observed which was both dose and time dependent and reached a maximum induction of 1,700-fold with 10 μM BESpm at 48 hr.

Clearly, such a large and rapid increase in activity of the back conversion pathway could lead to a very rapid decrease in intracellular polyamines. It is not yet known if there is a direct correlation with the above SSAT induction and cell sensitivity to the bis(ethyl)polyamines. However, the differential induction of this important polyamine metabolizing enzyme between two lung cancer phenotypes suggests at least one approach to defining the events of underlying differential drug sensitivity.

The molecular mechanisms mediating the induction of SSAT by the bis(ethyl) compounds are already under study. When the sensitive large cell line was treated with actinomycin D in addition to BESpm, 75% of the SSAT induction was inhibited, suggesting that a large portion of the induced activity resulted from increased transcription of SSAT mRNA. Additionally, when cells were treated with cycloheximide to block new protein synthesis 95% of the induction was prevented. These data are consistent with the hypothesis that the polyamine analogues can directly or indirectly effect the expression of the SSAT gene.

That the polyamine analogues (and the polyamines themselves) could directly effect the process of specific gene transcription is an important possibility for which there is growing body of data (Celano *et al.*, 1988; Celano *et al.*, 1989). We have recently shown in a colon cancer cell line that inhibition of polyamine biosynthesis by DFMO leads to specific decreases in expression of important growth related genes, specifically c-myc and c-fos. *In vitro* run-off transcription assays indicate that this decrease in expression occurs at the level of gene transcription (Celano *et al.*, 1988; Celano *et al.*, 1989). Whether the observed differential induction of SSAT in SCLC and non-SCLC cells actually results from differences in transcriptional activation by the bis(ethyl)polyamines will be more readily determined once the human SSAT gene has been cloned.

C. *In vitro* Lung Tumor Progression Model

We have also used the models of human lung cancer in culture to study the relationships between tumor progression events and changes in sensitivity of cells to inhibition of polyamine biosynthesis. These relationships are most important to the clinical use of the inhibitors under study and bear a special link to events in the lung cancer cells we are studying. As discussed previously, SCLC is a cancer which is initially quite sensitive to radiation and chemotherapy, but becomes treatment resistant with time. The non-SCLC lung cancers are relatively refractive to therapy from the outset. Studies in cell culture and in patients (Baylin and Gazdar, 1987) suggest that one component of developing therapy resistance in SCLC cells may be the previously discussed phenotypic shift of SCLC towards the non-SCLC lung cancer phenotypes and, particularly, towards large cell undifferentiated lung cancer. We have studied this possibility, with responses to inhibitors of polyamine biosynthesis, by using our model of *ras* gene induced conversion of SCLC to the large cell phenotype (Mabry *et al.*, 1988).

Although the previously discussed phenotypic differences between "classic" and "variant" SCLC have been well documented, the underlying molecular events responsible for the observed transitions are not known. Since it has been shown that *ras* family gene mutations and increased *ras* expression occur with greater frequency in non-SCLC tumors (Milici *et al.*, 1986; Valenzuela and Groffen, 1986; Nakano *et al.*, 1984; Rodenhuis *et al.*, 1987), we examined the effects of high viral Harvey *ras* gene (v-Ha-ras) expression in SCLC. Using an amphotropic helper-virus system to introduce the v-Ha-ras gene into SCLC, the effects of *ras* expression were examined in two lines of "classic" SCLC and the biochemical "variant" SCLC line, NCI H82, which possesses a highly expressed c-myc amplification (Gazdar *et al.*, 1985). Neither of the classic SCLC lines exhibited changes in growth rate, cloning efficiency or biochemical markers following integration and high expression of the v-Ha-ras gene. Most importantly for the current discussion, the high expression of v-Ha-ras in the "classic" lines had no effect on their extreme sensitivity to polyamine depletion by DFMO.

By direct contrast, the variant line, NCI H82, exhibited profound changes in morphology following expression of v-Ha-ras. The cells developed the typical anchorage dependent growth pattern characteristic of large cell undifferentiated carcinoma and histologic features of this non-SCLC tumor including desmosome formation. Further, there was a significant increase in growth rate of the cells and soft agar cloning efficiency. Most dramatically, the v-Ha-ras expressing variant SCLC cells became entirely refractory to the cytotoxic effects of DFMO, behaving more similar to the non-SCLC lines which actually grow in the presence of this drug (Luk *et al.*, 1982). Furthermore, we have determined that the v-Ha-ras expressing SCLC cells develop a sensitivity to the bis(ethyl)polyamine, BESpm, which is characteristic of large cell lung carcinoma cells as described above (Casero *et al.*, 1987; Casero *et al.*, 1989).

The findings with v-Ha-ras induced changes in SCLC cells emphasize how time dependent changes in cancer cell populations during tumor progression can profoundly alter treatment results. This system with *ras* gene insertion provides a model to manipulate and characterize, further, the precise molecular mechanisms underlying differences between human cell types in regulation of the polyamine biosynthetic pathway and in responding to inhibition of the component steps.

V. SUMMARY

The polyamine metabolic pathway represents a highly regulated system in which the biosynthesis and intracellular concentrations of these biologically important amines are tightly controlled. Although the exact molecular roles in which the polyamines are involved are not precisely known, these compounds are absolutely required for growth in most cell systems and for survival in others. Recent studies indicate that the polyamines may have an important function in the expression of specific growth related genes. As the availability of inhibitors of each of the biochemical steps becomes available it is hoped even more information will be gained with regards to other functions in which Spd and Spm participate.

The polyamine biosynthesis pathway continues to provide a logical target for chemotherapeutic intervention in diseases, and especially in the treatment of cancer. Recent studies of human lung cancer cells have revealed striking cell-type specific sensitivities to various compounds which interfere with polyamine metabolism. That small cell lung carcinoma cells respond cytotoxically to DFMO but not bis(ethyl) polyamine treatment, while large cell undifferentiated lung carcinomas respond cytotoxically to the bis(ethyl)polyamines but not DFMO, underscore the cell specific effects of polyamine depletion and inhibition of individual steps in the synthetic pathway. Additionally, these results suggest the potential for designing polyamine biosynthesis inhibitors which have therapeutic specificity. This important goal will be facilitated by understanding the molecular events which lead to differential sensitivity of human tumor cell types to interference with specific polyamine metabolism steps and by elucidating the specific roles of the polyamines themselves.

REFERENCES

Abeloff, M.D., Rosen, S.T., Luk, G.D., Baylin, S.B., Zeltzman, M., and Sjoerdsma, A. (1986). *Cancer Treat. Rep.* **70**, 843-845.

Artamonova, E.Yu., Zavalova, L.L., Khomutov, R.M. and Khomutov, A.R. (1986). *Bioorg. Khim.* **12**, 206-212.

Baylin, S.B. (1985). In "Peptide hormones as mediators in immunology and oncology" (R.D. Hesch and M.J. Atkinson, eds.), pp. 1-12. New York, Raven Press.

Baylin, S.B. and Gazdar, A.F. (1987). In "Lung carcinomas" (E.M. McDowell, ed.), pp. 346-354. New York, Churchhill Livingstone.

Berger, F.G., Szymanski, P., Read, E., and Watson, G. (1984). *J. Biol. Chem.* **259**, 7941-7946.

Bergeron, R.J., Stolowich, N.J., and Porter, C.W. (1982). *Synthesis* **39**, 689-692.

Bergeron, R.J., Garlich, J.R., and Stolowich, N.J. (1984). *J. Org. Chem.* **49**, 2997-3001.

Bergeron, R.J., Neims, A.H., McManis, J.S., Hawthorne, T.R., Vinson, J.R.T., Bortell, R., and Ingeno, M.J. (1988). *J. Med. Chem.* **31**, 1183-1190.

Carney, D.N., Gazdar, A.F., Bepler, G., Guccion, J.G., Marangos, P.J., Moody, T.W., Zweig, M.H., and Minna, J.D. (1985). *Cancer Res.* **45**, 2913-2923.
Carthew, R.W., Chodosh, L.A., and Sharp, P.A. (1985). *Cell* **43**, 439-448.
Casero, R.A., Jr., Bergeron, R.J., and Porter, C.W. (1984). *J. Cell. Physiol.* **121**, 476-482.
Casero, R.A., Jr., Baylin, S.B., Nelkin, B.D., and Luk, G.D. (1986). *Biochem. Biophys. Res. Commun.* **134**, 572-579.
Casero, R.A., Jr., Go, B., Theiss, H.W., Smith, J., Baylin, S.B., and Luk, G.D. (1987). *Cancer Res.* **47**, 3964-3967.
Casero, R.A., Jr., Ervin, S.J., Celano, P., Baylin, S.B., and Bergeron, R.J. (1989). *Cancer Res.* **49**, 639-643.
Casero, R.A., Jr., Celano, P., Ervin, S.J., Porter, C.W., Bergeron, R.J., and Libby, P.R. (1989). *Cancer Res.* **49,** 3829-3833.
Celano, P., Baylin, S.B., Giardiello, F., Nelkin, B.D., and Casero, R.A., Jr. (1988). *J. Biol. Chem.* **263**, 5491-5494.
Celano, P., Baylin, S.B., and Casero, R.A., Jr. (1989). *J. Biol. Chem.* **264**, 8922-8927.
Coward, J.K. and Pegg, A.E. (1987). *Adv. Enzyme Regul.* **26**, 107-113.
Danzin, C., Bey, P., Schirlin, D., and Claverie, N. (1982). *Biochem. Pharmacol.* **31**, 3871-3878.
Danzin, C., Casara, P., Claverie, N., Metcalf, B.W., and Jung, M.J. (1983). *Biochem. Biophys. Res. Commun.* **116**, 237-243.
Della Ragione, F. and Pegg, A.E. (1982). *Biochemistry.* **21**, 6152-6158.
Della Ragione, F. and Pegg, A.E. (1983). *Biochem. J.* **213**, 701-706.
Erwin, B.G. and Pegg, A.E. (1986). *Biochem. J.* **238**, 581-587.
Gazdar, A.F., Carney, D.N., Nau, M.M., and Minna, J.D. (1985). *Cancer Res.* **45**, 2924-2930.
Ghoda, L., van Daalen Wetters, T.D., Macrae, M., Ascherman, D., and Caffino, P. (1989). *Science* **243**, 1493-1495.
Goodwin, G. and Baylin, S.B. (1982). *Cancer Res.* **42**, 1361-1367.
Goodwin, G., Shaper, J.H, Abeloff, M.D., Mendelsohn, G., and Baylin, S.B. (1983). *Proc. Natl. Acad. Sci. U.S.A.* **80**, 3807-3811.
Hibasami, H., Tsukada, T., Maekawa, S., and Nakashima, K. (1986). *Biochem. Pharmacol.* **35**, 4031-4035.
Jänne, J. and Raina, A. (1968). *Acta Chem. Scand.* **22**, 1349-1351.
Jorstad, C.M., Harada, J.J., and Morris, D.R. (1980). *J. Bacteriol.* **141**, 456-463.
Kahana, C. and Nathans, D. (1984). *Proc. Natl. Acad. Sci. U.S.A.* **81**, 3645-3649.
Kameji, T. and Pegg, A.E. (1987). *J. Biol. Chem.* **262**, 2427-2430.
Kontula, K.K., Torkkeli, T.K., Bardin, C.W., and Jänne, O.A. (1984). *Proc. Natl. Acad. Sci. U.S.A.* **81**, 731-735.
Libby, P.R. (1980). *Arch. Biochem. Biophys.* **203**, 384-389.
Libby, P.R., Bergeron, R.J., and Porter, C.W. (1989). *Biochem. Pharmacol.* **38**, 1435-1442.
Luk, G.D. and Casero, R.A., Jr. (1987). *Adv. Enzyme Reg.* **26**, 91-105.

Luk, G.D., Goodwin, G., Marton, L.J., and Baylin, S.B. (1981). *Proc. Natl. Acad. Sci. U.S.A.* **78**, 2355-2358.

Luk, G.D., Goodwin, G., Gazdar, A.F., and Baylin, S.B. (1982). *Cancer Res.* **42**, 3070-3073.

Luk, G.D., Abeloff, M.D., Griffin, C.A., and Baylin, S.B. (1983). *Cancer Res.* **43**, 4239-4243.

Luk, G.D., Abeloff, M.D., McCann, P.P., Sjoerdsma, A., and Baylin, S.B. (1986). *Cancer Res.* **46**, 1849-1853.

Mabry, M., Nakagawa, T., Nelkin, B.D., McDowell, E., Gesell, M., Eggleston, J.C., Casero, R.A., Jr., and Baylin, S.B. (1988). *Proc. Natl. Acad. Sci. U.S.A.* **85**, 6523-6527.

McCann, P.P. (1980). In "Polyamines in biomedical research" (J.M. Gaugas, ed.), pp. 109-123. New York, Wiley.

McConlogue, L., Gupta, M., Wu, L., and Coffino, P. (1984). *Proc. Natl. Acad. Sci. U.S.A.* **81**, 540-544.

Mamont, P.S., Duchesne, M-C., Grove, J., and Bey, P. (1978). *Biochem. Biophys. Res. Commun.* **81**, 58-66.

Mathews, M.J. and Gazdar, A.F. (1981). In "Lung cancer" (R.B. Livingston, ed.), pp. 283-306. The Hague, Martinus Nijoff.

Matsui, I. and Pegg, A.E. (1982). *Cancer Res.* **42**, 2990-2995.

Metcalf, B.W., Bey, P., Danzin, C., Jung, M.J., Casara, P., and Vevert, J.P. (1978). *J. Am. Chem. Soc.* **100**, 2551-2553.

Milici, A., Blick, M., Murphy, E., and Gutterman, J.U. (1986). *Biochem. Biophys. Res.Commun.* **140**, 699-705.

Nakano, H., Yamamoto, F., Neville, C., Evans, D., Mizuno, T., and Perucho, M. (1984). *Proc. Natl. Acad. Sci. U.S.A.* **81**, 71-75.

Pajunen, A., Crozat, A., Jänne, O.A., Ihalainen, R., Laitinen, P.H., Stanley, B., Madhubala, R., and Pegg, A.E. (1988). *J. Biol. Chem.* **263**, 17040-17049.

Pegg, A.E. (1986). *Biochem. J.* **234**, 249-262.

Pegg, A.E. (1988). *Cancer Res.* **48**, 759-774.

Pegg, A.E. and McCann, P.P. (1982). *Am. J. Physiol.* **243**, C212-221.

Pegg, A.E., Erwin, B.G., and Persson, L. (1985). *Biochem. Biophys. Acta* **842**, 111-118.

Pegg, A.E., Jones, D.B., and Secrist, J.A. (1988). *Biochemistry* **27**, 1408-1415.

Porter, C.W. and Bergeron, R.J. (1983). *Science* **219**, 1083-1085.

Porter, C.W. and Sufrin, J.R. (1986). *Anticancer Res.* **6**, 525-542.

Porter, C.W., Miller, J., and Bergeron, R.J. (1984). *Cancer Res.* **44**, 126-128.

Porter, C.W., Cavanaugh, P.F., Jr., Stolowich, N., Ganis, B., Kelly, E., and Bergeron, R.J. (1985). *Cancer Res.* **45**, 2050-2057.

Porter, C.W., Ganis, B., Vinson, T., Marton, L.J., Kramer, D.L., and Bergeron, R.J. (1986). *Cancer Res.* **46**, 6279-6285.

Porter, C.W., McManis, J., Casero, R.A., and Bergeron, R.J. (1987). *Cancer Res.* **47**, 2821-2825.

Rodenhuis, S., Van de Wetering, M.L., Mooi, W.J., Evers, S.G., Van Zandwijk, N., and Bos, J.L. (1987). *N. Engl. J. Med.* **317**, 929-935.

Rogers, S., Wells, R., and Rechsteiner, M. (1986). *Science* **234**, 364-368.
Russell, D.H and Snyder, S.H. (1969). *Mol. Pharmacol.* **5**, 253-262.
Schechter, P.J., Barlow, J.L.R., and Sjoerdsma, A. (1987). In "Inhibition of polyamine metabolism: biological significance and basis for new therapies" (P.P. McCann, A.E. Pegg, and A. Sjoerdsma, eds.), pp. 345-364. San Diego, Academic Press.
Secrist, J.A., III. (1987). *Nucleosides nucleotides* **6**, 73-83.
Shinki, T., Takahashi, N., Kadofuku, T., Sato, T., and Suda, T. (1985). *J. Biol. Chem.* **260**, 2185-2190.
Shirahata, A. and Pegg, A.E. (1986). *J. Biol. Chem.* **261**, 13833-13837.
Silverberg, E. and Lubera, J. (1987). *CA* **37**, 2-19.
Sjoerdsma, A. and Schechter, P.J. (1984). *Clin. Pharmacol. Ther.* **35**, 287-300.
Tabor, C.W. and Tabor, H. (1976). *Annu. Rev. Biochem.* **45**, 285-306.
Tabor, C.W. and Tabor, H. (1984). *Annu. Rev. Biochem.* **53**, 749-790.
Takenoshita, S., Matsuzaki, S., Nakano, G., Kimura, H., Hoshi, H., Shoda, H., and Nakamura, T. (1984). *Cancer Res.* **44**, 845-847.
Tang, K-C., Pegg, A.E., and Coward, J.K. (1980). *Biochem. Biophys. Res. Commun.* **96**, 1371-1377.
Valenzuela, D.M. and Groffen, J. (1986). *Nucleic Acids Res.* **14**, 843-852.
Williams-Ashman, H.G. and Schenone, A. (1972). *Biochem. Biophys. Res. Commun.* **46**, 288-295.
Winqvist, R., Makela, T.P., Seppanen, P., Janne, O.A., Alhonen-Hongisto, L., Jänne, J., Grzeschik, K.H., and Alitalo, K. (1986). *Cytogenet. Cell. Genet.* **42**, 133-140.

14

THERAPEUTIC UTILITY OF SELECTED ENZYME-ACTIVATED IRREVERSIBLE INHIBITORS

Paul J. Schechter
Albert Sjoerdsma

Merrell Dow Research Institute
Cincinnati, OH 45215

Over the last two decades a number of enzyme-activated irreversible inhibitors have been designed and their biological activity confirmed (Seiler *et al.*, 1978; Sandler, 1980; Sjoerdsma, 1981; Silverman and Hoffman, 1984). This paper will focus on two such inhibitors, chosen because both have demonstrated unique therapeutic properties in clinical studies.

I. EFLORNITHINE

Although first discovered in human semen over 300 years ago (Leeuwenhoek, 1678), the importance of the polyamines, putrescine, spermidine and spermine, in biological function had not been appreciated until recently. The last 20 years have seen a vast expansion of interest in polyamine metabolism and the attribution of a host of functions to these ubiquitous compounds. This explosion of knowledge can be largely traced to the availability of inhibitors of polyamine biosynthesis, allowing depletion of polyamine concentrations and observation as to the physiological consequences of such depletion (Bachrach, *et al.*, 1983; McCann *et al.*, 1987).

Eflornithine (DL-α-difluoromethylornithine; DFMO; Ornidyl®) was designed and synthesized as a specific enzyme-activated inhibitor of ornithine decarboxylase, the enzyme catalyzing the first step in polyamine biosynthesis (Metcalf *et al.*, 1978). The hundreds of publications in which eflornithine has been used testify to the pivotal role its discovery has played in the polyamine field.

The recognition that the polyamines served as growth factors in mammalian cells (Tabor and Tabor, 1964) and reports of elevated polyamine concentrations in body fluids of cancer patients (Russell, 1971), led to the development of eflornithine as a potential antitumor therapy. Despite impressive antitumor effects in a wide range of laboratory *in vitro* and *in vivo* models (Koch-Weser *et al.*, 1981; Pegg and McCann, 1982; Sjoerdsma and Schechter, 1984; Sunkara *et al.*, 1987), clinical studies using eflornithine therapy in cancer patients have, to date, yielded disappointing results (Schechter *et al.*, 1987).

Enzymes as Targets for Drug Design

An alternative potential therapeutic target was presented with the finding by Bacchi *et al.*, (1980) that eflornithine blocks the formation of putrescine from ornithine in trypanosomes and cures rodents infected with *Trypanosoma brucei brucei*. Human African trypanosomiasis, or sleeping sickness, remains endemic in 36 countries of sub-Saharan Africa where the vector, the tsetse fly, occurs (World Health Organization, 1986). The presence of this disease has had dire economic consequences for much of Africa and has led to the depopulation of many parts of this continent since the beginning of the century. Fifty million people are at risk of acquiring this disease which, if untreated, is invariably fatal. Approximately 20,000 new cases are reported annually, however, this number is certainly a gross underestimation. There are two forms of human African trypanosomiasis, the West African or gambiense form and the more virulent East African or rhodesiense form. Two phases distinguish the clinical picture of each of the forms of African trypanosomiasis; an early (hemolymphatic) stage (stage I) and a late (encephalomeningeal) stage (stage II) in which the central nervous system (CNS) is affected.

Current therapy for *T.b. gambiense* trypanosomiasis consists of suramin or pentamidine for the early stage and melarsoprol when the CNS is involved (World Health Organization, 1986). All three drugs must be administered intravenously and are associated with significant side effects. The most severe side effects are encountered with the organic arsenical, melarsoprol, and consist of reactive encephalopathy in about 10% and death in 1-5% of all cases treated (Gutteridge, 1985; World Health Organization, 1986). Of even greater concern is the emergence of trypanosoma strains resistant to melarsoprol. There is, therefore, an obvious need for new, effective and safe treatments for this dread disease.

Treatment with eflornithine of the first patient with *T.b. gambiense* trypanosomiasis occurred in July, 1981 in the Southern Sudan (Van Nieuwenhove *et al.*, 1985). As of December 15, 1988, data on a total of 403 cases have been collected (this total includes all cases treated with any dose for any length of time; Table I). Of these 403 patients, 12 had early stage disease while 391 had evidence of CNS involvement (i.e., trypanosomes found in CSF and/or >5 WBC/mm^3 CSF and/or obvious CNS symptoms and signs present). Two hundred eighty-three of the 391 late-stage cases were considered to be refractory to arsenical therapy, having previously been treated with between one and five full courses of organic arsenical; five were intolerant to arsenicals, having experienced encephalopathy or dermatitis before completing a treatment course, and 103 had not been previously treated with arsenicals. About half of the patients were treated with intravenous and oral eflornithine (usually 400mg/kg/day intravenously for two weeks followed by 300 mg/kg/day orally for three to four weeks); the remaining patients were treated either with intravenous only or oral only eflornithine.

Of the 403 patients treated, 380 could be evaluated (i.e., pre- and post-treatment CSF was obtained and examined). In all evaluable cases, eflornithine treatment was associated with a disappearance of trypanosomes from body fluids (when present prior to treatment) and a decrease in CSF WBC count to or towards normal. These biological changes were generally accompanied by a dramatic amelioration of the characteristic signs and symptoms of sleeping sickness. Several patients were comatose or pre-comatose prior to eflornithine administration and the rapid reversal of this state within a few days of treatment was among the most dramatic responses seen. Figure 1 presents the cumulative percentage of patients relative to CSF WBC count (the most sensitive index of active disease) prior to and immediately after eflornithine treatment as well as at last follow-up. A clear leftward shift associated with treatment is obvious indicating a marked decrease in CSF WBC count in the population. A further smaller shift to the left occurs after the end of treatment.

Twenty-eight patients died during or shortly after treatment. Sixteen of these cases could be evaluated prior to death and all 16 had responded to eflornithine treatment with a disappearance of trypanosomes from CSF and/or decreases in CSF WBC counts. The causes of death were complex and included severe cachexia, intercurrent infections and status epilepticus. Many of these patients were near-terminal prior to starting treatment.

Sixteen cases relapsed (or were re-infected) after eflornithine treatment following an initial positive response. Twelve of these 16 relapsed cases were treated exclusively with oral eflornithine. It is of interest that 12 of these 16 cases were below the age of 15. In three relapsed childhood cases, eflornithine treatment was repeated at double

Table I. Patients with T.b. Gambiense Trypanosomiasis treated with Eflornithine

Total Patients			403
Early-stage disease (Stage I)		12	
Late-stage disease		391	
Arsenical-refractory (Stage II_{AR})	283		
Arsenical-intolerant (Stage II_{AI})	5		
Not previously treated with arsenicals (Stage II)	103		
Route of Eflornithine Administration			
Intravenous plus oral		202	
Intravenous only		76	
Oral only		125	
Evaluable Patients			380
Responders			380
Deaths			
Responders (prior to death)	16	23	
Unevaluable	12		
Relapses (or re-infections)		16	
Following oral only treatment	12		

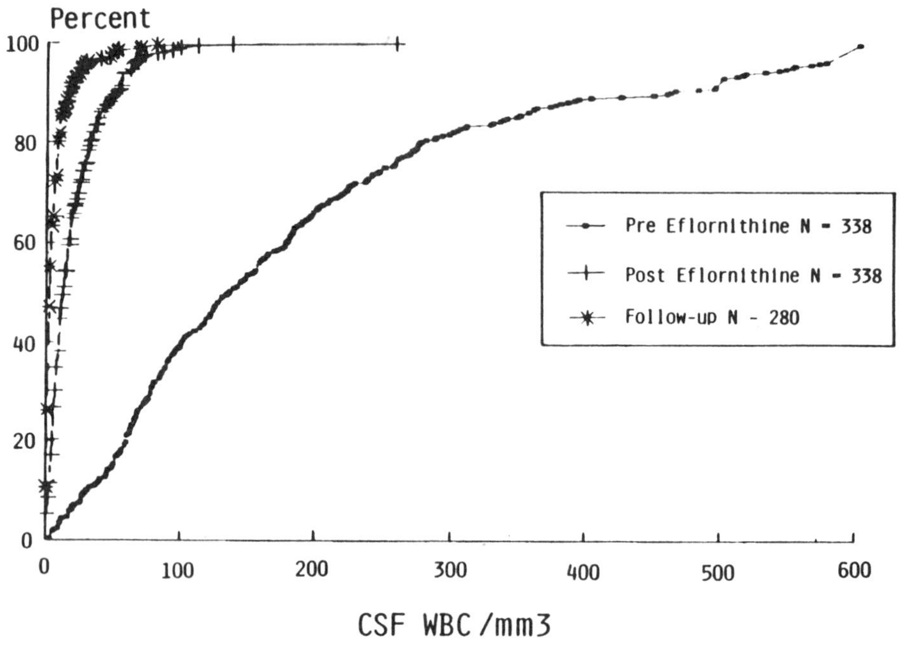

Fig. 1. *Cumulative percentage of patients with CSF counts less than or equal to that shown on X axis.*

the usual intravenous dose (i.e., at 800 mg/kg/day for 14 days) with a repeat positive response in all three.

Side effects noted during eflornithine treatment, or within one week after discontinuation, included diarrhea (usually during oral treatment), anemia, leucopenia, thrombocytopenia, seizures, abdominal pain, patchy hair loss, emesis and dizziness. All side effects were reversible upon discontinuing therapy or decreasing the dose. As most of these effects have been described in association with trypanosomiasis, the relative contributions of the disease and the treatment are conjectural.

To establish a curative effect in trypanosomasis, two-year follow-up of each treated case is necessary (World Health Organization, 1986). As of December 15, 1988, 72 patients have been followed up ≥24 months post-treatment without evidence of relapse, as determined by CSF examination (Table II). An additional 110 cases have been followed for at least one year without relapse.

Thus, eflornithine represents a safe and remarkably effective new treatment for human gambiense trypanosomiasis, including cases refractory to arsenicals, and appears to be curative in this disease. Experience in treating rhodesiense trypanosomiasis is very limited at this time, but it is clear that, as with other trypanocidal agents, eflornithine is less effective against this variety of trypanosomiasis than against the gambiense type. Based on laboratory experiments demonstrating synergism between

Table II. Follow-up (without relapse) of patients with T.b. Gambiense Trypanosomiasis treated with Eflornithine

Time After End of Treatment	Stage of Disease			Totals
	I	II*	II_{AR}	
≥ 36	2	2	9	13
24 - 35.5 mo.	3	8	48	59
18 - 23.5 mo.	0	18	30	48
12 - 17.5 mo.	2	21	39	62
6 - 11.5 mo.	1	11	32	44
3 - 5.5 mo.	2	5	15	22
	10	65	173	248

**Includes IIAI cases (n = 5).*

eflornithine and other agents (Clarkson *et al.*, 1984), clinical studies are currently ongoing in various East African countries to test the clinical efficacy of eflornithine in combination with other agents in rhodesiense trypanosomiasis.

II. VIGABATRIN

γ-aminobutyric-acid (GABA) has become generally accepted as one of the major inhibitory neurotransmitters in the vertebrate nervous system. Pathological alterations in the content or metabolism of GABA in various brain areas has been claimed in a variety of neurological and psychiatric syndromes. Hence, a pharmacological method of specifically altering CNS GABA concentrations would be of considerable interest to neurobiologists and could serve as a potential therapeutic agent. With this in mind, a program was launched in the mid-1970's to rationally design an inhibitor of GABA catabolism, based on the known mechanism of action of the enzyme involved, GABA-transaminase (GABA-T). From this effort emerged vigabatrin (γ-vinyl GABA; DL-4-amino-hex-5-enoic acid; SABRIL®) as an enzyme-activated irreversible inhibitor of GABA-T (Lippert *et al.*, 1977). Vigabatrin proved to be highly specific for this enzyme and demonstrated the necessary *in vitro* and *in vivo* biochemical properties in laboratory studies to increase brain GABA concentrations (Jung *et al.*, 1977; Schechter *et al.*, 1977). These biochemical changes were associated with various behavioral and pharmacological effects, including anticonvulsant activity in laboratory animals (Schechter *et al.*, 1977; Schechter and Trainer, 1978; Palfreyman *et al.*, 1981).

To determine if vigabatrin was biologically active in man, CSF was examined serially before and after oral vigabatrin treatment. From laboratory studies it was evident that increases in vigabatrin-induced concentrations of GABA in CSF correlate with similar increases in brain (Bohlen *et al.*, 1979). Vigabatrin produces dose-related increases in human CSF concentrations of GABA (both free and total), homocarnosine

(GABA-histidine dipeptide) and β-alanine (an alternate substrate for GABA-T); these changes being consistent with inhibition of brain GABA-T activity (Grove *et al.*, 1981; 1983; Schechter *et al.*, 1984).

Vigabatrin therapy was studied in a range of neurological and psychiatric diseases in which a theoretical rationale existed to assume that an increase in CNS GABAergic function could be beneficial. In trials of patients with Huntington's disease (Schechter, 1983; Scigliano *et al.*, 1984), cerebellar ataxia (Bonnet *et al.*, 1986) and dystonia (Carella *et al.*, 1986) vigabatrin treatment, at doses which increased CNS GABA concentrations, was without noticeable effect. On the other hand, in preliminary studies in neuroleptic-induced tardive dyskinesia (Tell *et al.*, 1981; Lambert *et al.*, 1982; Korsgaard *et al.*, 1983; Tamminga *et al.*, 1979; Danion *et al.*, 1984; Stahl *et al.*, 1984), neurogenic spasticity (Rinne, personal communication) and hebophrenic-catatonic schizophrenia (Lambert *et al.*, 1983), vigabatrin appears to have beneficial activity. The clinical efficacy of vigabatrin therapy is clearest in the treatment of epilepsy. The rationale for a role of GABA in epilepsy has been reviewed (Roberts, 1980; Spero, 1982; Schechter, 1986).

Initially, several single-blind, placebo-controlled studies were conducted adding oral vigabatrin to standard therapy in refractory epileptic patients, primarily experiencing complex partial seizures (Browne *et al.*, 1983; Gram *et al.*, 1983; Schechter *et al.*, 1984). In each of these studies vigabatrin was associated with a marked decrease in seizure frequency with excellent drug tolerance. Subsequently, six double-blind, placebo-controlled crossover studies were carried out in Great Britain, Denmark, France and Italy (Rimmer and Richens, 1984; Gram *et al.*, 1985; Loiseau *et al.*, 1986; Remy *et al.*, 1986; Tartara *et al.*, 1986; Tassinari *et al.*, 1987). In five of these studies outpatients with drug-refractory epilepsy, primarily of the complex partial type, were studied. In one center (Remy *et al.*, 1986) very severe cases requiring hospitalization were recruited. The study design was similar across the six studies and consisted of a two period crossover, each period lasting 7-12 weeks, comparing placebo or vigabatrin

Table III. Responses to Vigabatrin in 98 patients with complex partial seizures (with and without secondary generalization) in double-blind studies

		Mean Weekly Seizures			Vigabatrin-induced Decrease in Seizures [n (%)]		
Study	n	Placebo	Vigabatrin	P	75-100%	50-74%	25-49%
Gram et al., 1983	18	2.1	1.3	<0.01	4 (22)	3 (17)	4 (22)
Rimmer & Richens, 1984	22	6.5	3.8	<0.01	6 (27)	8 (36)	5 (23)
Loiseau et al., 1986	17	10.6	6.5	<0.01	1 (6)	8 (47)	2 (12)
Remy et al., 1986	11	4.0	2.5	<0.05	0	1 (9)	7 (64)
Tartara et al., 1986	15	3.0	1.5	<0.02	3 (20)	6 (40)	2 (13)
Tassinary et al., 1987	15	3.6	1.8	<0.02	2 (13)	3 (20)	6 (40)
Totals	98			<0.02	16 (16)	29 (30)	26 (27)

added to pre-existing antiepileptic therapy (Figure 2). The dose of vigabatrin was 3 g/day (1.5 g b.i.d.) except in two studies (Tartara *et al.*, 1986; Tassinari *et al.*, 1987) where lower doses were permitted for patients with lower body weights.

Considering only those patients exhibiting complex partial epilepsy (with or without secondary generalized seizures), representing 98 of the 131 patients studied, vigabatrin therapy was associated with a significant decrease ($p<0.05$) in seizure frequency compared to placebo in each of the six centers (Table III). Forty-six percent of the patients (45 of 98) had at least a 50% decrease in seizure frequency on vigabatrin while 72% (71 of 98) had at least a 25% decrease. The least favorable responses were noted in the one center where severe hospitalized cases were treated. Response of other seizure types to vigabatrin were less impressive overall, but many individual patients had dramatic amelioration of seizure frequency and severity.

Side effects were encountered both with placebo and vigabatrin treatments with somnolence, fatigue, irritability and dizziness among those effects reported more frequently with active drug. In these studies, as well as in other more chronic trials (Pedersen *et al.*, 1985; Browne *et al.*, 1987), these side effects tended to decrease with continued therapy.

Thus, vigabatrin represents an effective and well tolerated new anti-epileptic agent for patients with complex partial seizures; effective even in cases of severe drug-refractory disease.

In summary, eflornithine and vigabatrin are two examples of enzyme inhibitors, rationally designed that are based on known catalytic mechanisms of the enzymes

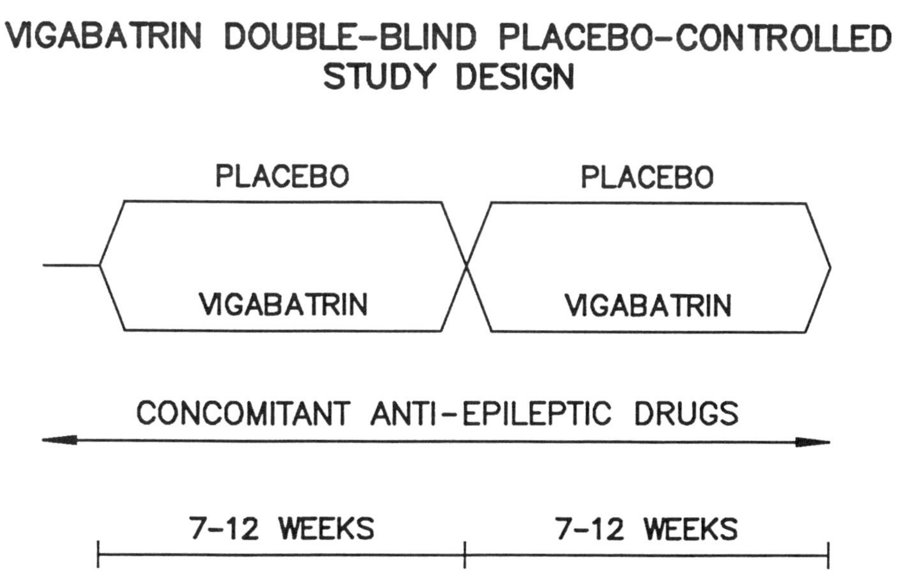

Fig. 2

targeted, which have been demonstrated to possess unique therapeutic properties in the treatment of serious, refractory diseases of man. These isolated examples serve to emphasize the promise of targeting enzymes to design new and useful therapeutic agents for the benefit of man.

REFERENCES

Bacchi, C.J., Nathan, H.C., Hutner, S.H., McCann, P.P., and Sjoerdsma, A. (1980). *Science* **210**, 332-334.

Bachrach, U., Kaye, A., and Chayen, R., eds. (1983). Advances in Polyamine Research, Vol. 4, New York, Raven Press.

Bohlen, P., Hout, S., and Palfreyman, M.G. (1979). *Brain Res.* **167**, 297-305.

Bonnet, A.M., Esteguy, M., Tell, G., Schechter, P.J., Hardenberg, J., and Agid, Y. (1986). *Can. J. Neurol. Sci.* **13**, 331-333.

Browne, T.R., Mattson, R.H., Napoliello, M.J., Penry, J.K., Smith, D.B., Treiman, D.M., and Wilder, B.J. (1983). *Epilepsia* **24**, 521.

Browne, T.R., Mattson, R.H., Penry, J.K., Smith, D.B., Treiman, D.M., Wilder, B.J., Ben-Menachem, E., Napoliello, M.J., Sherry, K.M., and Szabo, G.K. (1987). *Neurology* **37**, 184-189.

Carella, F., Girotti, F., Scigliano, G., Caraceni, T., Joder-Ohlenbusch, A.-M., and Schechter, P.J. (1986). *Neurology* **36**, 98-100.

Clarkson, A.B. Jr., Bienen, E.J., Bacchi, C.J., McCann, P.P., Nathan, H.C., Hutner, S.H., and Sjoerdsma, A. (1984). *Am. J. Trop. Med. Hyg.* **33**, 1073-1077.

Danion, J.-M., Singer, L., Tell, G., and Schechter, P. (1984). *Ann. Med. Psychol.* (Paris) **142**, 101-109.

Gram, L., Klosterskov, P., and Dam, M. (1985). *Ann. Neurol.* **17**, 262-266.

Gram, L., Lyon, B.B., and Dam, M. (1983). *Acta Neurol. Scand.* **68**, 34-39.

Grove, J., Palfreyman, M.G., and Schechter, P.J. (1983). *Clin. Neuropharmacol.* **6**, 223-229.

Grove, J., Schechter, P.J., Tell, G., Koch-Weser, J., Sjoerdsma, A., Warter, J.-M., Marescaux, C., and Rumbach, L. (1981). *Life Sci.* **28**, 2431-2439.

Gutteridge, W.E. (1985). *Br. Med. Bull.* **41**, 162-168.

Jung, M.J., Lippert, B., Metcalf, B.W., Bohlen, P., and Schechter, P.J. (1977). *J. Neurochem.* **29**, 797-802.

Koch-Weser, J., Schechter, P.J., Bey, P., Danzin, C., Fozard, J.R., Jung, M.J., Mamont, P.S., Prakash, N.J., Seiler, N., and Sjoerdsma, A. (1981). In "Polyamines in biology and medicine" (D.R. Morris and L.J. Marton, eds.) pp. 437-453. New York, Marcel Dekker.

Korsgaard, S., Casey, D.E., and Gerlach, J. (1983). *Psychiatry Res.* **8**, 261-269.

Lambert, P.A., Cantiniaux, P., Chabannes, J.-P., Tell, G.P., Schechter, P.J., and Koch-Weser, J. (1982). *Encephale* **8**, 371-376.

Lambert, P.A., Chabannes, J.-P., Cantiniaux, P., Schechter, P.J., and Tell, G.P. (1983). *Encephale* **9**, 145-149.

Leeuwenhoek, A. (1678). *Philos. Trans. R. Soc.* London **12**, 1040.

Lippert, B., Metcalf, B.W., Jung, M.J., and Casara, P. (1977). *Eur. J. Biochem.* **74**, 441-445.

Loiseau, P., Hardenberg, J.P., Pestre, M., Guyot, M., Schechter, P.J., and Tell, G.P. (1986). *Epilepsia* **27**, 115-120.

Metcalf, B.W., Bey, P., Danzin, C., Jung, M.J., Casara, P., and Vevert, J.P. (1978). *J. Am. Chem. Soc.* **100**, 2551-2553.

McCann, P.P., Pegg, A.E., and Sjoerdsma, A., eds. (1987). "Inhibition of polyamine metabolism. Biological significance and basis for new therapies," Orlando, Academic Press.

Palfreyman, M.G., Schechter, P.J., Buckett, W.R., Tell, G.P., and Koch-Weser, J. (1981). *Biochem. Pharmacol.* **30**, 817-824.

Pedersen, S.A., Klosterskov, P., Gram, L., and Dam, M. (1985). *Acta Neurol. Scand.* **72**, 295-298.

Pegg, A.E. and McCann, P.P. (1982). *Am. J. Physiol.* **243**, C212-C221.

Remy, C., Favel, P., Tell, G., Hardenberg, J., and Schehter, P.J. (1986). *Boll. Leg. Ital. Epilepssia* **54/55**, 241-243.

Rimmer, E.M. and Richens, A. (1984). *Lancet* **1**, 189-190.

Roberts, E. (1980). In "Antiepileptic drugs: mechanisms of actions" (G.H. Glaser, J.K. Penry, and D.M. Woodbury, eds.), pp. 667-713. New York, Raven Press.

Russell, D.H. (1971). *Nature* [New Biol.] **233**, 144-145.

Sandler, M., ed. (1980). Enzyme inhibitors as drugs, London, MacMillan Press.

Schechter, P.J. (1983). *J. Pharmacologie* **14**, 258-259.

Schechter, P.J. (1986). In "New anticonvulsant drugs" (B.S. Meldrum and R.J. Porter, eds.), pp. 265-275. London, John Libbey.

Schechter, P.J. and Trainer, Y. (1978). In "Enzyme-activated irreversible inhibitors" (N. Seiler, M.J. Jung, and J. Koch-Weser, eds.), pp. 149-162. Amsterdam, Elsevier.

Schechter, P.J., Trainer, Y., Jung, M.J., and Bohlen, P. (1977). *Eur. J. Pharmacol.* **45**, 319-328.

Schechter, P.J., Hanke, N.F.J., Grove, J., Huebert, N., and Sjoerdsma, A. (1984). *Neurology* **34**, 182-186.

Schechter, P.J., Barlow, J.L.R., and Sjoerdsma, A. (1987). In "Inhibition of polyamine metabolism. Biological significance and basis for new therapies" (P.P. McCann, A.E. Pegg, and A. Sjoerdsma, eds.), pp. 345-364. Orlando, Academic Press.

Scigliano, G., Giovannini, P., Girotti, F., Grassi, M.P., Caraceni, T., and Schechter, P.J. (1984). *Neurology* **34**, 94-96.

Seiler, N., Jung, M.J., and Koch-Weser, J., eds. (1978). "Enzyme-activated irreversible inhibitors," Amsterdam, Elsevier.

Silverman, R.B. and Hoffman, S.J. (1984). *Med. Res. Rev.* **4**, 415-447.

Sjoerdsma, A. (1981). *Clin. Pharmacol. Ther.* **30**, 3-22.

Sjoerdsma, A. and Schechter, P.J. (1984). *Clin. Pharmacol. Ther.* **35**, 287-300.

Spero, L. (1982). *Lancet* **2**, 1319-1322.

Stahl, S.M., Thornton, J.E., Simpson, M.L., and Berger, P.A. (1984). *Neurology* **34** (Suppl. 1), 85-86.

Sunkara, P.S., Baylin, S.B., and Luk, G.D. (1987). In "Inhibition of polyamine metabolism. Biological significance and basis for new therapies" (P.P. McCann, A.E. Pegg, and A. Sjoerdsma, eds.), pp. 121-140. Orlando, Academic Press.
Tabor, H. and Tabor, C.W. (1964). *Pharmacol. Rev.* **16**, 245-300.
Tamminga, C.A., Crayton, J.W., and Chase, T.N. (1979). *Arch. Gen. Psychiatry* **36**, 595-598.
Tartara, A., Manni, R., Galimberti, C.A., Hardenberg, J., Orwin, J., and Perucca, E. (1986). *Epilepsia* **27**, 717-723.
Tassinari, C.A., Michelucci, R., Ambrosetto, G., and Salvi, F. (1987). *Arch. Neurol.* **44**, 907-910.
Tell, G.P., Schechter, P.J., Koch-Weser, J., Cantiniaux, P., Chabannes, J.-P., and Lambert, P.A. (1981). *New Engl. J. Med.* **305**, 581-582.
Van Nieuwenhove, S., Schechter, P.J., Declercq, J., Bone, G., Burke, J., and Sjoerdsma, A. (1985). *Trans. R. Soc. Trop. Med. Hyg.* **79**, 692-698.
World Health Organization. (1986). Epidemiology and control of African trypanosomiasis. Report of a WHO expert committee. *WHO Tech. Rep. Ser.* **739**, 1-127.

15

INTRA-BLOOD-BRAIN-BARRIER IgG SYNTHESIS IN HIV[1]

Wallace W. Tourtellotte
Elyse Singer
Karl Syndulko

Neurology Service
VA Medical Center - West Los Angeles
Wadsworth Division
and
Department of Neurology
School of Medicine
University of California at Los Angeles
Los Angeles, CA 90024

Paul Shapshak
Lionel Resnick

Departments of Dermatology and Pathology
Mt. Sinai Medical Center
Miami, FL 33140

I. HIV NEUROLOGICAL DISEASE

HIV is a recently recognized pathogenic retrovirus which causes progressive deterioration of the immune system, and is associated with a wide variety of neurologic and neuropsychiatric syndromes (Dix and Bredesen, 1988; Helweg-Larsen *et al.*, 1986; Levy *et al.*, 1985; Navia *et al.*, 1986; Snider *et al.*, 1983; So *et al.*, 1986), either as primary manifestations of HIV-1 infection or as secondary consequences of immune activation or suppression. HIV infection has reached epidemic proportions in many parts of the world, and subsequently HIV-related neurologic disease has assumed increasing importance.

[1]*Supported by USAMRDC Log Number 6132002 (86PP6854), University AIDS Task Force (86LA,015), and the Research Service of the Veterans Administration Central Office*

The most common neurologic manifestation of HIV-1 is reported to be a subacute encephalopathy characterized by cognitive deficits, motor slowing, and behavioral changes, the end stage of which has been termed the "AIDS Dementia Complex (ADC)" (Levy and Bredesen, 1988; Navia *et al.*, 1986; Price *et al.*, 1986). The spectrum of HIV-1 encephalopathy ranges from mild cognitive abnormalities on neuropsychologic testing in otherwise asymptomatic individuals (Grant *et al.*, 1987) to a bedridden state marked by global dementia, severe hypokinesis, mutism, incontinence, and seizure (Levy and Bredesen, 1988; Navia *et al.*, 1986; Price *et al.*, 1986). Evidence of CNS disease has been reported in up to 90% of HIV-1 infected individuals at autopsy series (Levy and Bredesen, 1988). The exact percentage of infected persons who have more subtle manifestations of neuropsychiatric illness early in the course of their infection is unknown, and is the subject of a number of on-going longitudinal studies (Claypoole *et al.*, 1989; Grant *et al.*, 1989; Miller *et al.*, 1989; Riccio *et al.*, 1989; Selnes *et al.*, 1989a; Selnes *et al.*, 1989b; Singer *et al.*, 1989; Stern *et al.*, 1989; Syndulko *et al.*, 1989).

The detailed description and search for the etiopathogenesis of neurologic features of HIV-1 infection has been reviewed in a number of current papers (Johnson *et al.*, 1988; Levy and Bredesen, 1988; Michaels *et al.*, 1988; Perry and Marotta, 1987; Price and Brew, 1988; Price *et al.*, 1988). In this chapter we will focus on research on the use of cerebrospinal fluid evaluation of HIV-1 infected patients at different stages of disease to evaluate and monitor the CNS effects of HIV-1. In particular, we will discuss current concepts of intra-blood-brain-barrier (I-BBB) IgG synthesis associated with HIV-1 infection, briefly review its scientific basis, and its potential use to initiate and monitor enzyme- and non-enzyme-based treatments for ADC.

II. INTRA-BLOOD-BRAIN-BARRIER IgG SYNTHESIS

A. Background Studies in MS

The CSF provides a unique, relatively non-invasive, window into the neurochemical and neuroimmunological by-products of brain function and disease. The major impetus for our use of CSF analysis in HIV-1 is the considerable experience we have gained in analyzing CSF in MS and we draw extensively on that experience in guiding our research in detecting and understanding HIV-1 effects on the CNS (Tourtellotte *et al.*, 1980a, 1984a, 1985a, 1985b).

The predominant characteristic of the CSF profile in MS is an elevated IgG value in the absence of evidence for abnormal leakage across the BBB (Tourtellotte *et al.*, 1980a). Various hypotheses have been postulated regarding the source of the increased CSF IgG. Evidence from studies done by my laboratory (Tourtellotte, 1970a, 1970b, 1971, 1972, 1985; Tourtellotte and Ma, 1978; Tourtellotte *et al.*, 1975, 1980b, 1980c, 1980d, 1984b) suggest that the CNS-CSF compartment, surrounded by its BBB to protein, can perform as an immunologic unit and can synthesize IgG independently of the systemic immune system (Tourtellotte and Parker, 1966, 1967). We have termed this I-BBB IgG synthesis (Tourtellotte and Ma, 1978).

In healthy individuals the source of CNS-CSF IgG is exclusively the blood. The blood is also a source of CNS-CSF IgG when there is damage to the BBB. In this case, the elevated IgG concentration is accompanied by proportional elevation in albumin, which can be used to compensate for BBB leakage. It is postulated that in MS extra- and per-vascular lymphoplasmyocytes synthesize IgG within the CNS in multifocal regions probably at edges of active plaques of demyelination (Esiri, 1977; Prineas and Wright, 1978; Simpson *et al.*, 1969; Tourtellotte *et al.*, 1971). After IgG is secreted into the extravascular space of the CNS by these immune cells, it sinks by diffusion into the CSF, a fast moving 500 dl/day lacuna of CNS extracellular space, to be mechanically transported into blood via macropinocytes of the arachnoid villi. With provisos, the presence of CSF oligoclonal IgG may serve as evidence of *de novo* CNS IgG synthesis (Tourtellotte and Ma, 1978). In general, the same arguments and measurements may be applied to any source of elevated IgG within the CNS, including HIV-1.

B. I-BBB IgG Synthesis in HIV-1 Infection

Our group provided the first evidence that I-BBB IgG synthesis was elevated in AIDS and ARC (Aids Related Complex) (Resnick *et al.*, 1985). Using the Tourtellotte formula for I-BBB IgG synthesis rate (Tourtellotte, 1970b; Tourtellotte and Ma, 1978; Tourtellotte *et al.*, 1980a) and isoelectric focusing to detect oligoclonal IgG bands in CSF (Delmotte and Gonsette, 1977; Kjellen and Vesterberg, 1974; Laurenzi and Link, 1978), a group of patients with AIDS and ARC were identified who had abnormal I-BBB IgG synthesis. A high percentage of these patients showed the presence of HIV-1 specific antibody levels that were higher in CSF than in autologous serum. These results were compatible with local infection and active viral replication within the BBB. It was speculated that quantitative analysis of HIV-1 specific IgG within the BBB may become the basis of a diagnostic test for HIV-1 infection in the CNS and also the basis for objective evaluation of potential therapies of HIV-1 induced neurologic diseases (Resnick *et al.*, 1985).

A subsequent paper from our group confirmed and extended the findings in a broader range of HIV-1 infected individuals that included asymptomatic HIV-1 seropositives (ASP), ARC and AIDS (Resnick *et al.*, 1988). In this study, the I-BBB IgG synthesis rate was elevated in 39/52 (78%) of HIV-1 seropositive individuals. The frequency of abnormal I-BBB IgG synthesis rate was higher in non-AIDS HIV-1 infected individuals (28/32 or 88%) than in AIDS (11/20 or 55%). The frequencies of abnormal I-BBB IgG synthesis rates were substantially the same in those patients with (72%) and without (78%) neurologic manifestations. Although our laboratory performed the CSF analyses in both of the above studies, the patients were obtained from a variety of other centers.

The findings of elevated I-BBB IgG synthesis rate and thus early penetration of HIV of the CNS have been confirmed in a number of laboratories, although the exact frequency of abnormal CSF has varied (Appleman *et al.*, 1988; Elovaara *et al.*, 1988; Luer *et al.*, 1988; Marshall *et al.*, 1988). Elevated I-BBB IgG synthesis rate can be found in even newly seroconverted individuals, and at a relatively high rate in early

HIV-1 infection (Marshall *et al.*, 1988), further supporting the view that HIV-1 penetrates the CNS very early in the course of infection.

We have initiated a longitudinal study of CSF changes in HIV-1 infected individuals to further explore the time course and relationship to clinical and performance changes of I-BBB IgG synthesis by rate and oligoclonal bands. Preliminary reports of the baseline evaluations have fully confirmed our earlier findings (Singer *et al*, 1988, 1989). Our most recent unpublished results show that I-BBB IgG synthesis rate by the Tourtellotte formula was abnormal in 73% of ASPs and 80% of ARCs. Abnormal I-BBB IgG synthesis calculated by the IgG index (Tibbling *et al.*, 1977), Reiber and Felgenhauer formula (Reiber, 1980; Reiber and Felgenhauer, 1987), and Schuller and Sagar formula (Schuller and Sagar, 1983) showed almost identical results. The frequency of unique CSF IgG (oligoclonal) bands was 59% for ASP and 57% for ARC in these groups, with slightly lower frequencies for bands more intense in the CSF than in the serum. When evidence for abnormal I-BBB IgG synthesis was combined from the rate formula and bands, the percentage of abnormal CSFs went up to 88% for ASPs and ARCs combined. In contrast to the very high percentage of abnormal IgG findings in the CSF, only 7% of ASPs and 30% of ARCs showed direct evidence for presence of HIV-1 in the CNS by culture. 20% of the ASPs and 37% of the ARCs showed p24 antigen (≥10 pg/ml) in the CSF. Furthermore, we have found statistically significant ($p<0.01$) higher mean I-BBB IgG synthesis rate in HIV-1 infected individuals with neurologic dysfunction.

C. Potential Use of I-BBB IgG Synthesis to Monitor Treatment of HIV-1

The presence and expected increase in frequency of HIV-1 related neurological disease has three important implications for drug development and the design of clinical trials. First, an effective antiviral will probably have to penetrate the CNS, which may otherwise act as a reservoir for infection and may permit the progression of neurological sequalae even in the absence of continued systemic progression. Second, evaluation of HIV-related neurologic disease, and particularly performance based evaluations, will increasingly become an important end point in evaluating efficacy of putative treatments for HIV-1 infection (Price *et al.*, 1988). Many of those new drugs inhibit reverse transcriptase and other key enzymes needed for viral replication. Of considerable importance will be their ability to penetrate the blood brain barrier to attack this important viral reservoir. Third, since damage in the CNS is often not reversible, the treatment of CNS consequences may have to be initiated before clinical signs and symptoms of neurologic disease are evident. In the absence of clinical and neuropsychological abnormalities, a rational basis for initiating and following early treatment will most likely fall on CSF markers of HIV-1 penetration of the CNS. Of these, I-BBB IgG synthesis is a prime candidate. It has been used successfully in MS to monitor immunosuppressive treatments (Baumhefner *et al.*, 1979, 1986; Tourtellotte *et al.*, 1976, 1980b, 1980c, 1980d, 1986). The past experience with other treatable CNS infections, such as neurosyphilis, also provide ample precedent for using laboratory indices of infection to initiate and monitor treatment strategies that will certainly make use of powerful enzyme inhibitors such as AZT and its derivatives and inhibitors of glycoprotein processing enzymes.

REFERENCES

Appleman, M.E., Marshall, D.W., Brey, R.L., Houk, R.W., Beatty, D.C., Winn, R.E., Melcher, G.P., Wise, M.G., Sumaya, C.V., and Boswell, R.N. (1988). *J. Infect. Dis.* **158**, 193-199.

Baumhefner, R.W., Mendez, M., Ma, B.T., and Tourtellotte, W.W. (1979). *Neurology* **29**, 549 (Abstr. MSI-8).

Baumhefner, R.W., Tourtellotte, W.W., Syndulko, K., and Shapshak, P. (1986). In "Immunotherapies in multiple sclerosis" (O.R. Hommes, J. Mertin, and W.W. Tourtellotte, eds.), pp. 226-236. Suffolk, Stuart Phillips Publ.

Claypoole, K.H.J., White, D., Townes, D.B., Collier, A.C. Handsfield, H.H., Longstreth, W., Maravilla, K., and Murphy, V. (1989). *J. Clin. Exp. Neuropsychol.* **11**, 77.

Delmotte, P. and Gonsette, R. (1977). *J. Neurol.* **215**, 27-37.

Dix, R.D. and Bredesen, D.E. (1988). In "AIDS and the nervous system" (M.L. Rosenblum, R.M. Levy, and D.E. Bredesen, eds.), pp. 221-262. New York, Raven Press.

Elovaara, I., Seppala, I., Poutiainen, E., Sun, I.J., and Valle, S-L. (1988). *Neurology* **38**, 1451-1456.

Esiri, M.M. (1977). *Lancet* **2**, 478-480.

Grant, I., Atkinson, J.H., Hesselink, J.R., Kennedy, C.J., Richman, D.D., Spector, S.A., and McCutchan, J.A. (1987). *Ann. Intern. Med.* **107**, 828-836.

Grant, I., Pinching, A.J., Kingsley, D., Weiner, J., Kocsis, A., McDonald, W.I., and Trimble, M.R. (1989). *Neurology* **39** (Suppl. 1), 427.

Helweg-Larsen, S., Jakobsen, J., Boesen, F., and Arlien-Soborg, P. (1986). *Acta Neurol. Scand.* **74**, 467-474.

Johnson, R.T., McArthur, J.C., and Narayan, O. (1988). *FASEB J.* **2**, 2970-2981.

Kjellen, K.G. and Vesterberg, O. (1974). *J. Neurol. Sci.* **23**, 199-213.

Laurenzi, M.A. and Link, H. (1978). *Acta Neurol. Scand.* **58**, 148-156.

Levy, R.M. and Bredesen, D.E. (1988). *J. AIDS* **1**, 41-64.

Levy, R.M., Bredeson, D.E., and Rosenblum, M.L. (1985). *J. Neurosurg.* **62**, 475-495.

Luer, W., Poser, S., Weber, T., Jurgens, S., Eichenlaub, D., Pohle, H.D., and Felgenhauer, K. (1988). *Klin. Wochenschr.* **66**, 21-25.

Marshall, D.W., Brey, R.L., Cahill, W.T., Houk, R.W., Zajac, R.A., and Boswell, R.N. (1988). *Arch. Neurol.* **45**, 945-958.

Michaels, J., Sharer, L.R., and Epstein, L.G. (1988). *Immunodefic. Rev.* **1**, 71-104.

Miller, E., Satz, P., and Visscher, B. (1989). *J. Clin. Exp. Neuropsychol.* **11**, 34-35.

Navia, B.A., Jordan, B.D., and Price, R.W. (1986). *Ann. Neurol.* **19**, 517-524.

Perry, S. and Marotta, R.F. (1987). *Alzheimer Dis. Assoc. Discord.* **1**, 221-235.

Price, R.W. and Brew, B.J. (1988). *J. Infect. Dis.* **158**, 1079-1083.

Price, R.W., Navia, B.A., and Cho, E.S. (1986). *Neurologic Clin.* **4**, 285-301.

Price, R.W., Sidtis, J., and Rosenblum, M. (1988). *Ann. Neurol.* **23** (Suppl.), s27-s33.

Prineas, J.W. and Wright, R.G. (1978). *Lab. Invest.* **38**, 409-421.
Reiber, H. (1980). In "Progress in multiple sclerosis research" (H.J. Bauer, S. Poser, and G. Ritter, eds.), pp. 154-156. Berlin, Springer-Verlag.
Reiber, H. and Felgenhauer, K. (1987). *Clin. Chim. Acta* **163**, 319-328.
Resnick, L., DiMarzo-Veronese, F., Schupbach, J., Tourtellotte, W.W., Ho, D.D., Muller, F., Shapshak, P., Vogt, M., Groopman, J.E., Markham, P.D., and Gallo, R.C. (1985). *N. Engl. J. Med.* **313**, 1498-1504.
Resnick, L., Berger, J.R., Shapshak, P., and Tourtellotte, W.W. (1988). *Neurology* **38**, 9-14.
Riccio, M., Wilson, B., Hawkins, D., and Thompson, C. (1989). Preliminary findings of a prospective study.
Schuller, E. and Sagar, H.J. (1983). *Acta Neurol. Scand.* **67**, 365-371.
Selnes, O.A., McArthur, J.C., Miller, E., Sheridan, K., Gordon, B., and Polk, B.F. (1989a). *J. Clin. Exp. Neuropsychol.* **11**, 78.
Selnes, O.A., Munoz, A., Miller, E., Cohen, B., Sheridan, K., Gordon, B., and Mcarthur, J.C. (1989b). *Neurology* **39** (Suppl. 1), 426.
Simpson, J.F., Tourtellotte, W.W., Kokmen, E., Parker, J.A., and Itabashi, H.H. (1969). *Arch. Neurol.* **20**, 373-377.
Singer, E., Ruane, P., Resnick, L., Shapshak, P., Syndulko, K., Osborne, M., Rubinshtein, G., Rosario, I., Schmid, P., and Tourtellotte, W.W. (1988). *Ann. Neurol.* **24**, 166.
Singer, E.J., Ruane, P., Syndulko, K., Fahy-Chandon, B., Resnick, L., and Tourtellotte, W.W. (1989). *Neurology* **39** (Suppl. 1), 420.
Snider, W.D., Simpson, D.M., Nielson, S., Gold, J.W.M., Metroka, C., and Posner, J.B. (1983). *Ann. Neurol.* **14**, 403-418.
So., Y.T., Beckstead, J., and Davis, R.L. (1986). *Ann. Neurol.* **20**, 566-572.
Stern, Y., Sano, M., Williams, J.B., and Gorman, J. (1989). *J. Clin. Exp. Neuropsychol.* **11**, 78.
Syndulko, K., Singer, E., Ruane, P., Chandon, B., Shapshak, P., Tourtellotte, W.W., and Resnick, L. (1989). *J. Clin. Exp. Neuropsychol.* **11**, 35.
Tibbling, G., Link, H., and Ohman, S. (1977). *J. Clin. Lab. Invest.* **37**, 385-390.
Tourtellotte, W.W. (1970a). In "Handbook of clinical neurology, Vol. 9: Multiple sclerosis and other demyelinating diseases" (P.J. Vinken and G.W. Bruyn, eds.), pp. 324-382. Amsterdam, North Holland.
Tourtellotte, W.W. (1970b). *J. Neurol. Sci.* **10**, 279-304.
Tourtellotte, W.W. (1971). *Res. Publ. Assoc. Res. Nerv. Ment. Dis.* **49**, 112-155.
Tourtellotte, W.W. (1972). In "Symposium on multiple sclerosis: immunology, virology and ultrastructure" (UCLA Forum in Medical Sciences, Vol. 16), (F. Wolfgram, F.W. Ellison, J.G. Stevens, and J.M. Andrew, eds.), pp. 285-332. New York, Academic Press.
Tourtellotte, W.W. (1985). In "Handbook of clinical neurology, rev. ser. Vol. 3 (Vol. 47): Demyelinating diseases" (P.J. Vinken, G.W. Bruyn, H.L. Klawans, and J.C. Koetsier, eds.), pp. 79-130. Amsterdam, Elsevier.
Tourtellotte, W.W. and Ma, B.I. (1978). *Neurology* **28** (Suppl.), 76-83.

Tourtellotte, W. and Parker, J.A. (1966). *Science* **154**, 1044-1046.

Tourtellotte, W.W. and Parker, J.A. (1967). *Nature* **214**, 683-686.

Tourtellotte, W.W., Kokmen, E., Tavolato, B., and Kruger, P.S. (1971). *Trans. Am. Neurol. Assoc.* **96**, 318-320.

Tourtellotte, W.W., Murthy, K.N., Brandes, D.W., Jurkowitz, J., and Fleming, J.O. (1975). *Neurology* **25**, 362-363.

Tourtellotte, W.W., Murthy, K., Brandes, D., Sajben, N., Ma, B., Comiso, P., Potvin, A., Costanza, A., and Korelitz, J. (1976). *Neurology* **26**, 59-61.

Tourtellotte, W.W., Potvin, A.R., Fleming, J.O., Murthy, K.N., Levy, J., Syndulko, K., and Potvin, J.H. (1980a). *Neurology* **30**, 240-244.

Tourtellotte, W.W., Baumhefner, R.W., Potvin, A.R., Ma, B.I., Potvin, J.H., Mendez, M., and Syndulko, K. (1980b). *Neurology* **30**, 1155-1162.

Tourtellotte, W.W., Potvin, A.R., Baumhefner, R.W., Potvin, J.H., Ma, B.I., Syndulko, K., and Petrovich, Z. (1980c). *Arch. Neurol.* **37**, 620-624.

Tourtellotte, W.W., Potvin, A.R., Mendez, M., Baumhefner, R.W., Potvin, J.H., Ma, B.I., and Syndulko, K. (1980d). *Ann. Neurol.* **8**, 402-408.

Tourtellotte, W.W., Walsh, M.J., Baumhefner, R.W., Staugaitis, S.M., and Shapshak, P. (1984a). *Ann. NY Acad. Sci.* **436**, 52-67.

Tourtellotte, W.W., Shapshak, P., Staugaitis, S.M., and Saxton, R.E. (1984b). *Prog. Clin. Biol. Res.* **146**, 371-378.

Tourtellotte, W.W., Staugaitis, S., Walsh, M.J., Shapshak, P., Baumhefner, R.W, Potvin, A.R., and Syndulko, K. (1985a) *Ann. Neurol.* **17**, 21-27.

Tourtellotte, W.W., Staugaitis, S.M., Walsh, M.J., Shapshak, P., Baumhefner, R.W., and Potvin, A.R. (1985b). *Protides Biol. Fluids* **32**, 161-164.

Tourtellotte, W.W.,Syndulko, K.,Baumhefner, R.W. and Potvin, A.R. (1986). In "Immunotherapies in multiple sclerosis" (O.R. Hommes,J. Mertin, and W.W. Tourtellotte, eds.), pp. 219-225. Suffolk, Stuart Phillips Publ.

16

PROTEINASES IN THE FIBROSIS OF SYSTEMIC SCLEROSIS

E. Carwile LeRoy
M. Bashar Kahaleh

Division of Rheumatology and Immunology,
Department of Medicine
Medical University of South Carolina,
Charleston, SC 29425

Scleroderma is a group of afflictions which may be localized (morphea or linear) or generalized (systemic sclerosis) (Medsger, 1989). Systemic sclerosis is a generalized fibrosing disorder that leads to scarring of the skin, lungs, heart, gastrointestinal track, and kidneys with functional organ failure as an irreversible outcome (LeRoy, 1988). The scarring reaction stems from a state of activation/transformation of the interstitial fibroblast which expresses two major phenotypic abnormalities (LeRoy *et al.*, 1989):

1) an enhanced transcriptional activation of several genes which encode key proteins in the extracellular matrix, including collagens I and III, fibronectin, and proteoglycan core protein; and

2) a defect in cell growth regulation characterized by a failure to undergo growth arrest upon serum deprivation, an altered response and receptor level to several growth factors, and a persistent expression of the c-myc gene and its message, suggesting a transient but propagable, partial state of cellular transformation (Trojanowska *et al.*, 1988).

Ongoing studies are designed to characterize the activational state of the scleroderma fibroblast in molecular terms. Studies have focused on two growth factors: platelet-derived growth factor (PDGF) and transforming growth factor-β (TGF-β) and their interactions; on two cytokines: tumor necrosis factor (TNF-α) and lymphotoxin (TNF-β) and on the growth and matrix synthesis responses of the scleroderma fibroblast to a variety of structural and cytoadhesive proteins of the interstitial connective tissue matrix. At the moment our findings are consistent with the hypothesis that the cell ultimately responsible for the irreversible fibrosis of systemic sclerosis, the systemic sclerosis fibroblast, is temporarily transformed probably by an autocrine pathway. These cells provide a suitable model for cellular and molecular studies of fibrosis varying from atherosclerosis, pulmonary fibrosis, systemic sclerosis, keloid formation, to cirrhosis of the liver.

Fortunately, there is a prefibrotic stage of systemic sclerosis which involves abnormalities of the vascular and microvascular systems. An episodic vasospastic condition termed Raynaud's phenomenon antecedes systemic sclerosis in virtually 100% of patients, when sensitive measurements of the peripheral circulation, especially digital blood flow, are used to detect vasospasm.

Fewer than 5% of persons, male and female over 18 years of age, have Raynaud's phenomenon. Of these, about 15% can be shown to have early features of systemic sclerosis (Maricq *et al.*, 1989). These features are dilated and destroyed skin capillaries detected by nailfold capillary microscopy and a selective antinuclear antibody pattern called anti-centromere (Tan *et al.*, 1980). The use of these two techniques, nailfold capillary microscopy and antinuclear antibody determinations, can completely identify the disorder, systemic sclerosis, in its prefibrotic stage; if we knew what to do to prevent fibrosis, this population would be appropriate subjects for therapeutic trials.

Increasing evidence points to the vascular and microvascular systems as primary target organs for early injury in systemic sclerosis. The endothelium is continuous between arterial and capillary vessels, although some functional differences have been noted between arterial, venous, and capillary endothelial cells in culture. The vascular hypothesis of systemic sclerosis was placed on a firm footing by work showing that systemic sclerosis sera contained an activity that was injurious to all types of endothelial cells (Kahaleh *et al.*, 1979). These same sera which showed varying titers of endothelial cytotoxic activity also showed a deficient ability to inhibit serine proteinases, especially trypsin (Kahaleh and LeRoy, 1983). This was the first suggestion that a proteinase, presumably a serine proteinase, or alternately a deficiency in the elaborate system of proteinase inhibitors, was implicated in the endothelial injury of systemic sclerosis. To strengthen the association, a highly significant relationship was shown between endothelial cytotoxicity and loss of proteinase inhibitor, functional activity despite normal concentrations of several specific proteinase inhibitor proteins. At least two possibilities exist to explain these observations: 1) proteinases are released by cells which inactivate the soluble proteinase inhibitors in the surrounding microenvironment or 2) proteinase inhibitors are incapacitated by oxygen free radicals or other non-proteinase molecules which bind and prevent proteinase binding but do not cleave the active site of the proteinase inhibitor.

We have considered several putative proteinases as potentially responsible for the injury to endothelial cells. So far, we have not found any which fit all of our criteria. The search for a proteinase in scleroderma has led us to the connective tissue mast cell which is rich in one or more proteinases. Chymase is a serine proteinase of the carboxypeptidase family which is selectively present in connective tissue mast cells, in contrast to the tryptase proteinase which is predominantly in the mucosal subset of mast cell (Siraganian, 1988).

The mast cell has been associated with wound healing, scar formation, and fibrosis since it was first discovered using the early dyes of Ehrlich. The three major scleroderma-like fibrotic states show increased numbers and enhanced degranulation of mast cells: these are scleroderma itself, the scleroderma-like skin disease of chronic graft vs host disease and the mouse model of skin fibrosis, the tight skin mouse.

We first noted increased numbers and markedly increased states of degranulation of mast cells in the autosomally dominant inherited skin fibrosis model, the tight skin mouse (tsk/+, on a C57 black mouse strain) (Walker *et al.*, 1985). We elected to see if known inhibitors of mast cell degranulation could prevent skin fibrosis if animals were chronically treated beginning before the histologically detectable appearance of skin fibrosis. Despite its poor solubility, we chose cromolyn and gave it in the drinking water beginning at 4 weeks of age and continuing well into the period of maximal skin fibrosis seen in untreated tsk/+ mice. The cromolyn has a substantial but incomplete effect on mast cell degranulation, reducing extracellular granules significantly but not completely toward normal. Despite this incomplete effect on mast cells, the effects on prevention of skin fibrosis were remarkable. Tsk mice treated with cromolyn showed skin thickness measurements equal to those of unaffected syngeneic C57 mice; i.e. cromolyn completely prevented the development of skin fibrosis in the tight skin mouse (Walker *et al.*, 1987).

To return to the endothelial cell-proteinase story, recent studies from several laboratories have demonstrated that this cell actively participates in a selective proteinase-antiproteinase system whose imbalance could have a major impact on the prominent endothelial injury and fibroblast activation-fibrosis of systemic sclerosis. Endothelial cells have been shown to synthesize and secrete a member of the proteinase family of sulfhydryl proteinase which is also calcium dependent. This Ca^{++} activatable proteinase is capable of clearing and activating coagulation factors; it could have a direct effect on endothelial cells and basement membranes as well. Further study is needed to determine if this newly described, endothelial cell derived calcium activatable proteinase is active in the endothelial cell injury pattern so prominent in the affected tissue of systemic sclerosis (Rodgers *et al.*, 1987).

Endothelial cells, especially when activated, have other capacities to express and secrete proteinases which could effect the structure of the microvascular system and beyond it the interstitial tissues which are the site of fibrosis in systemic sclerosis. Recent studies have described a factor isolated from human placenta (18 kilodaltons) which stimulates the production of plasminogen activator, a serine proteinase, and latent collagenase, a metalloproteinase, from cultured bovine capillary endothelial cells. This activator is active in concentrations between 0.1 ng to 10 ng protein/ml. Interestingly, the same molecule induces DNA synthesis (and presumably cell proliferation) and chemotaxis in capillary endothelial cells. These studies show further the capacities of the endothelial cell to produce a variety of proteinases which affect the interstitial connective tissue (Moscatelli *et al.*, 1986).

Since most of the endothelial cell studies referred to depend on endothelial cell growth in cell culture settings, it should be noted that *in vivo* in both arterial, capillary and venous endothelium, the endothelial cell divides very infrequently and even then usually in response to mechanical or chemical injury. Thus, endothelial cells in culture, where the conditions (plastic surface and serum mitogens) are designed for rapid proliferation, are in an environment foreign to their natural setting and may well be activated, stimulated, or injured even in the best *in vitro* conditions. Under these conditions, endothelial cells not only express and secrete proteinases of several types as discussed but they also express and secrete a major member of the antiproteinase (Serpin) family which is selectively inhibitory to plasminogen activator. This factor is called plasminogen activator inhibitor (PAI). This molecule is synthesized as a 402 amino acid precursor molecule containing a leader sequence of 23 amino acid residues. Its sequence includes three asparagine-linked glycosylation sites for hetero-polysaccharide attachment and the fully-glycosylated native molecule is 52,000 daltons while the unglycosylated recombinant molecule produced in *E. coli* is 43,000 daltons (Pannekoek *et al.*, 1986). This plasminogen activator inhibitor is the major protein expressed by stimulated endothelial cells, comprising up to 10% of the total protein synthesized by these cells. Any cessation or inhibition of the synthesis or secretion of plasminogen activator inhibitor by endothelial cells *in vivo* would shift the balance strongly in favor of proteinase action, which would have the patho-biologic effects of endothelial cell injury and fibroblast activation, the prevailing situation in systemic sclerosis. It is interesting that endothelial cells, platelets, plasma and serum all contain a similar plasminogen activator inhibitor, suggesting that endothelial cells may be the major source of this important antiproteinase *in vivo* (Erickson *et.al* , 1985).

Consequences of excessive proteinase activity in the tissues unbalanced by adequate antiproteinase capacity are not limited to endothelial cells, fibroblasts, basement membranes, or interstitial connective tissue, as important as these effects may be. Proteinases can also affect significant informational molecules as well. As an example, the cytokine γ-interferon will be used here. γ-Interferon was originally described as an antiviral molecule produced by a variety of cells, especially T lymphocytes and fibroblasts. It has two major effects on fibroblasts: the inhibition of collagen synthesis at the level of transcription (reduced mRNA levels) and the induction of expression on the cell membrane of HLA antigens which participate in cell immune events. An attractive hypothesis regarding the excessive collagen synthesis of systemic sclerosis fibroblasts is that some informational molecule such as γ-interferon is deficient in its natural role of inhibiting fibroblast collagen synthesis and secretion. So attractive is this hypothesis that recombinant γ-interferon is currently being used as an experimental therapy in systemic sclerosis patients. The interesting relationship between γ-interferon and proteinases is that proteinases selectively cleave γ-interferon between residue 129 and 130 removing 11 residues from the C-terminal end. Functionally, this proteinase altered γ-interferon molecule showed a 2,000-fold reduction in its capacity to bind γ-interferon receptors. Thus, increased proteinase activity in interstitial tissues could alter γ-interferon's ability to suppress fibroblast collagen secretion, resulting in excessive collagen and fibrosis (Leinikki *et al.*, 1987). It is appropriate to suggest studies of the type of γ-interferon produced in the microenvironment of systemic sclerosis lesions.

γ–Interferon may have a special relationship with endothelial cells in other ways as well. It is well known in immunological circles that lymphocyte trafficking plays a significant role in the *in vivo* immune response. Furthermore, the key signal in the regulation of lymphocyte trafficking stems from a morphologically specialized endothelial cell lining the lymph node blood vessels. This specialized endothelial cell has been termed high venular endothelium. Recently a monoclonal antibody has been developed which selectively and specifically recognizes determinants on high venular endothelium. In a search for factors which are capable of inducing the expression of this unique endothelial cell determinant on mouse lung or bone marrow derived endothelial cells, interleukin 1, endothelial cell growth factors (all heparin binding and all homologous with the basic fibroblast growth factor molecule), and interferon β were unable to induce high venular endothelium antigen whereas γ-interferon induced high levels of this epitope recognized by the monoclonal antibody, suggesting that γ-interferon may play a role in endothelial cell differentiation (Duijvestijn *et al.*, 1986).

What implications do these studies have for future drug design for the human fibrotic disorders? They seem to lead to one family of biologically active proteins, the serine proteinases, and they also permit the hypothesis that serine proteinases of mast cell origin may be of particular interest. Since fibrosis in all its forms is largely irreversible and usually unresponsive to pharmacologic intervention, the pursuit of specific, orally administrable, inhibitors of human connective tissue mast cell proteinases would seem to hold promise for the future therapy of this family of disorders.

REFERENCES

Duijvestijn, A.M., Schreiber, A.B., and Butcher, E.C. (1986).*Proc. Natl. Acad. Sci. U.S.A.* **83**, 9114-9118.

Erickson, L.A., Hekman, C.M., and Loskutoff, D.J. (1985). *Proc. Natl. Acad. Sci. U.S.A.* **82**, 8710-8714.

Kahaleh, M.B. and LeRoy, E.C. (1983). *J. Lab. Clin. Med.* **101**, 553-560.

Kahaleh, M.B., Sherer, G.K., and LeRoy, E.C. (1979). *J. Exp. Med.* **149**, 1326-1335.

Leinikki, P.O. Calderon, J., Luquette, M.H., and Schreiber, R.D. (1987). *J. Immunol.* **139**, 3360-3366.

LeRoy, E.C. (1988). In "Cecil textbook of medicine". 18th ed. (J.B. Wyngaarden and L.H. Smith, eds.), pp. 2018-2024. Philadelphia, W.B. Saunders.

LeRoy, E.C., Smith, E.A., Kahaleh, M.B., Trojanowska, T., and Silver, R.M. (1989). *Arthritis Rheum.* **32**, 817-825.

Maricq, H.R., Weinrich, M.C., Keil, J.E., Smith, E.A., Harper, F.E., Nussbaum, A.I., LeRoy, E.C., McGregor, A.R., Diat, F., and Rosal, E.J. (1989). *Arthritis Rheum.* In press.

Medsger, TA., Jr. (1989). In "Arthritis and allied conditions". 11th ed. (D.J. McCarty, ed.), pp. 1118-1165. Philadelphia, Lea and Febiger.

Moscatelli, D., Presta, M., and Rifkin, D.B. (1986). *Proc. Natl. Acad. Sci. U.S.A.* **83**, 2091-2095.

Pannekoek, H., Veerman, H., Lambers, H., Diergaarde, P., Verweij, C.L., van Zonneveld, A., and van Mourik, J.A. (1986). *EMBO J.* **5**, 2539-2544.

Rodgers, G.M., Cong, J.Y., Goll, D.E., and Kane, W.H. (1987). *Biochim. Biophys. Acta* **929**, 263-270.

Siraganian, R.P. (1988). In "Inflammation: basic principles and clinical correlates" (J.I. Gallin, I.M. Goldstein, and R. Snyderman, eds.), pp. 513-542. New York, Raven Press.

Tan, E.M., Rodnan, G.P., Garcia, I., Moroi, Y., Fritzler, M.J., and Peebles, C. (1980). *Arthritis Rheum.* **23**, 617-625.

Trojanowska, M., Wu. L., and LeRoy, E.C. (1988). *Oncogene* **3**, 477-481.

Walker, M., Harley, R., Maize, J., DeLustro, F., and LeRoy, E.C. (1985). *Proc. Soc. Exp. Biol. Med.* **180**, 323-328.

Walker, M., Harley, R., and LeRoy, E.C. (1987). *J. Rheumatol.* **14**, 299-301.

17

THERAPEUTIC OPPORTUNITIES IN EICOSANOID BIOSYNTHESIS

John A. Oates

Department of Medicine
Vanderbilt University School of Medicine
Nashville, TN 37232

The clinical syndromes elicited by myocardial ischemia include unstable angina, myocardial infarction, and sudden cardiac death. Considerations of pharmacologic interventions aimed at preventing these ischemic syndromes have focused primarily on the platelet. This is because a substantial body of evidence implicates the disruption of atheromatous plaques with attendant thrombus formation in the coronary artery as a cause of each of these cardiac ischemic syndromes (Davies and Thomas, 1985). Accordingly, there is a strong logic for drugs that inhibit platelet function and the formation of occlusive thrombi.

In addition to pharmacologic strategies that address coronary occlusion, the responses of the heart to ischemia must also be considered. One such consequence of ischemia is ventricular fibrillation, the predominant cause of sudden cardiac death. Thus, the pharmacologic scope is extended to encompass both occlusive and post occlusive events.

The principal current antiplatelet drug is aspirin. By blocking the cyclooxygenase enzyme in platelets, aspirin inhibits the biosynthesis of thromboxane A_2, which is a major pro-aggregatory mediator released by platelets.

Aspirin acetylates the cyclooxygenase enzyme, thus blocking it irreversibly. By blocking cyclooxygenase in the vessel wall, aspirin can also inhibit the biosynthesis of prostacyclin, which is a potent inhibitor of platelet activation. Because aspirin inhibits the cyclooxygenase enzyme irreversibly, and platelets do not synthesize new cyclooxygenase enzyme, very low doses of aspirin exhibit a relative selectivity for platelet cyclooxygenase. However, the doses of aspirin employed in most clinical trials also inhibit prostacyclin biosynthesis.

The evidence supporting plaque disruption with attendant thrombus formation as a cause of unstable angina is of several types. Folts and his colleagues developed an animal model of recurring myocardial ischemia that mimics unstable angina (Folts *et al.*, 1976). This model is created by constricting an epicardial coronary vessel by about 90% with an occluder and creating endothelial damage within the occluded lumen. It has been shown that the intermittent reductions in coronary flow seen in this model can be related to the formation and subsequent disruption of platelet aggregates at the site of endothelial disruption and stenosis of the vessel. The recurrent ischemia can be prevented by potent inhibitors of platelet activation such as prostacyclin (Aiken, 1985); aspirin is only partially effective in this model.

Several types of anatomic evidence in humans support the disruption of a coronary plaque as a cause of unstable angina. This includes pathological evidence of plaque disruption in patients who died following unstable angina (Davies *et al.*, 1986; Falk, 1985), angiographic evidence suggestive of mural thrombus in patients with unstable angina (Vetrovec *et al.*, 1982; Mandelkorn *et al.*, 1983; Ambrose *et al.*, 1986), and angioscopic demonstration of plaque disruption at the time of surgery for unstable angina (Sherman *et al.*, 1986). The anatomic findings have been supported by evidence for a state of platelet activation in patients with unstable angina. This has been demonstrated in several studies (Schwartz *et al.*, 1980; Hirsh *et al.*, 1981; Fitzgerald *et al.*, 1986). In order to avoid the artifactual release of platelet mediators associated with the *ex vivo* activation of platelets, Desmond Fitzgerald, *et al*, examined the urinary and plasma metabolites of thromboxane A_2 as a reflection of platelet activation during unstable angina (Fitzgerald *et al.*, 1986). The two major pathways of thromboxane A_2 metabolism in man are to 2,3-dinor-thromboxane B_2 which can be measured in urine and 11-dehydro-thromboxane B_2 which can be measured in both plasma and urine (Roberts *et al.*, 1981; Lawson *et al.*, 1986). Dinor-thromboxane B_2 was found to be elevated in the urine of patients with unstable angina and 11-dehydro-thromboxane B_2 was increased in their plasma. This evidence for *in vivo* platelet activation was greatest in conjunction with the episodes of ischemic pain. Although coronary spasm may also cause myocardial ischemia at rest, plaque disruption would appear to be the principal cause of unstable angina.

Pathologic as well as angiographic evidence implicates coronary thrombosis as the cause of myocardial infarction. This is the basis for the beneficial effects of thrombolytic therapy and the efficacy of prophylactically administered aspirin in reducing the incidence of myocardial infarction.

Sudden cardiac death in conjunction with coronary disease is associated with a rupture of lipid rich atheromatous plaques and mural thrombus formation in approximately 90% of cases (Davies and Thomas, 1984; El Fawal *et al.*, 1987). This suggests that most instances of sudden cardiac death are the consequences of ischemic ventricular fibrillation. As this is triggered by mural thrombus, one would anticipate a reduction in sudden death in conjunction with prophylactic administration of aspirin, just as has been the case with myocardial infarction (Antiplatelet Trialists' Collaboration, 1988). Accordingly, the effect of aspirin on sudden death was examined in trials conducted in the last decade in which aspirin was the only randomized pharmacologic intervention and in which only a minority of patients received β-adrenoceptor blocking drugs which have a well established action to prevent sudden death (Pedersen, 1981; Beta-Blocker Heart Attack Trial Research Group, 1982). The three trials in this category were the Aspirin Myocardial Infarction Study (1980) and the PARIS I Trial (The Persantin-Aspirin Reinfarction Study Research Group, 1980) which were conducted in myocardial infarction survivors, and the Physician's Health Study in which aspirin was evaluated as primary prevention (The Steering Committee of the Physician Health Study Group, 1989). Whereas myocardial infarction was significantly reduced in all of these studies in conjunction with aspirin treatment, there was no reduction in sudden cardiac death. Sudden cardiac death was increased by 35%, 27% and 83% respectively in these trials. Whereas these increases do not achieve statistical significance in the separate trials, it is clear from these results that aspirin in these three trials did not reduce the prevalence of sudden cardiac death despite the significant reduction in myocardial infarction. As plaque rupture is the common etiology for myocardial infarction and sudden cardiac death, aspirin might also have been expected to have reduced sudden cardiac death. Because aspirin did not decrease sudden cardiac death, it was considered that it might be enhancing coupling between thrombus formation (occlusion) and ventricular fibrillation. As aspirin blocks prostacyclin biosynthesis in the doses employed in these trials, the effect of prostacyclin on the coupling of ischemia to ventricular fibrillation has been examined in studies conducted with John Hammon.

Ischemic ventricular fibrillation was induced in the unanesthetized analgesic-treated dog by inflation of a balloon cuff on a circumflex coronary artery. In this model, any ventricular fibrillation that occurred usually developed within ten minutes of occlusion. To assess the effects of prostacyclin on ventricular fibrillation, a dose was selected that did not alter hemodynamics but inhibited platelet aggregation. This dose of prostacyclin, 100 ng/kg/min, reduced the frequency of ventricular fibrillation from 53% in the control dogs to 6% in the prostacyclin group (Starnes *et al.*, . 1982).

Further assessment of platelet-active eicosanoids on the coupling of ischemia to ventricular fibrillation was performed by examining effects of the thromboxane synthase inhibitor, RO 22-4679, a 1-substituted imidazole, on ischemic ventricular fibrillation. This inhibitor of thromboxane synthase reduced the frequency of ventricular fibrillation from 33.3% in the control animals to 6% in the RO 22-4679 treated group (Huddleston *et al.*, 1983). Inhibition of thromboxane synthase has two potential biochemical consequences; it reduces the biosynthesis of thromboxane A_2 and at the same time causes accumulation of the prostaglandin endoperoxide (PGH_2). Under certain *in vitro* conditions (Marcus *et al.*, 1980; Needleman *et al.*, 1980) as well as in normal humans (FitzGerald *et al.*, 1983a), the accumulation of PGH_2 leads to an increase in the biosynthesis of other prostaglandins including prostacyclin and PGE_2. To assess the possible contribution of prostaglandins formed from the accumulated endoperoxide to the beneficial effect of thromboxane synthase inhibition on ventricular fibrillation, the effects of a thromboxane synthase inhibitor alone and together with an inhibitor of endoperoxide biosynthesis, indomethacin, were compared. For this study the thromboxane synthase inhibitor, U-63557A (5-(3'-pyridinylmethyl) benzofuran-2-carboxylic acid), was infused in a dose of 4 mg/kg. Ventricular fibrillation was completely abolished by U-63557A, in contrast to a 38% occurrence in animals receiving vehicle (Hammon and Oates, 1986). This beneficial effect of the thromboxane synthase inhibitor, however, was reversed by the concurrent administration of indomethacin 8 mg/kg; ventricular fibrillation occurred in 46% of the animals receiving the cyclooxygenase inhibitor in conjunction with the inhibitor of thromboxane synthase.

From these experimental evaluations, it is concluded that ischemic ventricular fibrillation can be inhibited by the administration of prostacyclin and also by the elevation of prostacyclin and perhaps other prostaglandins during thromboxane synthase inhibition.

As prostacyclin inhibits ischemic ventricular fibrillation, two important questions regarding the administration of aspirin to patients at risk of sudden cardiac death are:

1. Do doses of aspirin used in the clinical trials block prostacyclin biosynthesis?
2. Is prostacyclin released during human myocardial infarction?

Evidence from several investigators has demonstrated that doses of aspirin greater than 80 mgs daily will block the biosynthesis of prostacyclin in man, as reflected by reduction in the excretion of a urinary metabolite of prostacyclin, 2,3-dinor-6-keto-$PGF_{1\alpha}$ (FitzGerald *et al.*, 1983b) and by inhibition of the biosynthesis of prostacyclin by arterial and venous tissue studied *ex vivo* following prior administration of aspirin (Preston *et al.*, 1981; Hanley *et al.*, 1981; Weksler *et al.*, 1983; Weksler *et al.*, 1985). Accordingly, as the dose of aspirin employed in the large cardiovascular trials cited above was greater than this, all of these studies examine the effects of inhibiting the biosynthesis of both thromboxane A_2 and prostacyclin.

The biosynthesis of prostacyclin during acute myocardial infarction has been evaluated by measuring the urinary excretion of 2,3-dinor-6-keto-$PGF_{1\alpha}$. Excretion of this prostacyclin metabolite was found to be significantly and markedly elevated in patients with myocardial infarction, an elevation that correlated with peak plasma levels of creatine kinase (Fitzgerald *et al.*, 1986). Thus, it is concluded that prostacyclin biosynthesis increased during myocardial ischemia.

The hypothesis that prostacyclin inhibits the coupling of ischemia to ventricular fibrillation is therefore supported by several lines of evidence. Prostacylin inhibits ischemic ventricular fibrillation as does thromboxane synthase inhibition acting via a prostaglandin dependent mechanism. Prostacyclin biosynthesis is increased in myocardial infarction and the doses of aspirin used in the clinical trials in coronary disease were in the range that inhibit prostacyclin biosynthesis. Also consistent with the hypothesis is the observation that aspirin employed in trials in which beta blocker use is minimal does not prevent sudden cardiac death despite a clear reduction in the incidence of myocardial infarction. This hypothesis encourages the development of antiplatelet therapy that enhances, or at least does not inhibit, prostacyclin biosynthesis.

The discovery that imidazole selectively inhibits thromboxane synthase (Raz and Needleman, 1977) initiated a line of research that holds considerable therapeutic potential. Because prostacyclin is a more potent inhibitor of platelet aggregation than aspirin, the increase in prostacyclin biosynthesis when thromboxane synthase is inhibited during intravascular aggregation affords a pharmacologic circumstance in which normal endothelium can antagonize the aggregation of platelets in its proximity. Thus, in the Folts model of intermittent coronary occlusion (Folts *et al.*, 1976), thromboxane synthase inhibition is more effective in retarding aggregation induced interruption of flow than is aspirin (Aiken, 1985). Similarly, in the model of canine coronary occlusion induced by electrical injury of the endothelium (Fitzgerald *et al.*, 1976) thromboxane synthase inhibition also is more potent than aspirin in retarding the formation of an occlusive thrombus. Thus, there is a basis for expecting that inhibition of occlusive coronary thrombi by thromboxane synthase inhibitors is superior to that achieved by aspirin. These observations regarding the inhibition of the occlusive process coupled with the inhibition of ischemic ventricular fibrillation by thromboxane synthase inhibitors together provide an attractive therapeutic hypothesis that thromboxane synthase inhibition should be more effective than aspirin in reducing cardiovascular mortality in patients with coronary heart disease.

Thromboxane synthase inhibition alone could be sufficient to achieve a reduction in occlusion and post occlusive coronary ischemic events with an acceptable risk of bleeding complications. A limitation on the potency of thromboxane synthase inhibition as antiplatelet therapy derives from the fact that the endoperoxide, PGH_2, accumulates during thromboxane synthase inhibition, and PGH_2 is an agonist for the thromboxane A_2 receptor (Needleman *et al.*, 1977; Grimm *et al.*, 1981; FitzGerald *et al.*, 1985; Mais *et al.*, 1985). Accordingly, the administration of a thromboxane synthase inhibitor *in vivo* leads to little or no inhibition of *ex vivo* aggregation. Thus, an antagonist of the thromboxane receptor, by blocking the aggregatory effect of accumulated PGH_2, can enhance the antiaggregatory effect of thromboxane synthase inhibition *in vivo* (Fitzgerald *et al.*, 1988) and *in vitro* (Bertele *et al.*, 1984; FitzGerald *et al.*, 1985). Whether greater therapeutic benefit from a combination of a thromboxane synthase inhibitor with a thromboxane receptor antagonist would exceed the greater probability of bleeding remains to be ascertained.

The introduction of thromboxane synthase inhibition into cardiovascular therapy awaits the development of an inhibitor that has an acceptable safety profile together with a clearance sufficiently low to enable a dosage interval compatible with its use in prophylaxis. The availability of such an inhibitor would afford an opportunity to assess whether prostacyclin enhancing antiplatelet therapy could yield a reduction in cardiovascular mortality exceeding that found with aspirin.

REFERENCES

Aiken, J.W. (1985). In "Interaction of platelets with the vessel wall" (J.A. Oates, J. Hawiger, and R. Ross, eds.), pp. 79-87. Baltimore, Williams & Wilkins.

Ambrose, J.A., Winters, S.L., Arora, R.R., Eng, A., Riccio, A., Gorlin, R., and Fuster, V. (1986). *J. Am. Coll. Cardiol.* **7**, 472-478.

Antiplatelet Trialists' Collaboration. (1988). *Br. Med. J.* **296**, 320-331.

Aspirin Myocardial Infarction Study Research Group. (1980). *J. Am. Med. Assoc.* **243**, 661-669.

Bertele, V., Falanga, A., Tomasiak, M., Chiabrando, C., Cerletti, C., and de Gaetano, G. (1984). *Blood* **63**, 1460-1466.

Beta-Blocker Heart Attack Trial Research Group. (1982). *J. Am. Med. Assoc.* **247**, 1707-1714.
Davies, M.J. and Thomas, A. (1984). *N. Engl. J. Med.* **310**, 1137-1140.
Davies, M.J. and Thomas, A.C. (1985). *Br. Heart J.* **53**, 363-373.
Davies, M.J., Thomas, A.C., Knapman, P.A., and Hangartner, J.R. (1986). *Circulation* **73**, 418-427.
El Fawal, M.A., Berg, G.A., Wheatley, D.J., and Harland, W.A. (1987). *Br. Heart J.* **57**, 329-335.
Falk, E. (1985). *Circulation* **71**, 699-708.
FitzGerald, G.A., Brash, A.R., Oates, J.A., and Pedersen, A.K. (1983a). *J. Clin. Invest.* **72**, 1336-1343.
FitzGerald, G.A., Oates, J.A., Hawiger, J.J., Maas, R.L., Roberts, L.J., II., Lawson, J.A., and Brash, A.R. (1983b). *J. Clin. Invest.* **71**, 676-688.
FitzGerald, G.A., Reilly, I.A.G., and Pedersen, A.K. (1985). *Circulation* **72**, 1194-1201.
Fitzgerald, D.J., Roy, L., Catella, F., and FitzGerald, G.A. (1986). *N. Engl. J. Med.* **315**, 983-989.
Fitzgerald, D.J., Fragetta, J., and FitzGerald, G.A. (1988). *J. Clin. Invest.* **82**, 1708-1713.
Folts, J.D., Crowell, E.B., and Rowe, G.G. (1976). *Circulation* **54**, 365-370.
Grimm, L.J., Knapp, D.R., Senator, D., and Halushka, P.V. (1981). *Thromb. Res.* **24**, 307-317.
Hammon, J.W. and Oates, J.A. (1986). *Circulation* **73**, 224-226.
Hanley, S.P., Bevan, J., Cockbill, S.R., and Heptinstall, S. (1981). *Lancet* **1**, 969-971.
Hirsh, P.D., Hillis, L.D., Campbell, W.B., Firth, B.G., and Willerson, J.T. (1981). *N. Engl. J. Med.* **304**, 685-691.
Huddleston, C.B., Lupinetti, F.M., Laws, K.H., Collins, J.C., Clanton, J.A., Hawiger, J.J., Oates, J.A., and Hammon, J.W., Jr. (1983). *Circ. Res.* **52**, 608-613.
Lawson, J.A., Patrono, C., Ciabattoni, G., and FitzGerald, G.A. (1986). *Anal. Biochem.* **155**, 198-205.
Mais, D.E., Burch, R.M., Saussy, D.L., Jr., Kochel, P.J., and Halushka, P.V. (1985). *J. Pharmacol. Exp. Ther.* **235**, 729-734.
Mandelkorn, J.B., Wolf, N.M., Singh, S., Shechter, J.A., Kersh, R.I., Rodgers, D.M., Workman, M.B., Bentivoglio, L.G., LaPorte, S.M., and Meister, S.G. (1983). *Am. J. Cardiol.* **52**, 1-6.
Marcus, A.J., Weksler, B.B., Jaffe, E.A., and Broekman, M.J. (1980). *J. Clin. Invest.* **66**, 979-986.
Needleman, P., Raz, A., Ferrendelli, J.A., and Minkes, M. (1977). *Proc. Natl. Acad. Sci. U.S.A.* **74**, 1716-1720.
Needleman, P., Wyche, A., and Raz, A. (1979). *J. Clin. Invest.* **63**, 345-349.
Pedersen, T.R. (1981). *N. Engl. J. Med.* **304**, 801-807.
The Persantin-Aspirin Reinfarction Study Research Group. (1980). *MMW* **122**, 1761-1771.

Preston, F.E., Whipps, S., Jackson, C.A., French, A.J., Wyld, P.J., and Stoddard, C.J. (1981). *N. Engl. J. Med.* **304**, 76-79.

Raz, A. and Needleman, P. (1977). *Prostaglandins* **13**, 611-618.

Roberts, L.J., II, Sweetman, B.J., and Oates, J.A. (1981). *J. Biol. Chem.* **256**, 8384-8393.

Schwartz, M.B., Hawiger, J., Timmons, S., and Friesinger, G.C. (1980). *Thromb. Hemost.* **43**, 185-188.

Sherman, C.T., Litvack, F., Grundfest, W., Lee, M., Hickey, A., Chaux, A., Kass, R., Blanche, C., Matloff, J., Morgenstern, L., Ganz, W., Swan, H.J.C., and Forrester, J. (1986). *N. Engl. J. Med.* **315**, 913-919.

Starnes, V.A., Primm, R.K., Woosley, R.L., Oates, J.A., and Hammon, J.W., Jr. (1982). *J. Cardiovasc. Pharmacol.* **4**, 765-769.

The Steering Committee of the Physician Health Study Group. (1989). *N. Engl. J. Med.* **321**, 129-135.

Vetrovec, G.W., Leinbach, R.C., Gold, H.K., and Cowley, M.J. (1982). *Am. Heart J.* **104**, 946-952.

Weksler, B.B., Pett, S.B., Alonso, D., Richter, R.C., Stelzer, P., Subramanian, V., Tack-Goldman, K., and Gay, W.A., Jr. (1983). *N. Engl. J. Med.* **308**, 800-805.

Weksler, B.B., Tack-Goldman, K., Subramanian, V.A., and Gay, W.A., Jr. (1985). *Circulation* **71**, 332-340.

18

MAO AS A TARGET FOR DRUG DESIGN: CLINICAL ISSUES

Donald S. Robinson[1]

Bristol-Myers Company
Pharmaceutical Research and Development Division
Wallingford, CT 064921

I. INTRODUCTION

A. History of the Discovery of Monoamine Oxidase Inhibitor Drugs

As with many of the drugs that issued in the modern era of psychopharmacology, discovery of the therapeutic activity of monoamine oxidase inhibitors (MAOIs) was due to the serendipitous observation that there were behavioral side effects of a drug developed for treating tuberculosis (Maxwell and Eckhardt, in press). Discovery of this important drug class and its relevance to psychiatry followed by some 25 years the isolation in 1928 of monoamine oxidase from rabbit liver (Hare, 1928). The first MAOI was iproniazid, which some five years after this fortuitous finding in tuberculosis patients was introduced into clinical medicine as the first antidepressant drug and, following electroconvulsive therapy (ECT), was only the second somatic treatment effective for depression.

As with many important drug discoveries, the history of iproniazid and monoamine oxidase inhibitors has many interesting perspectives. In this review I will focus on a few of these milestones. The late Dr. Nathan Kline and associates noted in 1957 that iproniazid was the first drug known to "energize" rather than sedate depressed patients, a property they regarded as being useful (Loomer et al., 1957). They considered this an important distinction when treating disorders that are associated with profound

[1]*Research cited in this chapter includes work conducted by the author under Public Health Service grants MH15533, MH27836, MH36179, MH37617 and MH32176. The author also wishes to acknowledge the Burroughs-Wellcome Scholar in Clinical Pharmacology Award and the Pharmaceutical Manufacturers Association Foundation Clinical Pharmacology Award for programmatic support.*

depressed mood, retardation, and diminished mood reactivity. One early report on this drug in the treatment of sanitorium patients with tuberculosis described patients as "dancing in the halls -- and it wasn't because they had seen their x-ray pictures!" (Maxwell and Eckhardt, in press). During the time it was being tried in psychiatric patients, Zeller and associates (Zeller *et al.*, 1952) showed that iproniazid inhibited monoamine oxidase activity. Crane (1956), in turn, suggested that the undesired side effects of this tuberculosis therapy might be beneficial for treating depression. This proved to be the case. Unfortunately iproniazid was suspected, perhaps incorrectly, of producing hepatotoxicity and was withdrawn from the U.S. market shortly after introduction in 1961, although this was not the case in the United Kingdom and a number of other countries (Kline and Cooper, 1980).

Armed with the knowledge that inhibition of the enzyme monoamine oxidase was a likely prerequisite for activity, several pharmaceutical houses developed alternative agents, including phenelzine, isocarboxazid, tranylcypromine, and nialamide, which were all marketed as antidepressants in the 1960's. These successors to iproniazid have been regarded by many experienced clinicians as being less effective than the original MAO inhibitor, which has remained popular and in continuous use in some countries for over 30 years for treatment of atypical depression and patients refractory to other drug therapies. A reputation for inferior efficacy of these MAOIs has proven to be undeserved as borne out by more recent research.

However, other difficulties with MAOIs emerged at the same time that tricyclic antidepressants (TCAs) were introduced into clinical medicine. The efficacy of this latter class of drugs was felt to approach that of ECT. When MAOIs were found to cause life-threatening hypertensive reactions, later shown to result from ingestion of foods with high tyramine content, this drug class fell into almost total disuse in the U.S. At this time in the 1960's MAOIs as a class were nearly removed from the market by the FDA (Schiele, 1965).

Despite an aura of negativity there remained continuing clinical interest in the MAOIs, particularly in the United Kingdom and Canada where some advocates supported the notion that they were particularly useful in treating neurotic depression and so-called "atypical depression," disorders for which tricyclic antidepressants seemed less beneficial. However, the climate in the U.S. remained such that for a period of over 10 years few U.S. psychiatrists had ever encountered a patient treated with a MAO inhibitor -- in retrospect, a rather remarkable circumstance! More recently, despite the fact that MAOIs are subject to generic competition in the U.S. and hence are not actively promoted by their pharmaceutical sponsors, they have sustained a constant annual growth in new prescriptions written, as documented over the 5 year period, 1983-1987.

B. Development of Platelet Monoamine Oxidase Assay

Our involvement with these drugs derived in 1966 from the ongoing research interest in the enzyme monoamine oxidase in Sjoerdsma's laboratory at the National Heart Institute. As Blaschko (1972) pointed out at a "festschrift" held in his own honor, one reason why the discovery of the action of MAO on adrenaline in 1937 attracted the interest of many pharmacologists was the search that had been going on for a catalyst that played a role in the adrenergic neurotransmitter system analogous to the catabolic enzyme acetylcholinesterase in the cholinergic system. While MAO was specifically associated with the adrenergic neurons at that time, important differences were recognized to exist between the metabolic fates of the adrenergic and cholinergic neuronal mediators. This realization proved to be well founded with the discovery by Axelrod in 1957 of the enzyme catechol-O-methyl transferase (Axelrod, 1957). This finding also added two new important substrates of MAO to those already known: metanephrine and normetanephrine.

There had been a pervading interest within Sjoerdsma's laboratory in the endogenous monoamines and their metabolism, especially as they related to cardiovascular phenomenology. Several significant discoveries relating to this area of investigation emanated from this laboratory in the 1960's; to cite some of these:

1. The cardiovascular effects of monomine oxidase inhibition in man (Gillespie and Sjoerdsma, 1961).
2. The clinical utility of the prodrug, methyldopa, as an antihypertensive agent (Oates *et al.*, 1960).
3. A sensitive fluorometric assay of monoamine oxidase activity utilizing measurement of the rate of indoleacetic acid conversion from tryptamine (Lovenberg *et al.*, 1962).
4. Measurement of *in vivo* MAO inhibition in man, utilizing changes in urinary excretion of monoamines and enzyme activity in tissue obtained by peroral jejunal mucosa biopsy (Levine and Sjoerdsma, 1963).

In the 1960's MAOIs were touted not only as psychiatric drugs but also as useful antihypertensive, antianginal and antimigraine agents. These clinical applications derived to a significant extent from the observation by Gillespie and Sjoerdsma (1961) that MAOIs produced orthostatic blood pressure changes, an effect proportional to the degree of enzyme inhibition. This finding was based on studies utilizing indirect measurement of *in vivo* MAO inhibition estimated by quantifying 24-hr excretion of tyramine, tryptamine and serotonin. Recognizing the advantages of direct measurement of MAO inhibition, Levine and Sjoerdsma (1962) the following year demonstrated the feasibility of assaying MAO activity in specimens of mucosa obtained by peroral biopsy of human jejunum. In a related development at the time, Lovenberg, Levine and Sjoerdsma (1962) described a sensitive *in vitro* fluorometric assay for MAO activity utilizing measurement of the rate of conversion of tryptamine to indoleacetic acid.

With McEwen's characterization in 1965 of plasma benzylamine oxidase, a soluble form of MAO, my task was to extend these findings by exploring whether sufficient enzyme activity resided in a human blood specimen to provide a useful index of *in vivo* MAO activity. Initial results showed that red cells lacked the enzyme MAO, while erratic levels of MAO activity were detectable in the plasma supernatant. Subsequently, a significant source of MAO of the mitochondrial type sedimenting in the plasma layer was identified in blood platelets (Robinson *et al.*, 1968). This report presented evidence that blood platelets are rich in MAO activity, and it described a convenient radiometric assay for measuring the enzyme activity in platelet rich plasma and in the washed platelet fraction.

Studies of enzyme kinetics in these tissues showed that platelet and plasma MAO were distinct forms of the enzyme, whereas the platelet and liver enzymes were quite similar, particularly in their response to inhibitor agents. It was apparent that platelet MAO content expressed as enzyme activity per gram of tissue approached that of liver, the tissue richest in this ubiquitous enzyme. On the other hand, plasma contains relatively sparse levels of MAO activity by comparison (Table I).

Substrate characteristics of the platelet, plasma and liver enzymes are summarized in Table II. In these experiments, values for V_{max} and K_m were calculated from double reciprocal plots of velocity and substrate concentration for the platelet and liver enzymes, according to the technique reported by McEwen for the plasma enzyme (1965). Benzylamine proved to be the substrate most actively deaminated by the enzyme in each of the tissues.

In this early work (Robinson *et al.*, 1968) the activity of a variety of inhibitors was compared utilizing enzyme from each of these sources (Table III). These *in vitro* inhibitor studies served to characterize the enzyme present in each of these tissues. Since plasma MAO is a pyridoxal-containing enzyme, it can be seen that it was effectively inhibited by carbonyl reagents, such as isoniazid and KCN, while the activities of the liver and platelet enzymes were not affected by these agents. The noncarbonyl compounds, pargyline and tranylcypromine, on the other hand potently inhibited the liver and platelet enzymes.

Although the A and B forms of MAO had not yet been described at that time, in retrospect it is of interest to note that the MAO-B specific inhibitor, pargyline, employed in these experiments, proved to be the most potent inhibitor of the mitochon-

Table I. Monoamine oxidase activity of human liver, platelets and plasma with ^{14}C-benzylamine substrate[a]

Enzyme source	Average enzyme activity[b]	
	(mμmole/g tissue/hr)	(mμmole/ml whole blood/hr)
Liver	33,000	
Platelets	5,000	75
Plasma	30	15

[a]*Adapted from Robinson et al. (1968).*

[b]*The results represent average values of assays of homogenates of liver, suspensions of platelets, and crude platelet-free plasma performed at optimal pH.*

Table II. Relative reaction rates and substrate affinity constants for monoamine oxidases of liver, platelets and plasma[a]

	Liver[b]		Platelets		Plasma[c]	
Substrate	Vmax[d]	Km(M)	Vmax[d]	Km(M)	Vmax[d]	Km(M)
Benzylamine	100	4.5×10^{-4}	100	1.5×10^{-4}	100	3.3×10^{-4}
Tyramine	92	6.3×10^{-5}	45	5×10^{-5}	12	
Tryptamine	62	4.5×10^{-4}	11	1.8×10^{-5}	10	
Serotonin	36	4.0×10^{-5}	10	2×10^{-3}	<1	

[a] *Adapted from Robinson et al. (1968).*
[b] *Enzyme source is mitochondrial fraction of rat liver.*
[c] *Plasma values calculated from the data of McEwen on purified enzyme (1965).*
[d] *Maximum reaction velocity rates estimated from double reciprocal plots relative to benzylamine (100%).*

drial enzymes subjected to study. The nonselective MAO inhibitor, tranylcypromine, also was observed to be a potent inhibitor.

C. Initial Clinical Pharmacologic Studies with MAO Inhibitor Drugs

In clinical pharmacologic studies with various MAO inhibitors, measurement of *in vivo* enzyme inhibition using the platelet assay was found to correlate closely with temporally associated increases in urinary tryptamine excretion. Both platelet MAO

Table III. Effects *in vitro* of various inhibitors on monoamine oxidase activity of liver, platelets, and plasma with benzylamine substrate[a]

Inhibitor	IC_{50}[b]			
	Human liver	Rat liver	Platelets	Plasma
Pargyline	1×10^{-7} M	2×10^{-8} M	2×10^{-8} M	$> 10^{-3}$ M
JB-516	1×10^{-5} M	3×10^{-6} M	2×10^{-7} M	5×10^{-6} M
Tranylcypromine	1×10^{-6} M	1×10^{-7} M	1×10^{-6} M	1×10^{-3} M
Iproniazid	5×10^{-5} M	1×10^{-6} M	2×10^{-6} M	1×10^{-3} M
Isoniazid	$> 10^{-3}$ M	$> 10^{-3}$ M	$> 10^{-3}$ M	$> 10^{-3}$ M
KCN	$> 10^{-3}$ M	$> 10^{-3}$ M	$> 10^{-3}$ M	5×10^{-4} M

[a] *Adapted from Robinson, et al. (1968).*
[b] *Inhibitor concentrations resulting in 50% enzyme inhibition.*

activity and tryptamine excretion during MAOI treatment were found to change in a dose-dependent manner (Figure 1).

Effects of five MAOIs on platelet enzyme and urinary tryptamine in normal subjects are presented in Table IV. It can be seen that all of the drugs significantly affected (inhibited) MAO levels *in vivo*. However, the relatively lesser degrees of inhibition with phenelzine in this clinical pharmacology study for the first time raised doubts about the adequacy of a 45 mg/day therapeutic dose. As we shall see later, this observation proved to be critical in selecting the proper dosage when finalizing a study design appropriate to establish efficacy and to define the clinical profile of MAOIs. As it later turned out, clinical pharmacology studies such as these did play a key role as forerunners of definitive efficacy trials that helped foster re-emergence of MAOIs into wider medical practice.

Another interesting finding emanating from these early clinical pharmacology studies of MAOIs was the observation that enzyme activity recovered slowly, with return to baseline levels of MAO activity and tryptamine excretion typically occurring two weeks or longer after discontinuing treatment with an irreversible MAO inhibitor (Figure 1). The clinical implications of this finding were important in clarifying the drug free interval necessary to safely switch a patient from an MAOI to another antidepressant, for example, tranylcypromine, imipramine, or fluoxetine. The need for a minimum 14-day hiatus, if possible, between drugs following MAOI therapy was generally unappreciated at the time. This has proven to be critical to good patient management as toxic drug-drug interactions with MAOIs became increasingly recognized. Animal studies have determined the half-life of brain monoamine oxidase to be of the order of 11 days (Goridis and Neff, 1971).

The discovery that platelets were an abundant source of MAO, coupled with the development of a convenient platelet MAO assay in man in the late 1960's, generated on the part of my colleague, Alexander Nies, and myself an interest in re-examining the clinical utility of MAO inhibiting drugs. We undertook a series of well-controlled double-blind trials to address several basic questions. These investigations involved primarily the MAOI, phenelzine, which was selected because of its apparent safety advantages over the other leading MAOI, tranylcypromine. At the time in the late 1960's, the FDA had placed all four clinically available MAOIs, phenelzine, tranylcypromine, nialamide, and isocarboxazid, in the "possibly effective" category because adequate efficacy data derived from well designed, placebo-controlled trials were lacking for all of them. As mentioned, up until then the clinical pharmacology of these drugs had been largely unstudied and undefined. Recommended therapeutic doses of these drugs at that time were empiric, and often erroneous.

Earlier, animal (Dubnick *et al.*, 1962; Gey and Pletcher, 1961) and human postmortem (Ganrot *et al.*, 1962) brain studies had shown that high levels of brain MAO inhibition, in the range of 75-85% or greater, were required to induce detectable increases in brain monoamines (dopamine, norepinephrine, or serotonin). The platelet MAO studies had implied that the marginal levels of MAO inhibition achieved with the then recommended doses for phenelzine would probably prove inadequate for significant clinical activity. A dose of phenelzine, 45 mg/day, seemed at best borderline

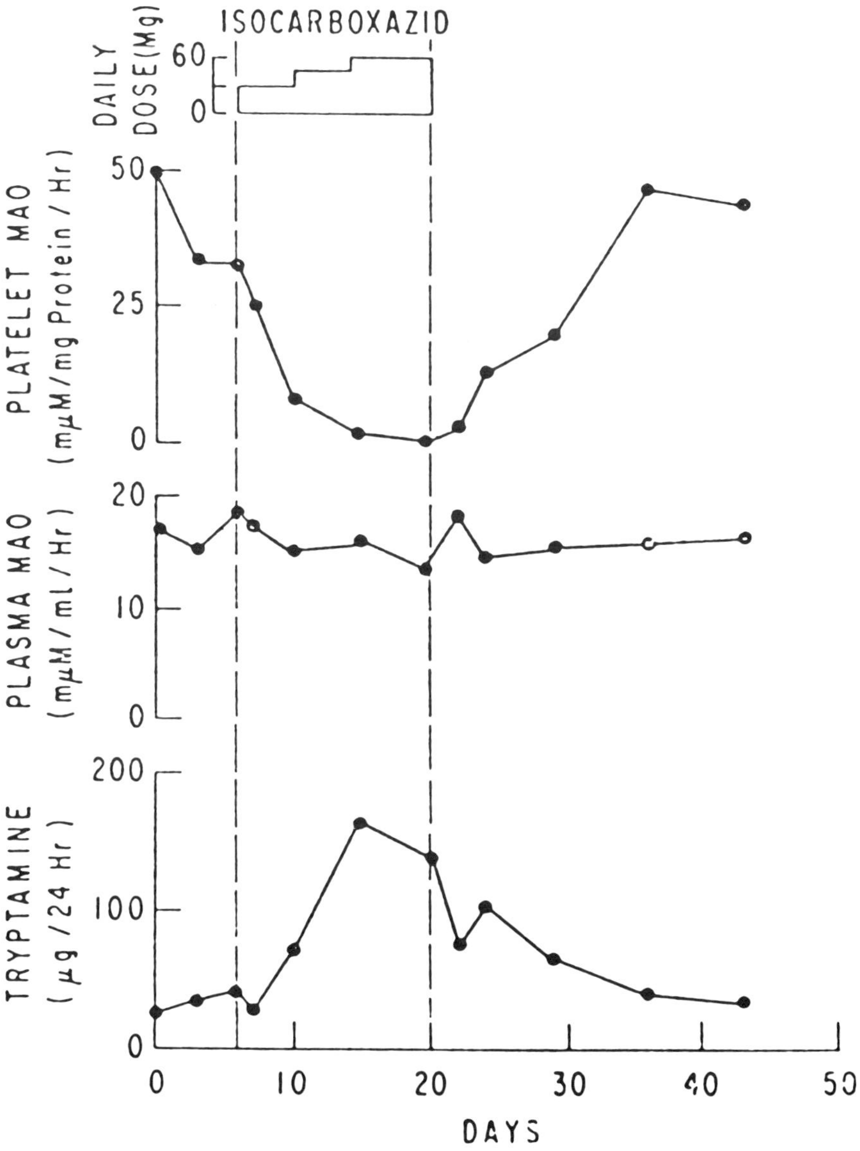

Fig. 1. *Changes in platelet and plasma MAO activity and tryptamine excretion in a subject during administration of the MAOI isocarboxazid, at doses ranging from 30 to 60 mg/day. Adapted from Robinson et al. (1968).*

Table IV. Effect of various drugs on platelet monoamine oxidase and urinary tryptamine excretion[a]

Subj. No.	Drug	Dose (mg/day)	Platelet MAO activity[b]			Urinary tryptamine[c]		
			Baseline	Rx	(% Change)	Baseline	Rx	(% Change)
1	Pargyline	100	32.7	0.0	(-100)	156	1000	(+541)
2	"	100	32.7	0.1	(- 99)	90	459	(+410)
3	"	100	17.7	0.5	(- 97)	264	1600	(+506)
4	Isocarboxazid	60	37.1	2.8	(- 93)	106	585	(+452)
5	"	60	38.0	3.9	(- 90)	34	122	(+259)
6	"	60	29.3	1.7	(- 94)	158	940	(+495)
7	Tranylcypromine	30	57.3	1.8	(- 97)	33	184	(+458)
8	"	20	39.4	2.5	(- 94)	165	1141	(+506)
9	Phenelzine	45	33.3	7.5	(- 78)	181	545	(+201)
10	"	45	34.0	10.5	(- 69)	95	174	(+ 83)
11	Nialamide	100	30.0	9.8	(- 66)	177	856	(+384)
12	"	100	23.7	12.5	(- 47)	293	944	(+222)

[a]*Adapted from Robinson, D.S., et al. (1968).*
[b]*Expressed as mu-Mole/mg protein/hr.*
[c]*Expressed as mg excreted/24 hr.*

even though this was the approved dosage in the package insert at the time. It became evident that more sophisticated dose response studies were needed to resolve the therapeutic dose issue.

II. CLINICAL EFFICACY STUDIES OF MAO INHIBITORS

In order for investigations of MAO inhibitor drugs to have any scientific relevance, it seemed essential to conclusively establish their efficacy, which had been impugned. It was important to show that phenelzine (and by inference other MAO inhibitor drugs) was indeed an efficacious antidepressant, a nagging concern at the time. The initial goal was the design and implementation of well-controlled efficacy trials with capacity to fully assess the therapeutic effects of these drugs in patients with primary depression. In these studies it was essential to utilize adequate, effective doses of drug predicated on extrapolations from the earlier clinical pharmacology findings. For these efficacy trials a higher dose of phenelzine, 60 mg/day, was employed outside the then recommended dose range. Treatment was given at a fixed dose for 6 weeks. As shown in Figure 2, significantly greater improvement in depressive symptoms was evident with phenelzine than placebo treatment. It should also be noted that antidepressant effect

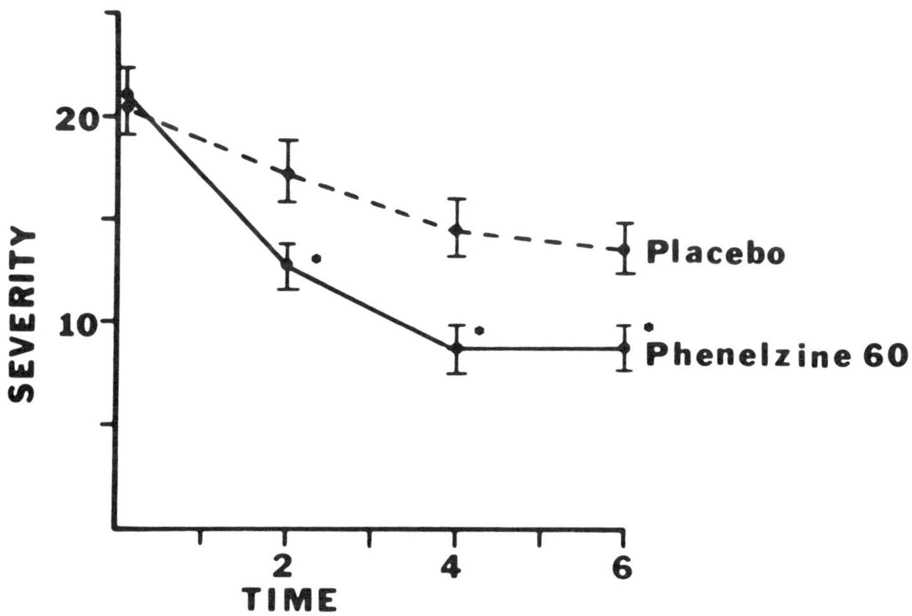

Fig. 2. *Improvement in Beck Depression Inventory rating scale scores during double-blind treatment with either phenelzine 60 mg/day (n=47) or placebo (n=46). *P<.05. Adapted from Robinson, et al. (1978).*

became apparent as early as 14 days after the beginning of drug therapy. Clinical lore at the time held that antidepressant drug therapy generally required 3 to 4 weeks before clinical improvement was to be expected. This study was a forerunner of subsequent reports confirming that antidepressant drugs, when used in adequate doses, have discernible therapeutic effects after only one or two weeks of treatment.

This placebo-controlled phenelzine study was one of the first to conclusively show an MAOI's efficacy in treating depression (Robinson *et al.*, 1978). This is somewhat ironic, given that the drug had been in clinical use for well over 10 years at the time. Indeed, this trial proved to be the positive efficacy study cited by the FDA in reaching its decision to grant an "effective" rating to phenelzine (Nardil). This was the only MAO inhibitor to receive this rating at the time. Tranylcypromine was not upgraded to effective status for several more years, and isocarboxazid, the remaining marketed MAOI, still retains its "probably effective" rating. The current isocarboxazid labeling also limits use to patients who have failed with other therapies.

With phenelzine's antidepressant efficacy established, subsequent studies were directed at defining its therapeutic dose range. A double-blind study was set-up, utilizing two different fixed doses of phenelzine, 60 mg/day and 30 mg/day, compared to placebo in treating outpatients with primary depression (Ravaris *et al.*, 1980). Mean improvement scores after 6 weeks of treatment clearly showed the higher phenelzine dose to be superior to the lower dose and placebo treatment (Table V). This study replicated the previous one in showing a daily dose of 60 mg phenelzine to be effective.

Table V. Improvement of depressed patients treated with two different doses of phenelzine or placebo[a]

Rating scale	Mean improvement (change from baseline)			
	Placebo (n=19)	Ph30 (n=16)	Ph60 (n=14)	P
Ham-D (17-item)	8.0	7.4	13.9	<.025
Beck depression scale	6.0	3.7	13.4	<.025
SDI scales[b]				
depressed mood	4.1	3.6	8.4	<.01
work and activity	0.7	1.2	2.0	<.05
retardation	3.7	2.6	5.6	.NS
anxiety	6.5	6.0	12.1	<.005
% MAO inhibition[c]	---	58 ± 15	83 ± 15	<.05

[a] *Double-blind treatment with phenelzine 60 mg/day (Ph60), phenelzine 30 mg/day (Ph30), or placebo was compared. Adapted from Ravaris, et al. (1976).*

[b] *Vermont structured depression interview.*

[c] *Mean (±S.D.) platelet MAO inhibition at treatment week 4.*

Somewhat surprisingly, the lower dose proved to be totally ineffective, not differing from placebo treatment.

All outcome measures (Ham-D, Beck Depression Inventory, Structured Depression Interview scales) were consistent in showing superiority of the 60 mg/day dosage regimen to placebo or the lower dose. This was also evidenced by the degree of improvement in individual symptom ratings, such as depressed mood, work and activity, and anxiety items.

In the high dose group mean platelet MAO inhibition exceeded 80% while it was below 60% in the low dose group. These results lent support to the potential investigative value of the *in vivo* platelet MAO inhibition assay as a research tool in clinical pharmacology studies.

To further define the therapeutic dose range of phenelzine, an analysis of all subjects receiving the fixed 60 mg/day dose was carried out stratifying by body weight. This showed that clinical improvement was inversely related to body weight. As evidenced in Figure 3, greatest improvement in SCL-90 checklist scores occurred in those patients who received a daily dosage of 1-mg phenelzine per kg of body weight.

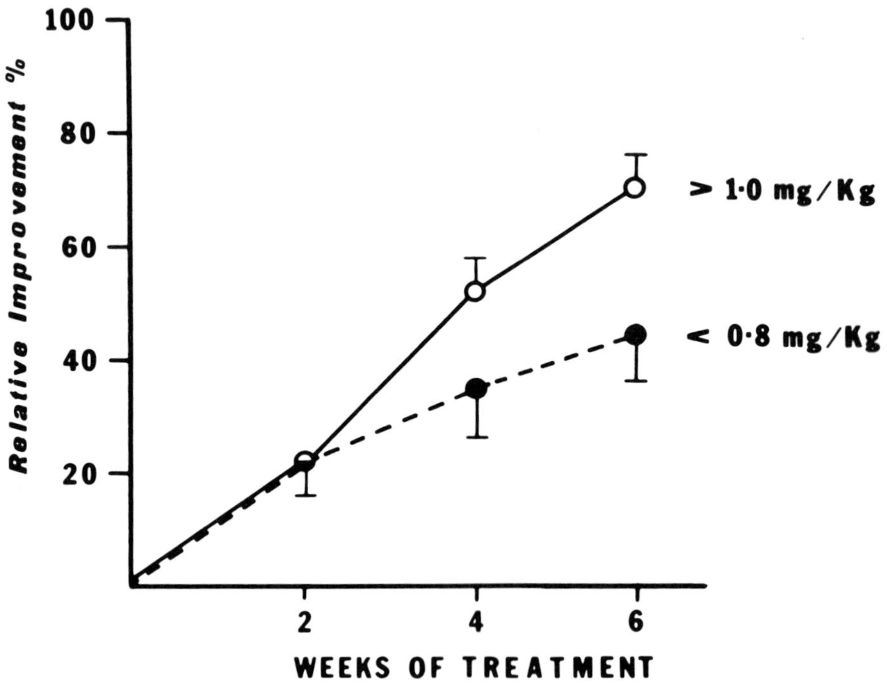

Fig. 3. *Mean improvement in Symptom Checklist-90 scores of depressed patients treated for 6 weeks with fixed dose of phenelzine 60 mg/day, stratified by dose per unit body weight. Adapted from Robinson, et al. (1978).*

The target dosage of 1 mg/kg/day was later replicated in a second double-blind controlled clinical trial also utilizing a fixed dose design (Ravaris *et al.*, 1980). Evidence thus emerged that the recommended dose of phenelzine in individual patients should be at least 1 mg/kg/day, given either in single or divided doses. These clinical pharmacology studies of phenelzine, conducted long after the drug was introduced into clinical practice, for the first time provided accurate dosing and efficacy data, information previously unavailable in the drug labeling or psychopharmacology literature.

III. PROGRESS IN MONOAMINE OXIDASE INHIBITOR RESEARCH DURING THE PRESENT DECADE

Over the past 10 years there has been a rebirth of scientific and clinical interest in the MAO inhibitor drugs. Space does not permit recounting all important advances and contributions to the literature; rather here I will cite selected findings that had a major impact on enhancing our understanding of this unique class of drugs and helped to spur further drug discovery and development efforts.

A. MAO Inhibitors: Preclinical Discoveries

The primary mechanism of action of all types of antidepressant drugs, not only the MAO inhibitors, remains an enigma despite intensive biological investigation of this question. The advent of selective serotonin uptake inhibitors and serotonin type-1A receptor agonists, both classes of drugs that show antidepressant activity, has recently focussed increasing attention on serotonergic mechanisms as being central to the pathophysiology of mood disorders. Serotonin neuronal systems have become a primary focus of new drug research in mood disorders. While TCAs profoundly affect both noradrenergic and serotonergic neuronal systems, the more selective 5HT-uptake inhibitors, the serotonin partial agonists (buspirone and gepirone), and the MAO inhibitors all appear to act through primary actions on serotonin pathways. Deranged or dysfunctional central serotonergic systems may be etiologic in the pathophysiology of both anxiety states and the affective disorders (Coppen, 1967; Willner, 1985; Iversen, 1984).

Recent work of deMontigny and coworkers (Blier *et al.*, 1988) employing electrophysiological techniques is particularly valuable in furthering our understanding of the neuronal effects of antidepressant drugs. Initially, MAOI administration inhibits neuronal firing of both noradrenergic and serotonergic neurons. With continued administration, however, there is progressive desensitization of the somatic autoreceptors of raphe serotonergic neurons, as evidenced by the decreased responsiveness of these neurons to systemically administered lysergic acid (LSD). Unlike serotonin systems, sensitivity of the somatic noradrenergic autoreceptors remains unmodified so that adrenergic firing continues to be suppressed with chronic treatment. Postsynaptically, serotonergically-enervated neurons exhibit enhanced responsiveness evidenced by hyperpolarization of hippocampal pyramidal neurons induced by electrostimulation of the ascending raphe 5HT pathway. A corresponding enhancement of

neurotransmission of the dorsal noradrenergic bundle does not occur. Therefore, it has been postulated that the ultimate effects of chronic MAOI antidepressant administration following at least 2 weeks of treatment is selective facilitation of serotonergic (5HT) neurotransmission (Blier *et al.*, 1986).

These same investigators recently reported that chronic imipramine treatment also increases the response to stimulation of the ascending 5HT pathway in the firing activity of dorsal hippocampal neurons (Blier *et al.*, 1988). This suggests that serotonin neurotransmission is enhanced not only by MAOIs, but also by other putative antidepressants, including the 5HT partial agonists and tricyclic antidepressants (imipramine), when given chronically. Whether all antidepressant therapies act similarly on central serotonergic systems requires confirmation (Blier *et al.*, 1988). Serotonergic pathways may be central to the action of antidepressant drugs. This appears to be the case for the MAOIs and the 5HT-1A partial agonist drugs. All known effective antidepressant drugs produce a compensatory decrease in cortical 5HT-2 receptors, suggesting a mechanistic linkage of antidepressant drugs to central serotonin systems.

B. Monoamine Oxidase Inhibitors: Development of New Agents

An important new direction in MAOI drug development has been the synthesis of a series of selective inhibitors of the MAO-A enzyme subtype. Unlike currently marketed drugs, some of these new chemical entities bind reversibly to the enzyme so that enzyme inhibition dissipates gradually over several hours to days rather than the period of weeks with present MAOIs before activity levels return to normal since new enzyme must be synthesized. New drug development strategy has focussed on alleviating unwanted side effects and minimizing the safety liabilities of MAOIs. The propensity of existing MAOIs to predispose to hypertensive reactions, including hypertensive crises, is particularly problematic and limits the use of these otherwise highly effective psychopharmacologic agents.

There is little convincing evidence that the MAO-B selective inhibitors possess significant antidepressant properties, except when given in high doses where selectivity is lost and MAO-A inhibition also occurs. Therefore, a major focus has been development of inhibitors of MAO-A, the form of the enzyme that degrades intraneuronal serotonin, norepinephrine, and certain other bioactive amines. Reversible MAO-A inhibitors may offer additional advantages. However, the antidepressant efficacy of reversible inhibitors, which remains to be established, may be diminished. It may not be possible to dissociate therapeutic effect from a requisite sustained level of neuronal MAO inhibition. Undesirable CNS-mediated side effects of MAOIs (appetite stimulation, excessive weight gain, vasomotor center inhibition, and sexual dysfunction) are particularly troublesome and may be linked to MAO-A rather than MAO-B inhibition. These side effects may not be amenable to the selective inhibitor approach. This could be the price one has to pay for clinical efficacy.

On the other hand, the potentially life-threatening, but fortunately rare, hypertensive reactions can seemingly be minimized via the selective MAO-A inhibition approach. A particularly ingenious strategy has been the synthesis of pharmacologically inactive precursors of MAO-A inhibitors that require decarboxylation by aromatic *L*-amino acid decarboxylase (AADC), an enzyme with high activity in monoaminergic neurons. Potent, selective inhibitors of the MAO-A enzyme subtype have been developed that are activated in this manner (McDonald *et al.*, 1980; Palfreyman *et al.*, 1986). One such agent, MDL 72,394, is an inactive pro-drug that on intraneuronal decarboxylation is converted to a fluoroallylamine that selectively and irreversibly inhibits MAO-A activity (Palfreyman *et al.*, 1986, Palfreyman *et al.*, chapter 11, 1989). Currently undergoing phase 2 trials, MDL 72,394 at a dose somewhat below that associated with enhanced tyramine sensitivity is showing promise of clinical antidepressant activity (W. Lovenberg, personal communication, 1989).

AADC activity of neuronal tissues may exceed that of liver, thereby resulting in a higher degree of neuronal MAO inhibition than that in other tissues. This could have obvious advantages for this prodrug approach. Critical levels of MAO inhibition in the central nervous system required for psychopharmacologic activity may be attained with MDL 72,394 while minimizing enzyme inhibition in the peripheral organs. The latter is responsible for vulnerability to an ingested tyramine load.

An observation by Campbell and coworkers (1980) could be relevant to the mechanism of action of MDL 72,394. Chronically administered phenelzine or tranylcypromine appears to enhance brain (neuronal) AADC activity. If this enhancement is a result of MAO inhibition in general, irrespective of the agent, then MDL 72,394 may induce its own activation within the neuron. This could further enhance the selectivity for neuronal tissues of this prodrug.

Conclusive evidence of its clinical efficacy from controlled trials is awaited with interest. The promising approach taken with MDL 72,394 and its analogues could offer significant advantages over existing MAOIs by selectively inhibiting intraneuronal MAO-A activity.

C. Clinical Pharmacology of Monoamine Oxidase Inhibitors: Recent Findings

The pharmacokinetics and metabolism in man of existing MAO inhibitor drugs received little attention until recently. Since these drugs were approved prior to the current legislation regulating drug development, studies conducted as a matter of course today were not done originally. At the time of its registration and approval, little was known about phenelzine's metabolic biotransformation or clinical pharmacokinetics. This situation continued despite years of therapeutic use. For many years it had been incorrectly assumed that acetylation formed the major metabolic pathway for phenelzine, governing its elimination half-life. Knowledge of tranylcypromine's fate was equally unstudied until relatively recently when its metabolism and clearance have been the subject of study. There remains a dearth of information about isocarboxazid's biotransformation and clearance. The pharmacokinetics and metabolism of MAOIs have been recently reviewed (Robinson and Kurtz, 1987).

The irreversible enzyme inhibitors in clinical use today, i.e., tranylcypromine, phenelzine, and isocarboxazid, are now known to act as substrates for monoamine oxidase. On binding to the enzyme a latent reactive group of the molecule activated by the normal catalytic activity of the enzyme forms an irreversible covalent bond. This results in irreversible inactivation of both the enzyme and the drug; hence these drugs have been termed "suicide" enzyme inhibitors (Paech *et al.*, 1979; Sjoerdsma, 1981). This property of MAOIs may account for their apparently unique pharmacokinetics. It could explain the very short plasma half-lives observed with phenelzine and tranylcypromine. Following single dose administration of phenelzine, its plasma elimination half-life ranges between 1 and 4 hours (Cooper *et al,*. 1978; Robinson *et al.*, 1985). Since irreversible MAO inhibitor drugs become substrates of the enzyme, it is also possible that with continued use MAOIs may impair their own clearance. This could be responsible for what appears to be the nonlinear pharmacokinetics of these irreversible enzyme inhibitors.

Available evidence suggests that there is gradual accumulation of phenelzine over a several week period before steady state is achieved (Cooper *et al.*, 1978; Robinson *et al.*, 1985). With longer term treatment profound levels of enzyme inhibition could develop in brain tissues. This phenomenon, which has received relatively little attention until now, deserves investigation. It is known that plasma concentrations of phenelzine progressively increase in depressed patients who have received several weeks or months of treatment (Robinson *et al.*, 1981).

For over a decade it was presumed that acetylation was the major metabolic pathway for degradation of MAOIs of the hydrazine class (such as phenelzine and isocarboxazid). Reports in the early literature suggested that appropriate clinical dosing of these drugs was dependent on acetylator phenotype, with slow acetylators more susceptible to MAOI side effects (Evans *et al.*, 1965). We investigated phenelzine's biotransformation, utilizing site-specific, stable isotope-labelled phenelzine. Labelled drug administered as a single 60 mg dose was given to drug naive and chronically treated subjects. Plasma and urine were collected over a 96 hour period. Using a highly sensitive GC/MS assay (Jindal *et al.*, 1980) both were analyzed for parent drug and potential metabolites. No acetylated products of parent drug or metabolites could be detected in plasma or urine. Over 80% of the administered dose was recovered as either phenylacetic acid or hydroxyphenylacetic acid, which constitute the major metabolites of phenelzine (Cooper *et al.*, 1978).

These stable isotope studies conclusively showed that acetylation is not a metabolic pathway for the degradation of phenelzine in man. These pharmacokinetic findings are consistent with other more recent clinical reports that failed to find association of acetylator phenotype and either side effects or clinical efficacy of isocarboxazid or phenelzine (Davidson *et al.*, 1978; Robinson *et al.*, 1980; Marshall *et al.*, 1978; Yates and Loudon, 1979).

Much more limited clinical pharmacologic data are available for the other clinically used MAOIs, tranylcypromine and isocarboxazid. Unlike phenelzine, for which a relationship between platelet MAO inhibition and clinical outcome has been shown (Robinson *et al.*, 1978; Ravaris *et al.*, 1976), these other MAOIs have been the subject of fewer studies. Utility of monitoring platelet MAO inhibition has not been shown in the case of these latter drugs. Only recently have studies of the kinetics and metabolism of tranylcypromine in man been reported. Youdim and coworkers detected tranylcypromine, amphetamine, metamphetamine, and phenylethylamine in the plasma of a patient who ingested a 250 mg dose of tranylcypromine (Youdim *et al.*, 1979). The plasma elimination half-life in this patient was estimated to be 3.5 hr. Phenylethylamine was observed to have an even shorter half-life, 2.8 hr. Amphetamine and methamphetamine had detectable although lower concentrations than parent drug or phenylethylamine following the overdose. However, their concentrations followed a pattern of increasing rather than declining levels over a 2-day period, suggesting that they represent tranylcypromine metabolites. The overdose syndrome exhibited by the patient was suggestive of mild psychotomimetic effects, attributed to either phenylethylamine or the amphetamine metabolites. These investigators also showed that when tranylcypromine was incubated with rat liver homogenate, both phenylethylamine and methamphetamine were generated. Amphetamines have also been reported to be metabolic products of the MAO-B inhibitor deprenyl (Reynolds *et al.*, 1978).

Simpson and coworkers found that following a single 10 mg challenge dose of tranylcypromine, peak plasma levels occurred at 1 to 2 hr, with a plasma elimination half-life of less than 2 hr (Simpson *et al.*, 1985). Significant platelet MAO inhibition persisted for at least 1 week following this single dose. Another study in normal subjects indicated that tranylcypromine kinetics exhibit a biexponential plasma elimination curve and may follow a two compartment model (Bailey and Barron, 1980). In a study to assess its biologic half-life, Bieck and Antonin estimated the pharmacodynamic half-life of tranylcypromine to be of the order of several days, based on evidence of residual sensitivity to tyramine infusion for some time after drug discontinuation (Bieck and Antonin, 1982). One subject treated for 4 weeks exhibited tyramine supersensitivity for a number of weeks after drug was stopped. In another recent report it was suggested that several cases of apparent autoinduction of hypertensive reactions during tranylcypromine treatment, which could not be attributed to dietary indiscretions, may involve the problem of conversion of tranylcypromine to metabolites with pressor activity (Keck *et al.*, 1989).

These clinical pharmacologic studies of the irreversible MAO inhibitor drugs all show that MAOIs are rapidly cleared from plasma. Nonetheless, enzyme inhibition and its biologic sequelae persist for extended periods after stopping these MAOIs. Whether the reversible MAO inhibitors currently under development will offer clinical advantage remains to be established in well-controlled studies. As might be predicted for drugs that act as suicide enzyme inhibitors, the pharmacokinetics and pharmacodynamics of MAOIs appear to be unique. Further investigation is needed to fully characterize the complex pharmacokinetics and pharmacodynamics of MAOIs in humans.

D. Clinical Efficacy of MAO Inhibiting Drugs: Recent Findings

Contrary to the earlier literature that suggested MAOIs play a limited role in the treatment of affective disorders (i.e., primarily for treating atypical depression) recent clinical investigations indicate these agents possess a broad therapeutic scope in the treatment of mood disorders. Numerous well-controlled trials now support the efficacy of the nonselective MAOIs, as well as the specific MAO-A inhibitor, clorgyline (investigational status in the U.S.), in the treatment of more typical as well as severe

Table VI. Clinical disorders for which monoamine oxidase inhibitors have evidenced clinical efficacy

Depression
major depression (including melancholic subtype)
dysthmic disorder
atypical depression
Panic disorder (with or without agoraphobia)
Social phobia
Eating disorders (bulimia)
Obsessive-complusive disorder
Migraine (prophylactic)
Post-traumatic stress disorder

forms of major depression. This includes treatment of endogenous and melancholic depression subtypes of major depression. The efficacy of MAOIs overall is at least equivalent to the tricyclic antidepressants (Robinson, 1985; Quitkin *et al.*, 1979; Janicak *et al.*, 1988), and their therapeutic applications may be broader. MAOIs may possess additional therapeutic properties beyond strictly antidepressant activity. They appear to have intrinsic anxiolytic activity. This might account for the apparent unique benefits of MAOIs in managing anxiety disorders as well as treating depression (Robinson and Kurtz, 1987; Larsen, 1988).

Disorders for which there is considerable evidence of the clinical utility of MAO inhibitors are listed in Table VI.

To summarize, MAOIs have gained widespread acceptance over the last decade as effective alternative therapy to more traditional antidepressant agents and as the treatment of choice for selected patients with a variety of disorders. Their therapeutic application in a spectrum of anxiety disorders is based on more recent evidence from both uncontrolled and controlled clinical trials. Further investigation is needed to fully define their clinical profile. This applies particularly to melancholic major depression, dysthymic disorder, and anxiety disorders where in recent reports MAOIs appear to show promise as effective alternative therapy.

MAOIs have proven useful in the treatment of depressive disorders of all types. They still remain a preferred treatment for atypical depression. Overall, MAOIs appear to be superior to tricyclic antidepressants in major depression based on both observer-rated and self-reported outcome measures. When placed on an MAOI for the first time patients frequently volunteer that they have never felt as well. Adverse effects tend to be the primary limiting factor to MAOI use. In spite of this limitation, an expanded future role for MAO inhibitors in clinical medicine is predicted. If the safety profile can be significantly improved by development of safer and more selective agents, MAOIs offer prospects of substantial therapeutic advantage in a number of psychiatric disorders.

E. The Safety Profile of MAO Inhibitor Antidepressants: Current Evidence

In addition to their liability for hypertensive reactions, MAOIs have other disturbing side effects that limit their clinical acceptability. These can lead to severe drug intolerance and ultimately force discontinuation of the MAOI despite an excellent initial therapeutic response. It is not uncommon for patients to volunteer that they have never felt better than when treated with an MAOI, yet over the longer term they cannot tolerate the adverse effects associated with long-term use of these drugs. One is frequently compelled to discontinue treatment after several weeks or months of therapy for this reason. Fortunately, many other patients seem to tolerate MAOIs very well, even when given for long periods of time. Studies conducted in recent years have enabled investigators to more accurately document the true incidence of MAOI side effects and to relate them to length of drug treatment.

Typical MAOI side effects of clinical importance are listed in Table VII.

One of the most problematic adverse effects of MAOIs, fortunately mild in most patients, is orthostatic hypertension. This can be severe in the occasional patient to the point of recurrent presyncope or syncope. This side effect reflects centrally-mediated vasomotor center inhibition, one of the pharmacologic effects of this drug class (Robinson *et al.*, 1982). Nearly all patients experience some orthostatic blood pressure changes as well as a general lowering of blood pressure. This is asymptomatic in most patients. Severe orthostatic hypotension is usually not amenable to treatment when it occurs, although the use of mineralocorticoids has been advocated to counter the hypotension.

Excessive weight gain and increased appetite, thought to be a drug-mediated effect on hypothalamic appetite centers, are perhaps the most common troublesome adverse effects of long term MAOI treatment. Weight gains over several months of MAOI treatment in excess of 40-50 lbs. are not uncommon, particularly in women. Sexual dysfunction, presumably also hypothalamically-mediated, can be a very distressing side effect in men. There seems to be no effective therapy for this adverse effect other than stopping the drug. Reducing the dose of MAO is occasionally helpful; however, typically dose reduction also results in lessened therapeutic benefit.

The recent advent of controlled clinical trials of maintenance treatment with MAOIs has afforded an opportunity to assess adverse experiences and risk-benefit of

Table VII. Clinically important and troublesome side effects of monoamine oxidase inhibitors.

Excessive weight gain (hyperphagia)
Sexual dysfunction
Insomnia
Symptomatic orthostatic hypotension
Edema (refractory to diuretics)
Myoclonic movements
Hypomania

Table VIII. Incidence of adverse experiences during 3 phases of phenelzine treatment of patients with major depression[a]

	Acute Rx (n=83)	Cont. Rx (n=73)	Maintenance Rx ≤ 6 mo. (n=34)	Maintenance Rx >6 mo. (n=17)
CV				
dizziness	18 (22%)	1 (1%)	0 (0%)	1 (6%)
fainting	6 (7%)	1 (1%)	0 (0%)	1 (6%)
palpitations	6 (7%)	4 (5%)	2 (6%)	1 (6%)
Neuro/Neuromusc.				
twitching	24 (29%)	25 (34%)	7 (21%)	4 (24%)
tremor	11 (13%)	7 (10%)	3 (9%)	2 (12%)
paresthesia	9 (11%)	16 (22%)	6 (18%)	4 (24%)
Behavioral				
drowsiness	16 (19%)	17 (23%)	5 (15%)	4 (24%)
dreaming	5 (6%)	12 (16%)	4 (12%)	3 (18%)
confusion	7 (8%)	10 (14%)	3 (9%)	2 (12%)
hypomania	8 (10%)	4 (5%)	2 (6%)	2 (12%)
GI				
hyperphagia	27 (32%)	27 (37%)	8 (24%)	2 (12%)
GU				
decreased libido	8 (10%)	5 (7%)	4 (12%)	4 (23%)
anorgasmia	16 (19%)	10 (14%)	4 (12%)	2 (12%)
impotence	2 (2%)	2 (3%)	0 (0%)	1 (6%)
ejac. disturb.	5 (6%)	6 (8%)	0 (0%)	0 (0%)
Autonomic				
blurred vision	14 (17%)	7 (10%)	5 (15%)	4 (23%)
dry mouth	44 (53%)	20 (27%)	8 (24%)	2 (12%)
ankle edema	16 (19%)	18 (25%)	8 (24%)	4 (23%)

[a]*Acute Rx = acute treatment period (6-10 wks); Cont. Rx = continuation treatment of responding patients (≥4 mos. of response); Maintenance Rx = double-blind maintenance treatment period (≤2 yrs.) Adapted from Robinson, et al. (1988).*

these agents over the longer term. The full impact of side effects during extended treatment of MAOI responders can now be better appreciated. In a study of benefits of 2-year maintenance treatment with phenelzine in preventing relapses, drug side effects were characterized as a function of treatment duration (Robinson et al., 1988; Robinson *et al.*, 1986). While adverse effects proved to be a relatively uncommon reason for discontinuation during the initial weeks of phenelzine treatment, with continuing therapy side effects often emerged and became problematic after therapeutic response had been attained. During continuation treatment of responding patients side effects proved to be the most common reason for treatment terminations (11/73 responding patients discontinued for reason of adverse drug experience). The most frequent treatment discontinuations for reason of adverse experience (intolerance) were excessive weight gain, sexual dysfunction and sleep disturbance. Most MAOI-treated patients report some changes in sleep patterns. Generally, there is an overall decreased need for sleep. In some instances, severe insomnia ensues, which is refractory to therapy as long as drug is continued. Again, reducing the dose of phenelzine is sometimes helpful in this situation, but often therapeutic benefit is also lost by so doing.

The incidence of all adverse experiences reported during long term treatment study with phenelzine are presented in Table VIII.

F. Efficacy of Maintenance Treatment of Depression with MAO Inhibitors

The efficacy, safety and optimal duration of drug treatment of depression are unresolved issues, currently the foci of several ongoing studies. The duration of a major depressive episode probably exceeds one year, based on evidence from the prepsychopharmacology era as well as on emerging data from maintenance treatment studies that allow estimates to be made based on relapse rates when placebo is substituted. It appears that all somatic treatments act as suppressive therapy only. The

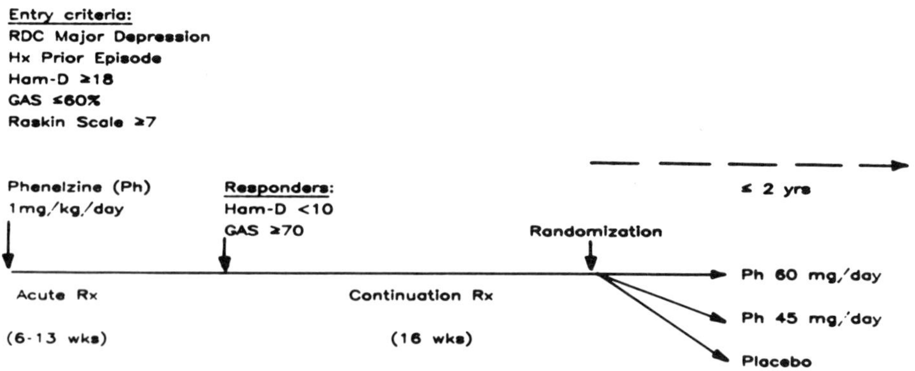

Fig. 4. *Long term maintenance treatment study design. Criteria for admission to study and for defining treatment response and relapse are shown. Patients who maintained therapeutic response for 16 weeks were then randomized to double-blind maintenance treatment for 2 years with one of two doses of phenelzine or placebo.*

depressive syndrome tends to return if treatment is stopped prematurely before the episode has run its coarse. A 5-year study recently completed at Marshall University assessed the clinical advantages of maintenance phenelzine treatment of patients with a major depression (Robinson *et al.*, 1988; Robinson *et al.*, 1986). The study design is shown in the following schematic (Figure 4).

After granting informed consent, 88 patients who met entry criteria were enrolled in the study. Seventy-three of these patients initially responded to phenelzine treatment and entered a continuation treatment phase for an additional 16 weeks. Forty-seven of these 73 patients successfully sustained a satisfactory therapeutic response to MAOI for the requisite 4 month period and became eligible for entry into double-blind maintenance treatment. These patients were randomized to two years of double-blind treatment with one of two doses of phenelzine or placebo. Comparative relapse rates for drug-treated and placebo-treated patients during this double-blind 2-year maintenance phase are presented in Figure 5.

Patients switched to double-blind placebo maintenance treatment underwent a 4-week dose reduction schedule of 15 mg/day every 7 days. These patients relapsed within 10 weeks of randomization at a high rate (80% of placebo patients). On the other hand, maintenance of active drug treatment was associated with significantly

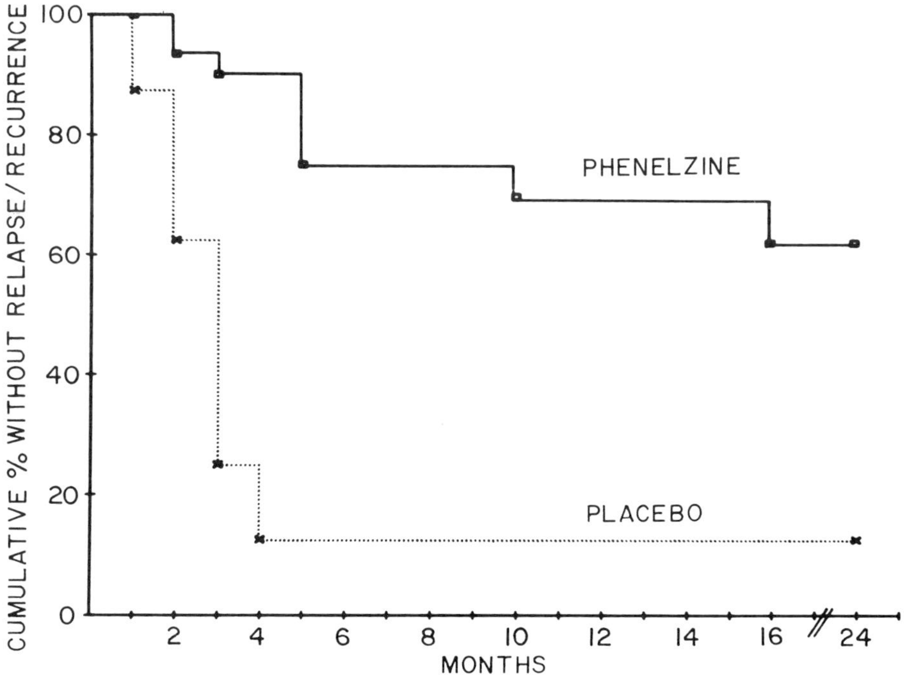

Fig. 5. *Survival curve analysis of patients randomized to phenelzine (45 or 60 mg/day) or placebo. Relapse (>2 mos.) and recurrence (<2 mos.) after assignment to double-blind maintenance treatments are indicated.*

Table IX. Maintenance treatment study of major depression: estimates of episode length[a]

Rx group	No. of relapsers	Mean duration of episode (months)	
Placebo	n=13	14.6 ± 2.2	
Phenelzine 45 mg/d	n= 2	20.2 ± 10.0	15 ± 2 mos.
Phenelzine 60 mg/d	n= 2	13.4 ± 6.5	

[a]*Episode length is estimated as the sum of the time from onset of the episode prior to entry into the study, and the total drug treatment time of patients relapsing within the first two months following randomization.*

lower relapse and recurrence rates. Both doses of phenelzine successfully prevented immediate relapse or delayed (late) recurrence of a depressive episode in most patients. Drug benefit was manifest over the two year period. The optimal duration of long term MAOI treatment remains in question, and resolution of this issue will require additional study.

Of interest to phenomenologists, extrapolation of these maintenance treatment data seems to indicate that the length of a major depressive episode may typically exceed 15 months (see Table IX).

Benefit of maintenance treatment with the tricyclic antidepressants has also been reported (Prien *et al.*, 1984). A preliminary study of maintenance treatment with the MAOI, isocarboxazid, indicates that responders should be continued on drug for at least 6 to 12 months (Davidson and Sullivan, 1983).

The findings of these and other controlled trials reaffirm the customary practice, based on longstanding clinical experience, of maintaining responding patients on antidepressant drugs for at least 6 to 12 months. In patients with recurrent or chronic depressions it is not uncommon to continue treatment even longer. Some patients may require antidepressant drug therapy for several years. Exacerbations often occur when attempts are made to withdraw medication in such patients. The emergence of a pharmacologic withdrawal syndrome following chronic antidepressant treatment does not account for these symptomatic relapses observed on stopping drug. Rather, there is typically a return of the full blown pretreatment depressive syndrome.

III. SUMMARY

Today monoamine oxidase remains a target for drug design, still a relevant issue some 30 years after introduction of the MAO inhibitors. Although MAOIs existed in near oblivion for over a decade in the U.S., scientific and clinical interest is now high. These drugs enjoy widespread acceptance in psychiatric practice because of unique therapeutic properties. While clinical applications in cardiovascular disorders have long since been abandoned, MAOIs remain extremely useful in the treatment of a variety of psychiatric disorders. Their re-emergence stemmed in part from clinical pharmacologic studies conducted at the National Heart Institute some two decades ago and subsequently continued at the University of Vermont. Recent studies have served to better define the clinical profile of MAO inhibitors.

As we move into the 1990s, MAO inhibiting drugs enjoy growing popularity because of a special place in the management of psychiatric disorders. The advent of safer and better tolerated drugs of this class, such as selective MAO-A inhibitors with neuronal specificity, offers the prospect of even greater safety without sacrificing efficacy. It is worth noting that among his many achievements, Dr. Albert Sjoerdsma has been associated with this evolution of MAO inhibitors in psychopharmacology.

REFERENCES

Axelrod, J.L. (1957). *Science* **126**, 400.

Bailey, E. and Barron, E.J. (1980). *J. Chromatogr.* **183**, 25-31.

Bieck, P.R. and Antonin, K.H. (1982). *Eur. J. Clin. Pharmacol.* **22**, 301-308.

Blaschko, H. (1972). *Adv. Biochem. Psychopharmacol.* **5**, 1-10.

Blier, P., de Montigny, C., and Azzaro, A.J. (1986). *J. Pharmacol. Exp. Ther.* **237**, 987-994.

Blier, P., de Montigny, C., and Chaput, Y. (1988). *Clin. Neuropharmacol.* **11** (Suppl.2), S1-S10.

Campbell, I.C., Murphy, D.L., Walker, M.N., Lovenberg, W., and Robinson, D.S. (1980). *Br. J. Clin. Pharmacol.* **9**, 431-432.

Cooper, T.B., Robinson, D.S., and Nies, A. (1978). *Commun. Psychopharmacol.* **2**, 505-512.

Coppen, A. (1967). *Br. J. Psychiatry* **113**, 1237-1264.

Crane, G.E. (1956). *Am. J. Psychiatry* **112**, 494-501.

Davidson, J.R.T. and Sullivan, J.L. (1983). *J. Clin. Psychopharmacol.* **3**, 42-43.

Davidson, J., McLeod, M.N., and Blum, M.R. (1978). *Am. J. Psychiatry* **135**, 467-469.

Dubnick, B., Leeson, G.A., and Phillips, G.E. (1962). *J. Neurochem.* **9**, 299-306.

Evans, D.A.P., Davison, K., and Pratt, R.T.C. (1965). *Clin. Pharmacol. Ther.* **6**, 430-435.

Ganrot, P.O., Rosengren, E., and Gottfries, C.G. (1962). *Experientia* **18**, 260.

Gey, K.F. and Pletscher, A. (1961). *J. Neurochem.* **6**, 239-243.

Gillespie, L. and Sjoerdsma, A. (1961). *Med. Clin. North Am.* **45**, 421-428.

Goridis, C. and Neff, N.H. (1971). *J. Neurochem.* **18**, 1673-1682.

Hare, M.L.C. (1928). *Biochem. J.* **22**, 968-979.

Iversen, S.D. (1984). *Neuropharmacology* **23**, 1553-1560.

Janicak, P.G., Pandey, G.N., Davis, J.M., Boshes, R., Bresnahan, D., and Sharma, R. (1988). *Am. J. Psychiatry* **145**, 93-95.

Jindal, S.P., Lutz, T., and Cooper, T.B. (1980). *J. Chromatogr.* **221**, 301-308.

Keck, P.E., Jr., Pope, H.G., Jr., and Nierenberg, A.A. (1989). *J. Clin. Psychopharmacol.* **9**, 48-51.

Kline, N.S. and Cooper, T.B. (1980). In "Psychotropic agents" (F. Hoffmeister and G. Stille, eds.) (Handbook of experimental pharmacology, v. 55/1), pp. 369-397. Berlin, Springer Verlag.

Larsen, J.K. (1988). *Acta Psychiatr. Scand.* **78** (Suppl. 345), 74-80.

Levine, R.J. and Sjoerdsma, A. (1962). *Proc. Soc. Exp. Biol. Med.* **109**, 225-227.

Levine, R.J. and Sjoerdsma, A. (1963). *Clin. Pharmacol. Ther.* **4**, 22-27.

Loomer, H.P., Saunders, J.C., and Kline, N.S. (1957). *Psychiatr. Res. Rep. Am. Psychiatr. Assoc.* **8**, 129-141.

Lovenberg, W., Levine, R.J., and Sjoerdsma, A. (1962). *J. Pharmacol. Exp. Ther.* **135**, 7-10.

McDonald, I.A., Palfreyman, M.G., Zreika, M., and Lacoste, J.M. (1984). Poster presented at the Eighth International Symposium of Medicinal Chemistry, Stockholm, Sweden.

McEwen, C.M., Jr. (1965). *J. Biol. Chem.* **240**, 2003-2010.

Marshall, E.F., Mountjoy, C.Q., Campbell, I.C., Garside, R.F., Leitch, I.M., and Roth, M. (1978). *Br. J. Clin. Pharmacol.* **6**, 247-254.

Maxwell, R. and Eckhardt, S. (1989). In "Handbook of drug discovery" (R. Maxwell and S. Eckhardt, eds.), Clifton, NJ, Humana Press. In press.

Oates, J.A., Gillespie, L., Udenfriend, S. and Sjoerdsma, A. (1960). *Science* **131**, 1890-1891.

Paech, C., Salach, J.I., and Singer, T.P. (1979). In "Monoamine oxidase: structure, function and altered functions" (T.P. Singer, R.W. VonKorff, D.L. Murphy, eds.), pp. 39-50. New York, Academic Press.

Palfreyman, M.G., McDonald, I.A., Bey, P., Schechter, P.J., and Sjoerdsma, A. (1986). Paper presented at the Fifteenth Congress, Collegium International Neuropsychopharmacologicum, San Juan, Puerto Rico.

Prien, R.F., Kupfer, D.J., Mansky, P.A., Small, J.G., Tuason, V.B., Voss, C.B., and Johnson, W.E. (1984). *Arch. Gen. Psychiatry* **41**, 1096-1104.

Quitkin, F., Rifkin, A., and Klein, D.F. (1979). *Arch. Gen. Psychiatry* **36**, 749-760.

Ravaris, C.L., Nies, A., Robinson, D.S., Ives, J.O., Lamborn, K.R., and Korson, L. (1976). *Arch. Gen. Psychiatry* **33**, 347-350.

Ravaris, C.L., Robinson, D.S., Ives, J.O., Nies, A., and Bartlett, D. (1980). *Arch. Gen. Psychiatry* **37**, 1075-1080.

Reynolds, G.P., Elsworth, J.D., Blau, K., Sandler, M., Lees, A.J., and Stern, G.M. (1978). *Br. J. Clin. Pharmacol.* **6**, 542-544.

Robinson, D.S. (1986). *Psychopharmacol. Bull.* **22**, 12-15.

Robinson, D.S. and Kurtz, N.M. (1987). In "Psychopharmacology: the third generation of progress" (H.Y. Meltzer, ed.), pp. 1297-1304. New York, Raven Press.

Robinson, D.S., Lovenberg, W., Keiser, H., and Sjoerdsma, A. (1968). *Biochem. Pharmacol.* **17**, 109-119.

Robinson, D.S., Nies, A., Ravaris, L., Ives, J.O., and Bartlett, D. (1978). *Arch. Gen. Psychiatry* **35**, 629-635.

Robinson, D.S., Nies, A., and Cooper, T.B. (1980). *Clin. Pharmacol. Ther.* **27**, 280.

Robinson, D.S., Nies, A., Corcella, J., and Cooper, T.B. (1981). *Psychopharmacol. Bull.* **17**, 154-157.

Robinson, D.S., Nies, A., Corcella, J., Cooper, T.B., Spencer, C., and Keefover, R. (1982). *J. Clin. Psychiatry* **43**, 8-15.

Robinson, D.S., Cooper, T.B., Jindal, S.P., Corcella, J., and Lutz, T. (1985). J. Clin. *Psychopharmacol.* **5**, 333-337.

Robinson, D.S., Kayser, A., Bennett, B., Devereaux, E., Lerfald, S.C., Albright, D., Laux, D., and Corcella, J. (1986).*Psychopharmacol. Bull.* **22**, 553-557.

Robinson, D.S., Lerfald, S., Laux, D., and Bennett, B. (1988). Paper presented at the Sixteenth Congress, Collegium Internationale Neuro-Psychopharmacologicum, August 15, 1988, Munich, Germany.

Schiele, B. (1965). *Minn. Med.* **48**, 356-357.

Simpson, G.M., Frederickson, E., Palmer, R., Pi, E., Sloane, R.B., and White, K. (1985). *Biol. Psychiatry* **20**, 684-687.

Sjoerdsma, A. (1981). *Clin. Pharmacol. Ther.* **30**, 3-22.

Willner, P. (1985). *Psychopharmacology* **85**, 387-404.

Yates, C.M. and Loudon, J.B. (1979). *Psychol. Med.* **9**, 777-779.

Youdim, M.B.H,. Aronson, J.K., Blau, K., Green, A.R., and Grahame-Smith, D.G. (1979). *Psychol. Med.* **9**, 377-382.

Zeller, E.A., Barsky, J., Fouts, J.R., Kirchheimer, W.F., and Van Orden, L.S. (1952). *Experientia* **8**, 349-350.

Index

F

G

H

I

N

O

P